Dedicated

To

Professor Y. R. Malhotra

Preface

An unbridled proliferation of human population, has generated interest in the areas of Limnology, Fish and Wildlife Ecology, as well as Biology. A corollary to this fact has come in the form of an increased pressure upon water and land resources. Further aggravation of this problem has been caused by the allied incidence of pollution, deforestation and the near extinction of wildlife. Consequently, the very survival of mankind is threatened. The only re-deeming factor is that among the people in general and scientists in particular, an increased awareness of the ecological problem is very much there now. This has generated a considerable volume of research work, in India as well as abroad. The present volume is a part of a series of such volumes that shall attempt to reflect the latest as well as original research work, related to these environmental issues.

The current inaugural volume is dedicated to Professor Y. R. Malhotra on his completion of 40 years of teaching and qualitative research, and his elevation as Vice-Chancellor of the University of Jammu. As a committed academician and teacher of infallible repute, Prof. Malhotra has conducted and guided research in limnology, fish ecology, entomology and wildlife. He is credited with about *200 publications* individually and collaboratively. This work is a symbol of gratitude and thanksgiving on part of a student, toward his teacher and inspirer.

The present volume includes articles by well known experts in their respective fields, many of whom have been students of Professor Malhotra. The editor is grateful to all the contributors who responded promptly by making available their valuable papers and it is this overwhelming response which had a catalytic impact upon me, to complete the task with an unflinching zeal and enthusiasm.

The editor also acknowledges the special and kind gesture of Prof. P. L. Duda, for agreeing to contribute a write - up in appreciation of Prof. Malhotra, with whom he has remained closely associated for almost four decades. I am also indebted to Dr. K. B. Razdan of the Department of English, University of Jammu, for looking after the stylistic aspect of the manuscript and the finer nuances of semantics. Thanks are also due to the publishers for their whole-hearted cooperation and goodwill in bringing out this volume, in the shortest possible time.

Bansi Lal Kaul

Advances in Fish and Wildlife Ecology and Biology

Vol. 1

Bansi Lal Kaul

2017
Daya Publishing House®
A Division of
Astral International (P) Ltd
New Delhi 110 002

First Published, 1996
Reprinted, 2017

ISBN: 978-93-5124-840-8 (Hardbound)

Published by : **Daya Publishing House®**
A Division of
Astral International Pvt. Ltd.
– ISO 9001:2008 Certified Company –
4760-61/23, Ansari Road, Darya Ganj
New Delhi-110 002
Ph. 011-43549197, 23278134
E-mail: info@astralint.com
Website: www.astralint.com

Digitally Printed at : Replika Press Pvt. Ltd

An Appreciation

PROFESSOR YOGRAJ MALHOTRA, the present Vice-Chancellor of the University of Jammu, remains essentially a teacher and an academician first and anything else later. Over four decades that I have known him in different capacities, I think that the way he influenced me is representative of the way he has influenced diverse branches of animal science that have been studied under his tutelage and guidance. His influence has been such that people around him have, I feel, achieved much more than they possibly could have without it. I pray that he continues to guide and influence his students and colleagues equally fervently as before for many many decades to come.

His contribution to animal science measured in terms of the number of publications, individual or in association (over 200 and well known) are signal and one cannot escape finding a continuous rise, starting from 60's (while at the university of Kashmir) to a staggering peak in 70's and 80's of this century at the university of Jammu. A glimpse into his academic and research activities, particularly during his tenure at the university of Jammu, reveals diversification in his scientific interest alongwith a crop of promising young men obviously inspired by him.

Professor Malhotra started initially as a student of Ichthyology, but eventually his sphere of research activity and influence diversified transcending ichthyology in such directions as systematics (entomology), limnology, reproductive biology, endocrinology, wild-life and ecology.

Professor Malhotra's classic work on feeding and reproduction of high altitude fishes of Central Asian Origin has eventually led to the formation of a well defined school of researchers on reproductive ecology of fishes in particular and vertebrates in general. In addition to his publications based mostly on field research have generated a wealth of knowledge on different aspects of the biology and ecology of fish and wild-life of J& K State. These works have provided ample support to some of the theoretical concepts on high altitude animal ecology.

I feel honoured to have been asked for this write up, and in doing so I may well have missed much more than I have provided here. The lapse surely is inadvertant.

P. L. Duda
Professor of Zoology,
Dept. of Biosciences,
University of Jammu,
Jammu

Contents

Preface v

An Appreciation vii

Contributors xiii

Section I : Fish and Limnology

1. Melanophore Occurrence in Early Life History Stages (Periodization) of Mahseer, *Tor tor* (Hamilton) and Its Role in Identification of the Larvae Inhabiting Jammu Waters of J and K State
Y. R. Malhotra and Subash Chander Gupta 3
2. Seasonal Variations in Biochemical Composition in Some Freshwater Fishes, Part – III, *Channa punctatus* (Bloch)
B. N. Pandey, Anupa Sharan, Rumana Perween and M. Kumar 31
3. Some Relationships Between Size Structure and Fertility of Rotifer Populations
S. S. S. Sarma 37
4. Effect of Accessory Pneumectomy on Some Haematological Values of Air Breathing Fish *Clarias batrachus*
B. D. Joshi 51
5. Food and Feeding Habits of *Heteropneustes fossilis* (Bloch) Inhabiting Gadigarh Stream, Jammu
S. P. S. Dutta 61
6. Observations on the Use of Ovaprim for Induced Spawning of Indian Major Carps
P. K. Roy 65
7. Aquatic Odonata and Hemiptera of Jammu and their Role in Aquaculture
Baldev Sharma, Neeru Dhalla and Renu Salaria 69
8. A Comparative Study of the Renal Organs of Freshwater Teleostean Fish, Part I – Morphology
B. L. Kaul 87
9. A Comparative Study of the Renal Organs of Freshwater Teleostean Fish, Part II – Histology
B. L. Kaul 105

10. Benthic – Macroinvertebrates as Indicators of Aquatic Environment
Usha Moza 135

11. Effect of Mechanical Stress on Early Embryonic Stages of *Tor tor*
Kuldeep K. Sharma 151

12. Icthyofauna of the Sector of Kaveri River in Head Region
M. N. Madyastha and S. Murugan 155

13. Biology of Indian Belone, *Xenentodon cancila* (Hamilton) – A Freshwater Fish from Jammu Waters of J & K State
I – *Periodization in Life History of Xenentodon cancila* (Hamilton)
Subash Chander Gupta and Kuldeep K. Sharma 159

14. Rotifer Fauna of Devikoppa Tank: Dharwad (Karnataka, India)
K. Vijay Kumar 187

15. Terminology of Various Developmental Stages of Fish Larvae Inhabiting Jammu Freshwaters
Subash Chander Gupta and Arun Kumar Gupta 193

16. Inter-specific Competition in Mixed Culture of Cladocera
Y. R. Malhotra and Seema Langer 203

17. Limnology of Farooq Nagar Pond, Jammu. Part II—Rotifera
S. P. S. Dutta and Jyoti Sharma 209

18. Relative Population Abundance of Ichthyo-fauna of Lake Mansar
Arun K. Gupta, Anil Khajuria, S. C. Gupta and Seema 221

19. On the Distribution and Ecology of Some Gastropod Molluscs of the Jammu Province in J & K State
Anil K. Verma and P. L. Duda 229

20. Macrobenthic Fauna in Relation to Some Environmental Factors in Eutrophicating Lake Mansar, Jammu
K. Gupta and Anil Khajuria 235

21. Population Structure and Seasonal Succession of Zooplankton of Lake Surinsar, Jammu (India)
M. K. Jyoti and H. S. Sehgal 245

Section II : Wildlife

22. On the Habitat and Behaviour, Maturity – Size and Sexual Dimorphism in a Population of Freshwater Emydid Turtles of Jammu, J & K State
P. L. Duda, Anil K. Verma and D. N. Sahi 265

23. Status and Management of Wildlife in Jammu & Kashmir State
B. L. Kaul and Indu Kanwal 279

24. Eco-geographical Distribution and Present Status of Herptiles in Kashmir Himalayas
D. N. Sahi and P. L. Duda 289

25. Shrinking Wetlands of India
S. K. Chadha 299

26. Ecology and Status of Wildlife in Ladakh
B. L. Kaul 303

27. On the Freshwater Chelonian Fauna of Jammu and Kashmir
D. N. Sahi and Anil K. Verma 309

28. Threatened Wildlife Habitat in Kashmir Himalayas
B. L. Kaul 317

29. Role of Gastropods in Trematode Transmission Among Herptiles – Part I Amphibia. A Numerical Analysis
Anil K. Verma and P. L. Duda 321

Index 327

Contributors

1. **Prof. Y. R. Malhotra**
   Vice–Chancellor,
   University of Jammu,
   Jammu – 180004.
2. **Prof. B. N. Pandey**
   P. G. Department of Zoology,
   Magadh University,
   Bodh-Gaya –824234.
3. **Prof. M. N. Madyastha**
   Department of Biosciences,
   Mangalore University,
   Mangalagangotri – 574199.
4. **Prof. P. L. Duda**
   Professor of Zoology,
   Department of Biosciences,
   University of Jammu,
   Jammu – 180004.
5. **Dr. S. S. S. Sarma**
   Laboratory of Animal Ecology,
   Unversity of Ghent,
   Ghent – Belgium
6. **Dr. S. K. Chadha**
   Controller of Examinations
   University of Jammu,
   Jammu – 180004.
7. **Dr. K. Vijay Kumar**
   Department of Zoology,
   Gulbarga University,
   Gulbarga – 585105.

8. **Dr. B. L. Kaul**
Department of Zoology,
Govt. College for Women,
Parade, Jammu – 180001.

9. **Dr. B. D. Joshi**
Department of Zoology,
Gurukul Kangri University,
Hardwar – 249404.

10. **Dr. Baldev Sharma**
Department of Biosicences,
University of Jammu,
Jammu - 180004.

11. **Dr. S. P. S. Dutta**
Department of Biosciences,
University of Jammu,
Jammu - 180004.

12. **Dr. P. K. Roy**
Indian Inland Fisheries Training Centre,
Berrackpore, West Bengal – 743101.

13. **Dr. Subash C. Gupta**
Department of Biosciences,
University of Jammu,
Jammu – 180004.

14. **Dr. Kuldeep Sharma**
Department of Biosciences,
University of Jammu,
Jammu – 180004.

15. **Dr. Arun Gupta**
Department of Zoology,
Govt. College for Women,
Parade, Jammu –180001.

16. **Dr. D. N. Sahi**
Department of Biosciences,
University of Jammu,
Jammu–180004.

17. **Dr. Usha Moza**
Central Inland Capture Fisheries Research Centre,
Session House Marg, Karnal (Haryana).

18. **Dr. Anil K. Verma**
Biotechnology Division,
Regional Research Laboratory (CSIR),
Canal Road, Jammu-180001.

19. **Dr. (Mrs.) K. Gupta**
Department of Biosciences,
University of Jammu,
Jammu-180004

20. **Dr. M. K. Jyoti**
Department of Biosciences,
University of Jammu,
Jammu-180004.

Section I

Fish and Limnology

Chapter 1

Melanophore Occurrence in Early Life History Stages (Periodization) of Mahseer, *Tor tor* (Hamilton) and Its Role in Identification of Larvae Inhabiting Jammu Waters of J and K State

Y. R. Malhotra* and Subash Chander Gupta**

Abstract

Periodization of early developmental stages of *Tor tor*, raised in the laboratory, from the eggs collected from nature has been discussed. Fertilized eggs are spherical, transparent, about 1.4-3.1 mm in diameter, with tough shells & characteristic rounded structure on the shell and float as air-bubbles in water. The newly hatched larvae have eyes near the anterior end of yolk sac, with head flexed ventrally but in line with the longitudinal axis of body. Yolk sac gets absorbed at 96 hours post hatching stage, when gills appear and alimentary canal becomes coiled. Protrusible mouth appears at 120 hours post hatching stage along with pectoral rays and dorsal & lateral line melanophores merge. An adult characteristic feature is the presence of a pair of maxillary barbels, with a spine in dorsal fin and appearance of ventral fin is achieved at 456 hours post hatching stage. 744 hours post hatching stage is characterized by the appearance of an adult characteristic feature of a pair of rostral barbels. Appearance of a black patch on the caudal peduncle is achieved at 1104 hours post hatching stage (which ultimately disappears in adults). At 1248 hours post-hatching stage, the fingerling is a small adult, with a "heartshaped" patch of melanophores on the head, and with full complement of fins.

Introduction

Development and life history studies of fish offer plenty of scope to anyone working on fish biology. Though these studies form one of the important branch of fisheries research, yet it is unfortunate that in India, the problem has not received due attention. Despite lot of work done on the eggs & larval stages of Indian freshwater fishes, the Mahseer *(Tor tor)* - an economically important food and game fish, has not been studied from this angle.

Among the cyprinids, identification of *Tor tor* larvae is a formidable & very difficult task because of their similarity in appearance in their early stages with other cyprinid larvae. The present knowledge on eggs and early developmental stages of *Tor tor* is very meagre & for lack of such knowledge, very little authentic information is available on this aspect of fishery biology. The only available information on

* Vice-Chancellor, University of Jammu, Jammu-180001.

** Department of Biosciences, University of Jammu, Jammu-180001.

Mahseer, *Tor tor* is on their breeding habitats & breeding season (Beavan, 1877; Thomas, 1897; Dunsford, 1911; Nevill, 1915; & Sir Reginald Spence & Prater, 1933 (c. f. Khan, 1939); Hora, 1940; Mac Donald, 1948; Codrington, 1946 & Karamchandani *et al.* 1967 (c.f. Desai, 1973). Desai (1973) did some work on maturity, fecundity & larval development of *Tor tor* from Narmada river.

Regarding the life history of Mahseer, there is no account except on the early development of *Barbus (Lissochilus) hexagonolepis*, the copper mahseer (Nazir Ahmed, 1948); larval stages of *Barbus (Tor) mosal mahanadicus*, the mosal mahseer (Devid, 1953); larval stages of *Tor khudree*, Deccan Mahseer (Kulkarni, 1971); and early larval stages of *Tor putitora*, the Putitor Mahseer (Desai, 1972), but none of these workers have tried to project a single character or a group of characters that could be useful in larval identification. Considering the lack of data available on the life history of this economically important fish, the present studies were intended for fish biologists in the identification of fish larvae in the field. In the present communication, an attempt has been made to project informations regarding the eggs & larval stages of *Tor tor* inhabiting freshwaters of Jammu (J & K), and their periodization based on morphological features and time sequence of appearance, merger and disappearance of melanophores during different developmental stages.

Materials and Methods

Eggs of *Tor tor* were collected from flooded river Tawi, Jammu below Bahu Fort during monsoon (July-August) with the aid of plastic sieve nets/hand nets made of mesh cloth. The eggs were then brought to the laboratory in plastic containers containing river water, a few however were fixed in formaldehyde/70 per cent alcohol in the field itself. In the laboratory, transparent fertilized eggs were separated from translucent unfertilized ones. The former were then maintained in different hatching troughs & subsequent developmental studies made at the room temperature. Most of the collected eggs were at varying stages of development & thus the time of their spawning & initiation of fertilization in the eggs could not be ascertained. The developing eggs were isolated according to their developmental stages, recording the intervening period between the successive stages until they hatched. The larvae were then transferred to Fish-hatching incubator (a modified version of B.O.D., specially fabricated for the studies, with water circulation arrangements alongwith cooling and heating mechanisms) maintained at 22-24°C & were examined every 24 hours. The periodization (time duration mentioned against various developmental stages) has been determined post-hatching. As the larvae exhaust, they were provided with different foods for exogenous feeding in the lab. during their post-larval stages. The larvae were fed twice a day (in the early morning before sun rise & in the late evening after sun set) with artificial and natural planktonic food.

Observations

I. Embryonic Development

From within the eggs collected, the following stages of development have been recognized :

(a) Cleavage Stages :

(i) Fertilized Egg. (Fig. 1).

Fertilized eggs are transparent, spherical, non-adhesive, with tough shells, measuring 1.45-3.15 mm. in diameter, with an inner vitelline membrane (covering the yolk) and an outer zona radiata or perivitelline membrane. The vitelline membrane is separated from the yolk, allowing the egg to rotate within the shell. The yolk is yellowish, transluscent & highly vacuolated.

(ii) 4-celled Stage (Fig. 2 A & B)

Meroblastic verticle cleavage separates the blastodisc into two equal halves; second cleavage at right angles to the first cleavage, divides two cells into four equal sized cells. Each blastomere with rounded outer & straight inner margin.

(iii) 16-celled Stage (Fig. 3)

Further verticle cleavages result in the formation of 16 - cells, smaller in size than the previous stage.

(iv) 32-celled Stage (Fig. 4)

Horizontal cleavage result in 32-cells forming a layer of upper and lower blastomeres, cell size further reduced over the 16-celled stage.

(v) 64-celled Stage (Fig. 5)

Further cleavages result in the formation of 64-cells, the size further decreases but blastoderm occupies the same space.

(vi) 128-celled Stage (Fig. 6)

128-cells are formed from the cleavages of 64-cells.

(vii) 256-celled Stage (Fig. 7)

This stage is attained by the cleavages of the previous stage. The cells are sufficiently smaller than the previous stage.

(viii) Morula Stage (Fig. 8)

Blastoderm cells increase in number but descrease in size; tightly packed and invade areas over the yolk. Blastodermal cells golden-yellow whereas uncovered yolk is yellowish green in colour & highly vacuolated.

(ix) Gastrulation Stage (Fig. 9)

Most of yolk is invaded by blastoderm cells; only a small portion of uncovered yolk still persists towards one pole of egg, extending slightly sidewards. The margins of blastoderm towards this pole & its side are thickened to form a germ ring, which are the embryonic rudiments.

(x) Yolk-plug Stage (Fig. 10)

Gastrulation complete, blastoderm just closing. A characteristic small knob-like yolk-plug is formed by a constricting mass of closely packed blastodermal cells.

(b) Post-Cleavage Stages :

(xi) Differentiation of Cephalic and caudal ends (Fig. 11)

Blastopore now closes. Embryo elongates on the periphery of yolk, distinguishing a slightly thickened anterior head region & an angular posterior tail region but without any sign of internal structures or massing of cells.

(xii) Initiation of Optic-vesicle and 6-12 myotomes (Fig. 12)

Embryo 'C' shaped with broad & anteriorly blunt head region, rudiments of optic-vesicles and 6-12 myotomes. Yolk enlarged further, closely adheres to elongated yolk-mass. Cephalic & caudal ends further differentiated & distinguishable.

(xiii) Appearance of brain, auditory-vesicle rudiments & 14-16 myotomes (Fig. 13).

Embryo with distinct head & tail region, inner thickening in the head indicates initiation of brain. Auditory vesicle rudiments without otoliths, appear posterior to brain. Embryo with 14-16 myotomes, makes occasional twitching movements within the egg & has narrow embryonic finfold surrounding the tail. Kupffer's vesicle has appeared (in one specimen only) as an oval area at the base of caudal region.

(xiv) 22-24 Myotome Stage (Fig. 14)

Embryo much enlarged with more prominent embryonic finfold both dorsally & ventrally; 22-24 myotomes extend upto vent only. Yolk-sac elongated, broad anteriorly tapering posteriorly at the vent. Embryo with prominent cephalic end and free caudal end from yolk-mass, makes frequent movements within the egg membrane.

(xv) Differentiation of brain, eyes, appearance of pectoral bud and 30-35 myotomes (Fig. 15)

Embryo has now taken up the shape of larva, much elongated, broad anteriorly tapering posteriorly, with 30-33 myotomes (22-23 pre-anal, 8-10 post-anals). Yolk-sac elongates further, tapers posteriorly towards vent (tapering more acute than in previous stage), yellowish-green, more than twice post anal length of body; some clearing of yolk between body & yolk-sac in the pectoral fin region. Head broad, brain initiated, with a small pigmented dot in the eye. Pectoral fin bud like on the antero-dorsal surface of yolksac, well formed median finfold with circular caudal end extended upto anterior 1/3 of body both dorsally & ventrally. Pre-anal length of body about thrice post anal length, with clearly distinguished anus. Heart pulsates rapidly. Embryo executes active movements, changes its orientation from time to time within the egg membrane & almost fills up entire perivitelline space. These movements become rapid & forceful as the hatching time approaches, the embryo hatches out of egg by bursting of egg membrane.

The process of hatching occurs by tail piercing the vitelline membrane (Fig. 15 a, b) & hatchling comes out through already limpid perivitelline membrane. Hatchlings settle down on one side, come to surface flickering with great speed, again lie on their sides at the bottom of container. They occasionally dart up, move about for a while & then again lie at the bottom of container on one side. This may be due to the weight of yolk-sac which drags the larva downwards.

II. Larval Development

Morphometric data & distribution of melanophores during larval stages are presented in the Tables 1 and 2 respectively.

(a) Fry Stages :

(i) Larva just after hatching (Fig. 16)

This is the first larval stage. Body slender, elongated, transparent; broad anteriorly, tapering posteriorly & devoid of mouth. Eyes formed anterior near the anterior end of yolk-sac but without lens. Yolk-sac broad anteriorly, tapers posteriorly upto vent. Head flexed ventrally, in line with the longitudinal axis of body, slightly free from yolk-sac. Brain & auditory vesicles well formed, auditory vesicles still without otoliths. Larval finfold extends dorsally & ventrally to 2/3 of body, circular around caudal end; caudal fin diphycarcal. Pectoral fin bud like on the dorso-anterior surface of yolk-sac, posterior to auditory vesicle. Anus situated far back in posterior 1/4 of body. Pre-anal length about thrice post-anal length, but post anal length of body about 1/3 of yolk sac length. Myotomes 31-33 (23 pre-anal, 8-10 post anals).

(ii) 24 hours post hatching stage (Fig. 17)

Head & body elongate further, head now straight & has come to the level of yolk sac, free from much elongated yolk-sac. Eyes circular, lateral on head, with a pigmented spot on the lower margin. Alimentary tract appears, continuing upto anus. Larval finfold has wavy margins, pectoral fin distinct. Brain, auditory vesicles & anus distinct. Body without pigments.

(iii) 48 hours post hatching stage (Fig. 18)

Mouth cleft indicated, opercular cleft & swim bladder formed; yolk-sac much reduced. Eyes well formed bulging on broad head, brain lobes distinct. Actinotrichia appear in caudal fin. Pre-anal length of body almost twice post-anal length, later almost equals the predorsal length. Myotomes increase to 22-23 pre-anals and 12 post-anals.

Head pigmented dorsally and on the opercular cleft with stellate & faint melanophores. Melanophores noticeable on eitherside at the base of dorsal finfolds. Melanophores faint & stellate on the upper half of swim bladder, a few scattered towards lower side (not counted as they are grouped together); continue anteriorly with melanophores

Table 1: Morphometric data (in cms) by age of *Tor tor* larvae

Morphic Characters	Developmental stages (age) Hours post-hatching)																							
	Larvae just after hatching	*24*	*48*	*72*	*96*	*120*	*144*	*168*	*192*	*216*	*240*	*264*	*296*	*312*	*408*	*456*	*504*	*672*	*744*	*840*	*960*	*1104*	*1248*	*1440*
Size Interval	I2.3–	3.0-	3.5-	3.6-	3.6-	3.6-	4.2-	3.5-	4.5-	3.8-	4.5-	5.2-	6.1	6.0-	6.7-	6.7-	7.5	9.7	7.3-	9.4-	11.7-	12.4-	15.1-	21.7-
Length	2.6	3.4	4.8	4.2	4.9	4.4	4.7	4.4	4.8	4.8	5.3	6.2	6.2	7.4	7.4	8.0			8.4	10.4	12.5	15.6	16.3	28.4
Total	2.5	3.1	3.7	3.8	4.4	4.8	4.4	4.8	4.6	4.5	4.7	5.6	6.2	6.7	7.8	7.3	7.5	9.7	5.0	9.8	11.9	14.4	15.6	24.2
Standard	2.3	2.8	3.4	3.6	4.1	3.8	4.2	3.7	4.3	4.2	4.4	5.0	5.3	5.5	5.9	6.1	6.8	7.7	6.4	7.8	9.0	11.1	12.8	19.8
Pre-anal	1.8	2.6	2.5	2.6	3.1	2.8	3.1	2.7	3.2	3.2	3.4	3.3	4.2	4.3	4.5	4.7	4.9	5.6	4.8	5.7	6.5	8.1	8.9	14.3
Post-anal	0.5	0.9	1.2.	1.2	1.2	1.3	1.3	1.3	1.3	1.2	1.2	1.7	2.0	2.3	2.6	2.6	2.6	4.1	3.1	4.8	5.4	6.2	6.7	9.9
Snout	0.1	0.1	0.1	0.1	0.2	0.2	0.2	0.2	0.2	0.2	0.2	0.3	0.4	0.3	0.4	6.4	6.4	0.6	0.5	0.6	0.9	0.8	8.9	1.5
Head	0.2	0.4	0.7	8.8	1.8	8.9	1.0	0.9	1.0	1.1	1.1	1.4	1.5	1.6	1.7	1.7	1.9	2.3	2.8	2.5	2.9	3.2	3.5	5.1
Head depth	0.3	0.4	0.6	0.6	0.7	9.6	0.6	8.6	0.7	0.7	0.7	6.9	1.9	1.9	1.2	1.1	1.2	1.8	1.2	1.7	1.9	2.3	2.7	3.5
Eye diameter	0.5	0.1	0.2	0.2	0.2	0.2	0.2	0.2	0.2	0.2	0.2	0.3	0.4	0.4	0.4	0.4	0.4	0.5	0.5	0.6	0.7	0.9	1.8	1.3
Greatest Body depth	0.6	0.6	0.6	0.6	0.6	0.6	0.6	0.5	0.7	0.6	0.7	1.8	1.2	1.3	1.4	1.3	1.3	1.4	1.2	1.6	1.8	2.2	2.6	3.7
Yolk sac length	1.6	1.7	1.4	1.5	-	-	-	-	-	-	-	-	-	-	-	-	-	-	-	-	-	-	-	-
Yolk-sac depth	0.5	0.4	0.2	0.2	-	-	-	-	-	-	-	-	-	-	-	-	-	-	-	-	-	-	-	-
Pre-dorsal length	1.6	1.1	1.5	1.3	1.7	1.6	1.6	1.5	1.7	1.6	1.6	2.1	2.4	2.7	2.7	2.9	2.9	3.6	3.1	3.0	4.2	5.5	5.9	8.8
Anal fin to caudal peduncle	-	-	-	-	-	-	-	-	-	-	-	-	-	0.7	0.8	2.0	2.1	3.2	2.2	3.0	4.2	4.5	5.1	7.9
Eye to dorsal surface of head	0.00	0.1	0.2	0.1	0.1	0.1	0.1	0.1	0.1	0.1	0.1	0.1	0.1	0.1	0.1	0.1	0.1	0.2	0.1	0.2	0.2	0.3	0.2	0.2
Caudal peduncle depth	0.3	0.1	0.1	0.1	0.1	0.1	0.1	0.1	0.2	0.1	0.2	0.2	0.4	0.4	0.5	0.5	0.5	0.5	0.5	0.8	0.9	1.1	1.1	1.7
Swim-bladder length	-	-	0.4	0.6	0.7	0.6	0.7	0.6	0.7	0.7	0.8	1.1	1.1	1.2	1.2	1.3	1.3	1.5	1.4	2.0	2.3	-	-	-
Swim-bladder depth	-	-	0.1	0.2	0.2	0.2	0.1	0.2	0.2	0.3	0.2	0.3	0.4	0.4	0.4	0.5	0.5	0.5	0.6	0.6	0.6	-	-	-

Table 2: Number and position of melanophores during different developmental stages of *Tor tor* larvae

Developmental Stages (Age)	Swim bladder	Between swim bladder and pectoral fin	Head region	Dorsal line region: Pre-Dorsal	Dorsal line region: Fin-fold	Dorsal line region: Caudal Peduncle	Fins: Dorsal	Fins: Pectoral	Fins: Ventral	Fins: Anal	Fins: Caudal
1	2	3	4	5	6	7	8	9	10	11	12
Larvae Just after hatching	-	-	-	-	-	-	-	-	-	-	-
24 - hrs post-hatching	-	-	-	-	-	-	-	-	-	-	-
48 - hrs. post-hatching	Pigmented	-	8-10	-	3-4	-	-	-	-	-	-
72 - hrs. post-hatching	Pigmented	-	30-37	4-5	10-12	5-6	-	-	-	-	-
96 -hrs. post-hatching	Pigmented	-	Many	6-3 (in 3 rows)	22-24 (in 3 rows)	6-0	-	-	-	-	-
120 -hrs. post-hatching	40-50	-	Many	7 (in 3 rows)	20-22 (in 3 rows)	6-8	-	-	-	-	-
144 -hrs.post-hatching	45-55	-	Many	6-7 (in 3 rows)	22-24	6-8	-	-	-	-	18-20
168 -hrs.post-hatching	45-50	Pigmented	140-150	7-10 (in 2 rows)	24-28	4-6	-	-	-	-	8-10
192 -hrs. post-hatching	45-55	Pigmented	160-210	5-10	24-28	4-6	-	-	-	-	15-25
216 -hrs. post-hatching	40-50	Pigmented	160-210	8-10	25-30	5-6	-	-	-	-	30-40
240 -hrs. post-hatching	25-40	Pigmented	140-180	8-10	24-30	5-9	-	-	-	-	25-30
264 -hrs. post-hatching	25-40	Pigmented	160-190	10-12 (in 3 rows)	18-20	8-10	4-5	-	-	-	30-45
288 -hrs. post-hatching	40-50	Pigmented	140-150	7-8 (on the mid-dorsal line 4-5 on either side)	10-14	5-6	4-6	-	-	-	30-45
312 -hrs. post-hatching	40-60	Pigmented	140-160	8-10 & 6-8	18-22	4-5	8-10	-	-	-	40-50
408 -hrs. post-hatching	40-60	Pigmented	200-220	10-12 & 6-10	10-22	6-8	20-30	-	-	-	Many
456 -hrs. postt-hatching	Few distinct	Pigmented	Many	Highly	Pigmented	-	Pigmented	-	-	-	Pigmented
504 -hrs. post-hatching		Indistinct	55-57	Indistinct	101-2	Indistinct	Indistinct	-	-	-	Indistinct
672 -hrs. post-hatching	26-28	Pigmented	Highly Pigmented	Highly	Pigmented	-	Pigmented	Pigmented (base)	-	-	Pigmented
744 -hrs. post-hatching	Few distinct	Pigmented	Highly Pigmented	Highly	Pigmented	-	Pigmented	Pigmented (base)	Pigmented	Pigmented	Highly Pigmented
840 -hrs. post-hatching	Few distinct	Pigmented	Highly Pigmented	Highly	Pigmented	-	Pigmented	Pigmented (base)	Pigmented	Highly Pigmented	
960 -hrs. post-hatching	Indistinct	Pigmented	Highly Pigmented	Highly	Pigmented	-	Pigmented	Pigmented (base)	Pigmented	Highly Pigmented	
1104 -hrs. post-hatching	Indistinct		Highly				Pigmented				
1248 -hrs. post-hatching	Indistinct		Highly				Pigmented				
1440 -hrs. post-hatching	Indistinct		Highly Pigmented				Highly pigmented		Indistinct	Pigmented	Pigmented

Table 2: Number and position of melanophores during different developmental stages of *Tor tor* larvae. Contd.

Developmental Stages (Age)	Lateral-line	Ventro-visceral line		Caudal Peduncle	Throat region	Heart region	Gills	Media
		Pre-anal	Post-anal					Ventral-line
	13	14	15	16	17	18	19	20
Larve just after hatching	-	-	-	-	-	-	-	
24 - hrs. post-hatching	-	-	-	-	-	- -	-	
48 - hrs. post-hatching	-	-	2.4	-	-	Pigmented	-	-
72 -hrs. post-hatching	16-20	14-46	14-16	6-8	-	Pigmented	-	-
96 -hrs. post-hatching	24-26	22-26	14-16	10-12	10-18	Pigmented	Pigmented	-
120 -hrs. post-hatching	30-32	24-26	22-24	18-24	18-24	10-20 (corts ventralis) 8-10 (base of heart)	Pigmented	
144 -hrs. post-hatching	24-30	23-26	10-20	30-40	10-16	10-20 (corts ventralis), 8-10 (base of heart)	Pigmented	8-10
168 -hrs. post-hatching	35-42	35-45	25-30	24-35	18-20	24-32 (12-15 on either side of corts ventralis) 10-12 (base of heart)	Pigmented	10-16
192-hrs. post-hatching	30-35	25-35	25-35	30-45	12-16	14-16 (corts ventralis) 6-8 base of heart)	Pigmented	8-10
216 -hrs. post-hatching	32-40	35-40	22-28	30-35	12-16	Heart thickly pigmented on the sides	Pigmented	10-14
240 -hrs. post-hatching	35-42	35-40	20-24	20-24	14-18	14-16 in a row on either side of corts ventralis; 8–10 base of heart.	Pigmented	10-16
264 -hrs. post-hatching	28-40	16-20	16-20	16-24	16-22	14-16 in a row on either side of corta Ventralis; 8–10 base of heart	Pigmented	10-30
288 -hrs. post-hatching	30-40	16-20	15-25	15-25	6-8	4-5 on either side of corta ventralist; 4-6 at the base of heart	Pigmented	10-12
312 -hrs. post-hatching	32-36	26	20-25	20-25	6-8	4-6 on either side of corta ventralis; 4-6 at the base of heart	Pigmented	16-25
408 -hrs. post-hatching	35-40	20	30-40	30-40	8-10	6-8 on either side of corta ventralis; 6-8 at the base of heart	Pigmented	10-15
456 -hrs. post-hatching	35-45	20-25	30-40	30-40	15-25	Heart lined by pigmented streak	Pigmented	18-20
504 -hrs. post-hatching	14-15	Indistinct	Indistinct	Indistinct	Indistinct	Indistinct	Indistinct	Indistinct
672 -hrs. post-hatching	44-46	Indistinct	22-24	Pigmented	Indistinct	Pigmented	Pigmented	18-20
744 -hrs. post-hatching	40-42	Indistinct	Indistinct	Pigmented spot	Indistinct	Pigmented	Pigmented	8-12
840 - hrs. post-hatching	Pigmented	Few distinct	Few distinct	Pigmentedspot	Pigmented	Indistinct	Pigmented	Indistinct
960 -hrs. post-hatching	Pigmented	Indistinct	Few distinct	Pigmented spot	Indistinct	Indistinct	Pigmented	Indistinct
1104 -hrs. post-hatching	Pigmented	Indistinct	Indistinct	Pigmented spot	Indistinct	Indistinct	Pigmented	Indistinct
1248 -hrs. Post-hatching	Pigmented	Indistinct	Indistinct	Pigmented spot	Indistinct	Indistinct	Pigmented	Indistinct
1440 -hrs. Post-hatching	Pigmented	Indistinct	Indistinct	Pigjmented spot	Indistinct	Indistinct	Pigmented	Indistinct

around the margins of auditory vesicles. Heart demarcated by pigmented base. Ventro visceral melanophores very faint.

(iv) 72 hours post hatching stage (Fig. 19)

Jaws indicated, opercular cleft distinct with a flap. Nasal sacs appear. Tail no more diphycercal, heterocercy initiated· caudal fin with denticulate margins. Ventrally branchiostegal rays visible on throat. Massive eyes occupy about 1/3 of total length of broad and massive head.

An overall increase in pigmentation, melanophores faint & stellate. Head and swim-bladder melanophores prominent with same pattern, swim-bladder melanophores continue posteriorly with ventrovisceral melanophores. Pre-dorsal melanophores appear; dorsal melanophores (including pre-dorsals) are arranged in a row on either side of mid dorsal line at the base of dorsal finfold, approximating in the caudal region. Latral-line melanophores faint and stellate appear between the level of centre of swim-bladder upto little posterior of anus. Ventro-visceral melanophores are faint dots arranged in a row on either side of mid ventral line, merge with those present near the lower end of caudal peduncle. Ventrally, pigmented streaks demarcate the aorta ventralis & base of heart.

(b) Post-Fry Stages

(v) 96 hours post hatching stage (Fig. 20)

Yolk-sac completely absorbed. Coiling in alimentary canal indicated. Larval finfold under progressive absorption. Lower lobe of tail large. Swim-bladder distinct as an oval chamber, allows passage of light through it. Four pairs of gills demarcated.

Head highly pigmented with small stellate melanophores, scattered on skin following the shape of underlying brain i.e., distribution of melanophores above fore brain, mid brain & hind brain. Head and dorsal melanophores merge. Upper jaw thickly pigmented than the lower. Nasal sacs pigmented. Body highly pigmented dorsally with stellate melanophores, arranged in three rows (i.e. one row in mid dorsal line, one row on either side of mid-dorsal line, a few small dots scattered in the caudal region), dorsal melanophores started extending laterally downwards upto the lateral-line melanophores which are more than the previous stage. Few small dots appear below eye orbit & ventrally on throat. Pigmented streaks mark the gill filaments. Ventro-visceral melanophores increase in number. Fine stellate melanophores scattered laterally over the surface of alimentary canal.

(vi) 120 hours post hatching stage (Fig. 21)

Mouth protrusible, indicated by a fold between it & dorsum of Skull, lower jaw slightly longer than the upper; pre-orbital head long & narrow. Pectoral fin elongate, rays appear in it & caudal fin. Body opaque with clear myotomes except in the region of Swim bladder &

ventral coelom. Larvae show slight reduction in size over the previous stage.

Head melanophores more prominent, retain the same number & pattern, as those of dorsal melanophores which mix with lateral-line melanophores. Opercular cleft bears few small stellate melanophores; upper jaw pigments extend upto maxilla. Swim-bladder melanophores large & more in number on dorsal half, small & less in number on lower half, extended over the surface of alimentary canal. Pre-anal ventro-visceral melanophores retain almost same number; post-anal ventro-viscerals increase in number, merge with those on lower edge of caudal peduncle which are almost double than previous stage. Ventrally throat melanophores more in number, gill filaments pigmented ventrally; heart melanophores retain same pattern & are countable.

(vii) 144 hours post hatching stage (Fig. 22)

Gape of mouth narrow, head stubbed in front, 4-5 candal rays noticed for the first time in lower half of caudal fin. Bulging eyes shifted little dorsal with a flat inter-orbital bridge. Head and body long and elongated.

Melanophores size decrease over the previous stage. Predorsal melanophores arranged in three rows, whereas those at the base of dorsal fin fold are arranged in a row on either side. Internal melanophores visible on auditory vesicle, ventrally melanophores increase on throat. Melanophores appear for the first time in caudal fin. Few medio-ventral melanophores appear for the first time, scattered below the heart over the surface of alimentary canal.

(viii) 168 hours post hatching stage (Fig. 23)

Preorbital head (Snout) flattened, body slightly reduced in size 3-4 rays appear in pectoral fin. Myotome body wall opaque, visceral body wall transluscent.

Body pigments are of variable shapes & sizes. Small dotted melanophores more in number than stellate melanophores, occupying mostly snout, dorsal surface of body, lateral-line, ventro-visceral line, lower half of swim-bladder, lateral surface of alimentary canal below swim bladder and ventral surface of head (throat). Dorsal melanophores (pre-dorsals & those at the base of dorsal finfold) arranged in a row along each body side. Melanophores less in number on head & more in number on lateral-line. Opercular & auditory vesicle melanophores are small dots, less numerous and melanophores on lateral surface of alimentary canal are more numerous than the previous stage. Ventro-visceral melanophores increase in number, merging anteriorly with swim bladder & posteriorly with those on lower caudal lobe. Pectoral fin pigmented at base, caudal fin melanophores half than previous stage; ventrally lower jaw bears few black dots. Melanophores between dorso–lateral & lateral-line are faint & stellate, medio-ventral melanophores increase in number. Heart melanophores retain the same pattern but increase in number.

(ix) 192 hours post hatching stage (Fig. 24)

7 rays noticeable in denticulate caudal fin. Alimentary canal distinctly coiled. Gills now visible through transparent operculum. Pigments appear on caudal rays. Lateral line, pre-anal, ventro-visceral, throat, heart and medioventral melanophores show numerical decrease whereas those on head (including operculum) show numerical increase. Melanophores distinctly stellate on dorsal surface of head, auditory-vesicle, swim bladder, lateral surface of alimentary canal & some of dorso-laterals; the rest are small & dotted.

(x) 216 hours post hatching stage (Fig. 25)

Mouth anterior, imbibing; lower jaw longer than upper. Massive & bulging eyes occupy about half of total height of head, 4-6 pectoral rays formed, tail definitely heterocercal; caudal fin furcate with 13-14 rays. 2 additional myotomes to the pre-anal part by corresponding reduction to post-anal part. (Thus 25 pre anal, 10 post anals).

General pigmentation pattern continued with melanophores of variable shapes and sizes. Head melanophores stellate, more prominent than the rest of body, without any numerical variations; Snout densely pigmented, upper jaw melanophores reaching maxilla. Melanophores fine stellate on operculum, auditory vesicle, lateral side of coiled portion of alimentary canal and pre-anal ventro-visceral region but dotted on post-anal ventro-visceral region. Nasal sacs pigmented, few melanophores appear around eye orbit, swim bladder melanophores distinct. Upper half of myotomes well marked from their horizontal myoseptum upto dorsal edge through stellate melanophores. Heart lined by thick pigmented lining (thus melanophores not counted); internal melanophores in lower jaw also noticed.

(xi) 240 hours post-hatching stage (Fig. 26)

Dorsal fin indicated, larval finfold much reduced. Eyes dorsal, pre-orbital head long and narrow; opercular bones formed. Vertebral column & associated myotomes flexed in the region of swim-bladder.

General body pigmentation retain same pattern but number slightly decrease on head, swim bladder, caudal peduncle & caudal fin but increase on heart and lateral sides of alimentary canal.

(xii) 264 hours post hatching stage (Fig. 27)

5 rays indicated for the first time in dorsal fin which is still continuous with finfold; caudal fin forked, upper lobe larger than the lower with 19 branched rays formed (adult form). Head and body broader than the previous stage. Increase in pre-anal and post-anal part by two myotomes each (27 pre-anal and 12 post anals).

Head melanophores prominent & stellate, occupy the surface over brain lobes. Melanophores stellate on operculum (even extend near lower posterior aspect of orbit), caudal peduncle & lateral-line (starting from theanterior end of swim bladder upto caudal peduncle). Predorsal melanophores arranged in three rows in some specimens (one row on mid-dorsal line and one row on either side of mid-dorsal row), a row on either side at the base of dorsal finfold. Melanophores between dorso-lateral & lateral-line are arranged in two rows. Preanal vento-visceral melanophores stellate, post-anal ventro-viscerals dotted, scattered along the entire length of alimentary canal (from posterior end of heart upto vent).

(xiii) 288 hours post hatching stage (Fig. 28)

One more ray added to dorsal fin (d); 6 hypurals distinct in the caudal part of notochord, caudal fin forked with equal caudal lobes. Snout obtuse.

A general decrease in overall pigmentation of variable shapes & size. Few distinct melanophores on operculum form an arc. Pre-dorsal melanophores in three rows (a mid dorsal row of stellate and prominent melanophores and a row of small faint dots on either side of mid dorsal row). Melanophores small and less numerous at the base of dorsal finfold, a few scattered between it & lateral-line. Auditory-vesicle melanophores becoming indistinct.

(xiv) 312 hours post hatching stage (Fig. 29)

Dorsal fin with 8 rays, anal fin demarcated. Operculum distinct with opercular bones. Dorsal fin-fold still persistent.

Melanophores slightly increase in number on head, swim bladder, caudal fin and medio-ventral region but decrease on operculum; otherwise general number and pattern remains the same.

(xv) 408 hours post hatching stage (Fig. 30)

3-4 anal rays appear for the first time, one more ray added to dorsal fin (9), post-anal larval finfold still continuous with the anal fin.

Melanophores from hind brain region extend down & merge with lateral-line melanophores. Melanophores increase in number on operculum (still retaining opercular arc) and above lateral-line. Caudal rays highly pigmented on their sides, upturned caudal portion of vertebral column pigmented, a few also noticed below lateral-line. Dorsal lateral-line and ventro-visceral melanophores merge at the caudal peduncle.

No specific variations observed in larvae at 432 hours post hatching stage.

(xvi) 456 hours post hatching stage (Fig. 31)

Appearance of ventral fin almost in the middle of body, first spine in dorsal fin (1/9) and a pair of maxillary barbels for the first time; anal fin with 5 rays, ventral finfold still persists. Swim-bladder, auditory-vesicle, gills & alimentary canal becoming indistinct.

Body highly pigmented, melanophores mostly stellate. Both jaws densely pigmented. Head highly pigmented with fine stellate melanophores, pigments many on pre-orbital head. Head and dorsal melanophores continuous. Pre-dorsal melanophores arranged in 6 rows (3 rows on either side of mid-dorsal line); dorsal melanophores scattered laterally downwards, mix with lateral-line melanophores and extend below lateral-line (in the lower half of body). Dorsal & caudal fins highly pigmented. Few swim bladder, auditory vesicle & ventro-visceral melanophores visible; ventrally throat melanophores distinct. Laterally, few faint, stellate melanophores grouped together on the belly (alimentary canal). Myotomes posterior to dorsal fin also bear pigments on their upper half. All the characters in the larvae at 480 hours post hatching stage remain unchanged.

(xvii) 504 hours post hatching stage (Fig. 32)

One more spine added to dorsal fin (2/9), pectoral with 8-10 rays, anal with 1/5, caudal deeply forked. Dorsal & post anal ventral finfold completely absorbed, pre-anal ventral finfold still persists. Swim-bladder bilobed. Pre-anal myotomes reduced to 16-18, post-anals remain unchanged as 12.

General pigments on the body indistinct, except for those on dorsal surface of head (only on the mid & hind brain region), very few small dots left on operculum. Snout melanophores disappear. Pre-dorsal melanophores indistinct, but very few, faint & stellate melanophores at the base of dorsal fin. Lateral-line melanophores very few. Rest all the pigments disappeared.

No specific variations observed in the larvae at 528 hours, 552 hours, 576 hours, 600 hours, 624 hours & 648 hours post-hatching stages.

(xviii) 672 hours post hatching stage (Fig. 33)

One more spine added to anal fin (2/5), 5 rays appear in the ventral fin for the first time. Mouth narrow, anterior, imbibing. Posterior half of body upturned, more elongated & tapering than previous stage. Throat bones formed.

Head and dorsal melanophores merge & extend over highly pigmented operculum. Internal pigmentation of head & body noticed, lower margin of eyes bear few pigments. Dorsal fin well pigmented, pectoral fin pigmented at the base, caudal peduncle well pigmented with stellate melanophores. Melanophores few and distinct on swim-bladder and post-anal ventro-visceral regions, dense & stellate on the lateral surface of posterior half of alimentary canal upto vent; few and stellate on the medio-ventral region scattered along the entire length of alimentary canal and indistinct ventrally on throat. Heart region demarcated by a pigmented area.

No specific change in the larvae at 696 hours & 720 hours post hatching stages.

(xix) 744 hours post hatching stage (Fig. 34)

Anal fin now 2/6, ventral fin distinct with 6 rays. A pair of rostral barbels appear for the first time. Larvae decrease in total length.

General pigmentation same with stellate melanophores over entire body. Pre-orbital head densely pigmented with melanophores arranged in a semi-circle around the internal half of nasal sac. Few distinct melanophores visible on operculum & caudal peduncle.

No specific change in the larvae at 768 hours, 792 hours & 816 hours post hatching stages.

(xx) 840 hours post hatching stage (Fig. 35)

Dorsal fin 3/9, ventral fin with 8 rays, pectoral fin with 12 rays, ventral fin-fold now completely absorbed. Gape of mouth narrow; tail now homocercal, its fin furcate. Rostral & maxillary barbels much distinguished. Pre-anal vertebrae 14, post-anals 12. Scales appear for the first time in the middle of body, below the level of dorsal fin.

Eyes surrounded all round by pigmented arc, pigments extend below orbit. Nostrils well pigmented. Melanophores from hind brain region extend down, merge with those scattered on operculum; Opercular melanophores distinct & increase numerically. Melanophores appear for the first time on pectoral, ventral and anal fine. Swim-bladder & pre–anal ventro-visceral melanophores indistinct, post-anal ventro-viscerals persistent; medio-ventrals disappear. Ventrally jaw bones pigmented, with very few melanophores noticed on throat. Heart melanophores can still be made out.

The characters in the larvae at 864 hours, 888 hours, 912 hours and 936 hours post hatching stage remain the same.

(xxi) 960 hours post hatching stage (Fig. 36)

Two more rays added to pectoral fin (14), one ray to ventral fin (9). Scales extend along entire body. Mouth inferior, lower jaw now smaller than the upper.

Melanophores deeply lodged in skin rather than being superficial as in previous stage. Scales pigmented on their exposed margins. Another streak of melanophores appear on the posterior border of eye, behind the previous arc. Gills well pigmented as seen through operculum. Heart melanophores disappear, ventrally only lower jaw pigmented. Alimentary canal entirely covered with stellate melanophores.

No specific change in the larvae at 984 hours, 1008 hours, 1032 hours, 1056 hours & 1080 hours post hatching stages.

(xxii) 1104 hours post hatching stage (Fig. 37)

Head and trunk demarcated by a depression on nape, head quite massive & broad. Mouth clearly protrusible with thick lips. Eyes massive present in the anterior half of head. Scales with 4-5 circulii.

Nasal sac melanophores now arranged in a complete arc. Lower lip less pigmented than the upper. Dorsal & caudal rays densely pigmented at their bases. Lateral-line melanophores deep seated, preanal ventro-visceral melanophores becoming indistinct. Few small distinctly stellate melanophores form a small spot/patch on the caudal peduncle. No specific change in the larvae at 1128 hours, 1152 hours, 1176 hours, 1200 hours and 1224 hours post hatching stages.

(xxiii) 1248 hours post hatching stage (Fig. 38)

Dorsal fin complete (4/9), anal fin complete (3/5) and pectoral fin complete with 15 rays, acquiring the adult form. Labial fold of mouth continuous. Scales appear on the ventral side of body for the first time, 25-27 on the lateral-line, 9-11 pre-dorsals.

Nasal arc of melanophores compartmented. Head melanophores (mid-brain & hind brain) arrangement "heart-shaped". Snout melanophores extend posteriorly below ventral edge of eye orbit; scattered melanophores also persist behind the posterior & ventral edges of orbit. Jaws pigmented internally. Rostral barbels pigmented for the first time. Scales pigmented both on their surface & on their exposed margins. Caudal spot much more distinct than the previous stage.

No specific variations in the larvae at 1272 hours, 1296 hours, 1320 hours, 1344 hours, 1368 hours, 1392 hours, & 1416 hours post hatching stages.

(c) Fingerling Stage

(xxiv) 1440 hours post hatching stage (Fig. 39)

Mouth inferior, lower jaw smaller than the upper. Head much more elongated than previous stage; snout long, narrow, obtuse and tuberculated. Caudal deeply forked with equal lobes.

Pigmentation same to previous stage except for a prominent arc of melanophores seen through operculum. Head melanophores (heart-shaped) & caudal spot prominent. Head and trunk pigmented & clearly demarcated. Pectoral & ventral fin melanophores disappear, melanophores few on anal fin; vertebral column, dorsal and caudal fins well pigmented.

Discussion

The identification of larvae has been done on the basis of pigmentation which varies at different stages of development. Though its importance in larval identification was realized by Balinsky (1948) in case of cyprinid larvae, yet he has not been able to clarify the sequence of melanophore occurrence in the larvae.

Although dorsal melanophores appear in the larvae of *Tor tor* at 48 hours post hatching stage, yet they make their appearance in the newly hatched larvae of *Puntius conchonius* & at 48 hours post hatching stage of *Barilius bendelisis* (unpublished) while in *Aspidoparia morar* they appear in the larvae at 7 hours post hatching stage (Malhotra *et. al.* 1980). Similar occurrence of dorsal melanophores has also been reported by Desai (1973) in the larvae of *Tor tor* from Narmada water at 14.25 mm. stage but he has not mentioned the age of larvae and pattern of pigmentation.

The lateral-line melanophores appear in conjunction with head melanophores in the larvae of *Tor tor* at 72 hours post hatching stage while in *A. morar*, they appear at 304 hours post hatching stage (op. cit.), in the newly hatched larvae of *P. conchonius* & 144 hours post hatching stage of *B. bendelisis* (op. cit.) The appearance of ventro-visceral melanophores is observed at 48 hours post hatching stage of *Tor tor*, but they make their appearance in the newly hatched larvae & 48 hours old larvae of *P. conchonius* & *B. bendelisis* (op. cit.) respectively, and in 24 hours old larvae of *A. morar* (op. cit.). The medio-ventral melanophores do not appear until at 144 hours stage of *Tor tor* of present investigations & *B. bendelisis* (op.cit) but they appear very early at 48 hours stage in case of *P. conchonius* & *B. bendelisis* (op. cit).

Heart melanophores appear at 48 hours post hatching stage in case of *Tor tor* but they appear late at 72 hours stage in *P. conchonius* & *A. morar* & 120 hours stage in *B. bendelisis* (op. cit). The nasal sac melanophores form a semi-circle around internai half of nasal sacs of 744 hours stage in case of *Tor tor*, turning into a complete arc around the nasal sac at 1104 hours stage but this semicircle in case of *P. conchonius* & *B. bendelisis* is observed at 456 hours & 720 hours post hatching stages respectively forming a complete arc at 648 hours & 1008 hours stages respectively, thus forming a characteristic feature of these larvae helping in the determination of their age and identification.

The importance of caudal peduncle, gills, operculum, swim bladder & alimentary canal melanophores has not been realized by Balinsky (1948) but they assume great significance in larval aging and identification. Black patch at the caudal peduncle form a characteristic feature of most of carps, has been noticed to perfect in 1104 hours old larvae of *Tor tor* to disappear in adult; its occurrence though has also been reported by Desai (1973) in the same fish from Narmada river but in the larvae at 12.16 mm. length without mention of their age. Such a black patch appears in *P. conchonius* at 648 hours stage to persist as an adult character. Melanophores on gills appear for the first time in the larvae of *Tor tor* at 120 hours stage but in case of *B. bendelisis* & *A. morar*, they appear at a later stage of 216 hours & 304 hours respectively (op. cit). These patches are absent in *P. conchonius*. Opercular melanophores arrange themselves in the form of an arc on the operculum at 288 hours stage in *Tor tor* but in case of *P. conchonius*, a distinct row of melanophores is reported by Gupta (unpublished) at 288 hours stage. In case of *Tor tor*, swim bladder melanophores appear at 48 hours stage & occupy both upper & lower halves of swim bladder; the same stage is also characterized in case of *P. conchonius* in which they appear on the dorsal half of bladder, merge anteriorly with those at the base of pectoral fin & posteriorly with ventro-viscerals, and also in *A. morar* in which they appear on the dorsal half of bladder & a few appear between swim bladder & pectoral fin. Though *Tor tor* is characterized by the occurrence of melanophores on the lateral surface of alimentary canal at 120 hours stage, but *P. conchonius* (op. cit.) is characterized by their presence at 192 hours post hatching stage.

Head melanophores depict a typical "Heart Shaped" pattern in the larvae of *Tor tor* at their 1248 hours stage but such a pattern is not reported in other fish larvae. In *A. morar*, Malhotra *et. al* (1980) reported the simple occurrence of melanophores on brain region. In *B. bendelisis* & *P. conchonius* (op. cit.), they are lobate generally after the shape of underlying brain lobes. These head melanophores then merge with dorsal melanophores in the larvae of *B. bendelisis* & *P. conchonius* (op. cit.) at their 168 hours & 576 hours post hatching stages respectively. This character thus is an identifying feature for the larvae of *Tor tor* at their 96 hours post hatching stage. Yet another identifying feature of larvae of *Tor tor* is the merger of dorso-lateral, lateral-line & ventro-visceral melanophores at the caudal peduncle at their 672 hours post hatching stage. This character is not achieved in the larvae of *A. morar* (op. cit.) until as early as at 376 hours stage & as late as at 1128 hours stage in the larvae of *B. bendelisis* (op. cit). Thus the body melanophores of *Tor tor* investigated now agree with the general pattern given by Balinsky (1948), Kasansky (1915, 1925, 1928) and Schnakenbeck (1936, 1941) c.f. Balinsky (1948), yet in many features, the melanophore distribution at various developmental stages is unique.

Summary and Conclusions

The periodization in the larvae of *Tor tor* (Mahseer) inhabiting Jammu fresh-waters has been done on the basis of melanophore occurrence during different stages of development, which has formed the basis for the identification of its larvae. The pattern of pigmentation during different developmental stages has been discussed & described in line with those of Balinsky (1948).

On the basis of observations made during present investigations & from the study of earlier works, it has been concluded that the pattern of pigmentation is uniform within the same species but is different in different species. Thus it has become very clear that no two species agree in the sequential appearance of melanophores. The sequential appearance of melanophores is a useful instrument of aging and identification of fish larvae in the field.

Acknowledgements

The authors are highly grateful to CSIR and the University Grants Commission, New Delhi for financial assistance during the course of investigations.

Legend to Figures

Plate-I

Figs. 1-17: Fertilized egg, Cleavage, post cleavage and fry stages of *Tor tor*.

1: Fertilized egg.

2 A: 4-celled stage (polar view)

2 B: 4-celled stage (side view)

3: 16-celled stage.

4: 32-celled stage.

5: 64-celled stage.

6: 128-celled stage.

7: 256- celled stage.

8: Morula stage.

9: Gastrulation stage.

10: Yolk-Plug stage.

Post Cleavage Stages

11: Differentiation of cephalic and caudal ends.

12: Intiation of optic-vesicle and 6-12 myotomes stage.

13: Appearance of brain, auditory-vesicle rudiments and 14-16 myotomes stage.

14: 22-24 myotomes stage. Note the disappearance of Kupffer's vesicle.

15: Differentiation of brain, eyes, appearance of pectoral fin bud and 30-35 myotomes stage.

Fry Stages

16: Larvae just after hatching (A - Lateral view, B-Dorsal view, C-Ventral view)

(Scale = 1.00 mm.)

(GR-Germ ring; PVM-Perivitelline membrane; VM-Vitelline membrane; Y-Yolk; YP-Yolk-Plug; AV-Auditory vesicle; BR-Brain rudiments; FF-Finfold; KV-Kupffers vesicle; MT-Myotomes; NC-Notochord; OV-Optic vesicle; PB-Pectoral bud; V-Vent; YS-Yolk-sac; AT-Alimentary Canal; B-Brain; E-Eye; PF-Pectoral fin).

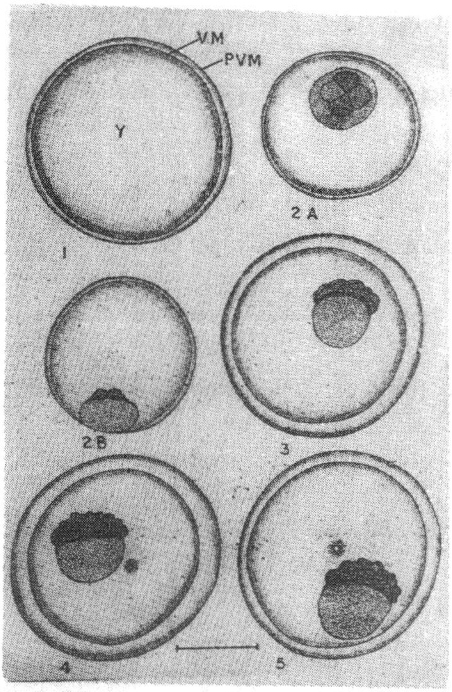
VM
PVM
Y
2 A
1
2 B
3
4
5

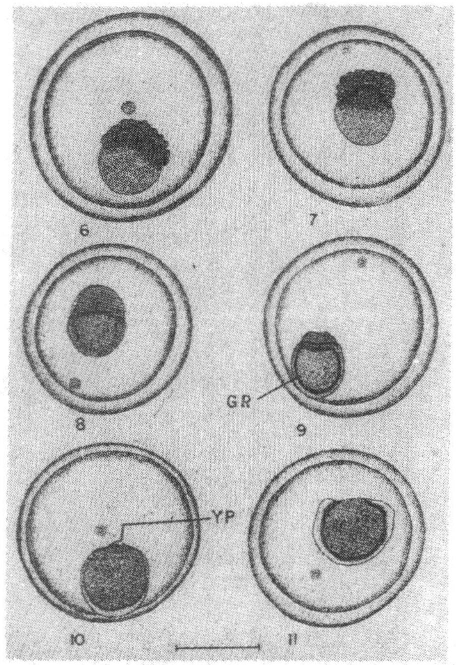
6
7
8
GR
9
YP
10
11

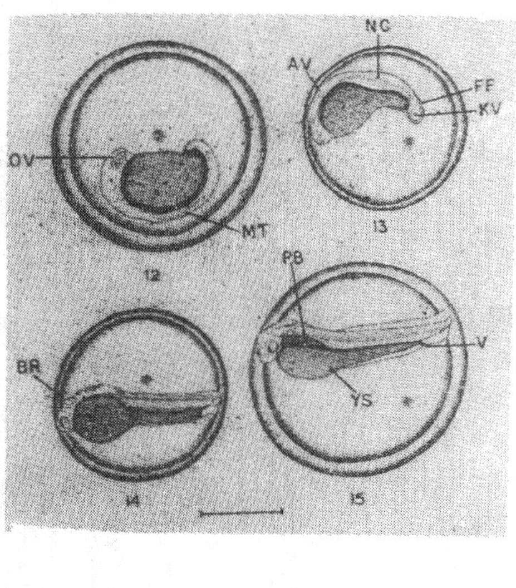
NC
AV
FF
KV
OV
MT
12
13
PB
BR
V
YS
14
15

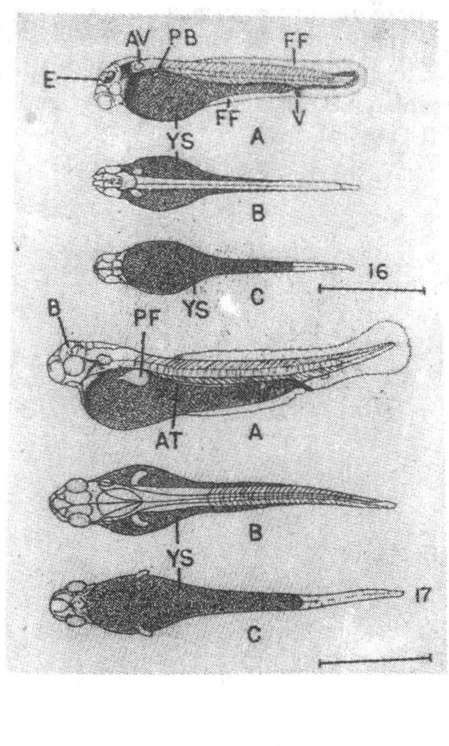
AV
PB
FF
E
FF
V
YS
A
B
C
16
YS
C
B
PF
A
AT
B
YS
17
C

Legend to Figures

Plate-II

Figs. 18-25: Fry and post-fry stages of *Tor tor.*

18-48: Hours post-hatching (A-lateral view, B-Dorsal view, C-ventral view).

19-72: hours post hatching (A-lateral view, B-Dorsal view, C-Ventral view).

20-96: hours post hatching-Post-fry stage (A-lateral view, B-Dorsal view, C-ventral view).

21: 120 hours post hatching (A-lateral view, B-Dorsal view, C-ventral view)

22: 144 hours post hatching (A-lateral view, B-Dorsal view, C-ventral view).

23: 168 hours post hatching (A-lateral view, B-Dorsal view, C-ventral view).

24: 192 hours post hatching (A-lateral view, B-Dorsal view, C-ventral view).

25: 216 hours post hatching (A -lateral view, B-Dorsal view, C-ventral view)

(Scale = 1.00 mm.)

(AC-Alimentary canal; AV-Aorta ventralis; BR-Branchiostegal rays; CP-Caudal peduncle; DM-Dorsal (dorso-lateral) melanophores; G-Gills; HM-head melanophores; HM—Heart melanophores; LLM-Lateral line melanophores; M-Mouth; NS-Nasal sacs; OC-Opercular cleft; OP-Operculum; SB-Swim bladder; SBM-Swim bladder melanophores; VVM-Ventro visceral menlanophores, YS-Yolk-sac; CR-Caudral rays; MVM-Medio-ventral melanophores; PR-Pectoral rays).

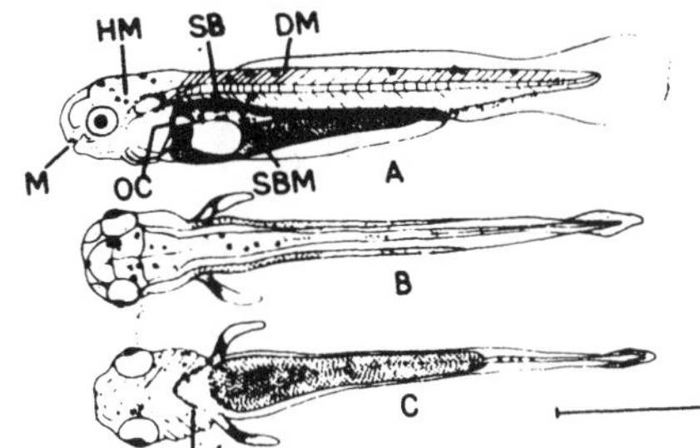

18

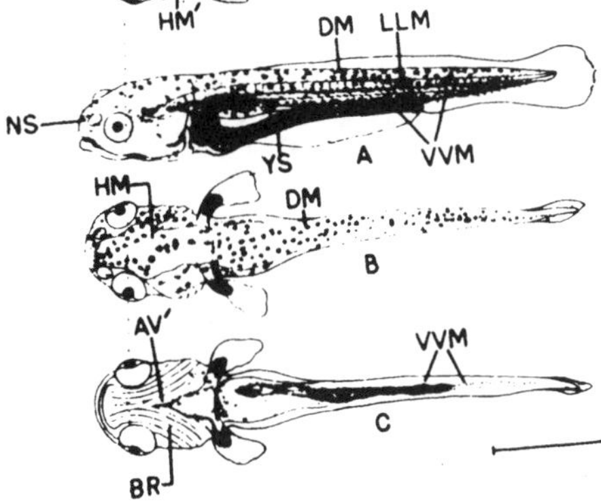

19

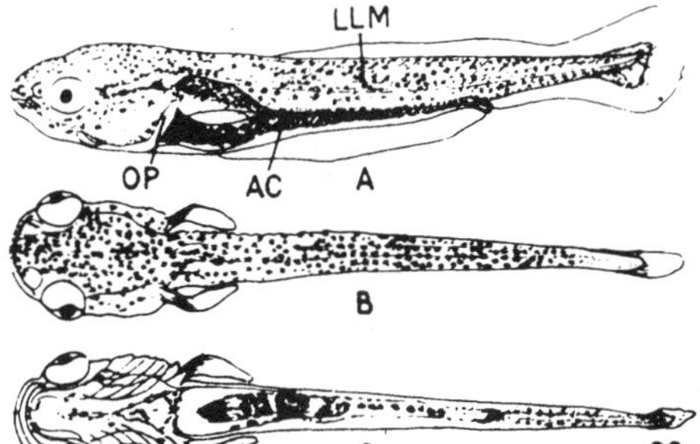

20

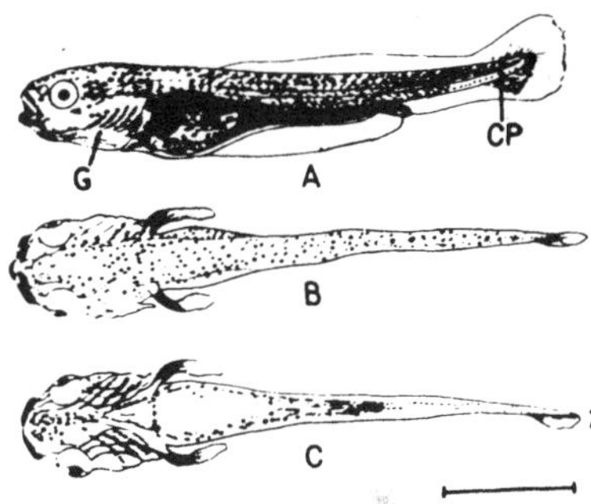

21

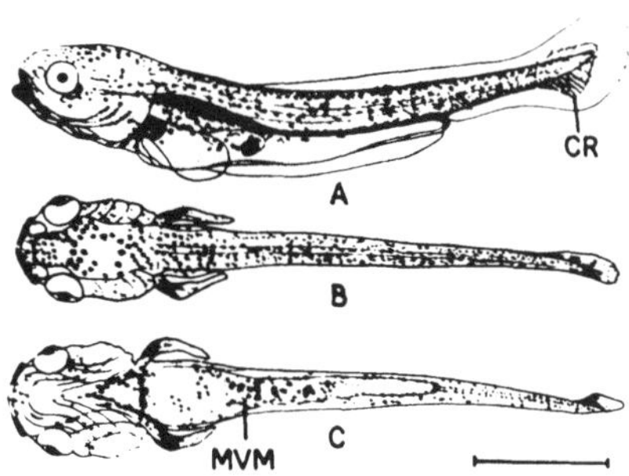

22

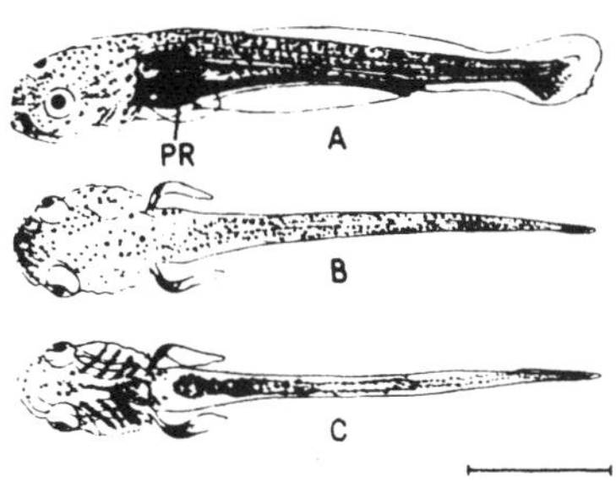

23

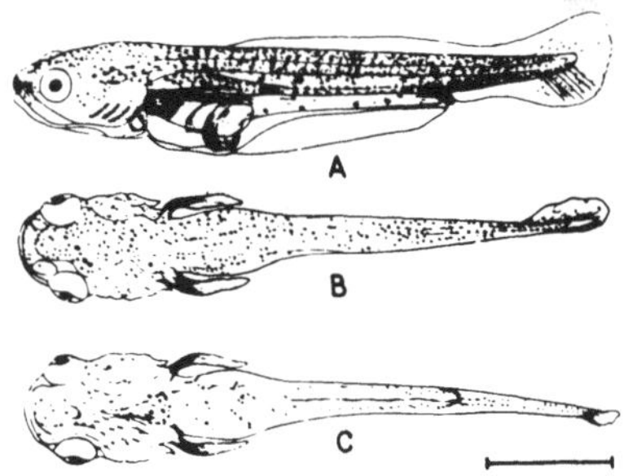

24

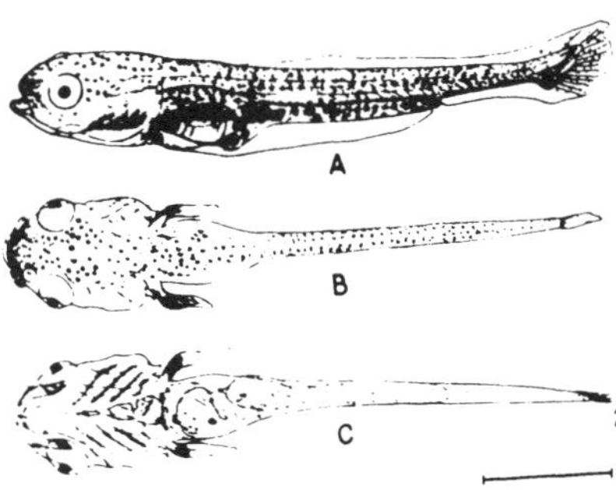

25

Legend to Figures

Plate-III

Figs. 26-30 : Post fry stages of *Tor tor*.

26-240 : hours post hatching (A-lateral view, B-Dorsal view, C-ventral view)

27-264 : hours post hatching (A-lateral view, B-Dorsal view, C-ventral view).

28-288 : hours post hatching (A-lateral view, B-Dorsal view, C-Ventral view).

29 -312 : hours post hatching (A-Lateral view, B-Dorsal view, C-Ventral view)

30-408 hours post hatching (A-Lateral view, B-Dorsal view, C-Ventral view).

(Scale = 1.00 mm.)

(AF-Anal fin; AR-Anal rays;

DF-Dorsal fin; DR-Dorsal rays; HY-Hypurals;

OPAM-Arc of melanophores on the operculum).

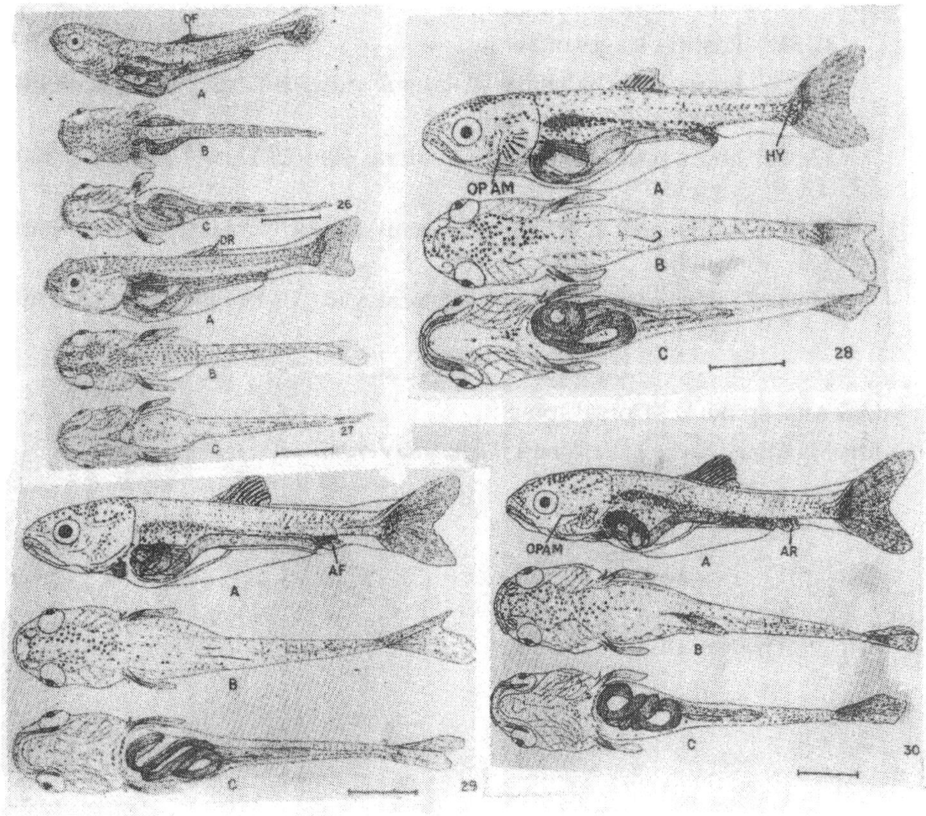
DF
A
B
C
26
DR
A
B
C
27
HY
OP AM
A
B
C
28
AF
A
B
C
29
OPAM
A
AR
B
C
30

Legend of Figures

Plate-IV

Figs. 31-34 : Post-fry stages of *Tor tor*.

31-456: hours post hatching (A-lateral view, B-Dorsal view, C-ventral view).

32-504: hours post - hatching (A-lateral view, B-Dorsal view, C-ventral view).

33-672: hours post hatching (A-lateral view, B-Dorsal view, C-ventral view).

34-744: hours post hatching (A-lateral view, B-Dorsal view, C-ventral view)

(Scale = 1.00 mm.)

(AS-Anal spine; DS-Dorsal spine;

MB-Maxillary barbels; VF-Ventral fin; VR-Ventral rays;

RB-Rostral barbels).

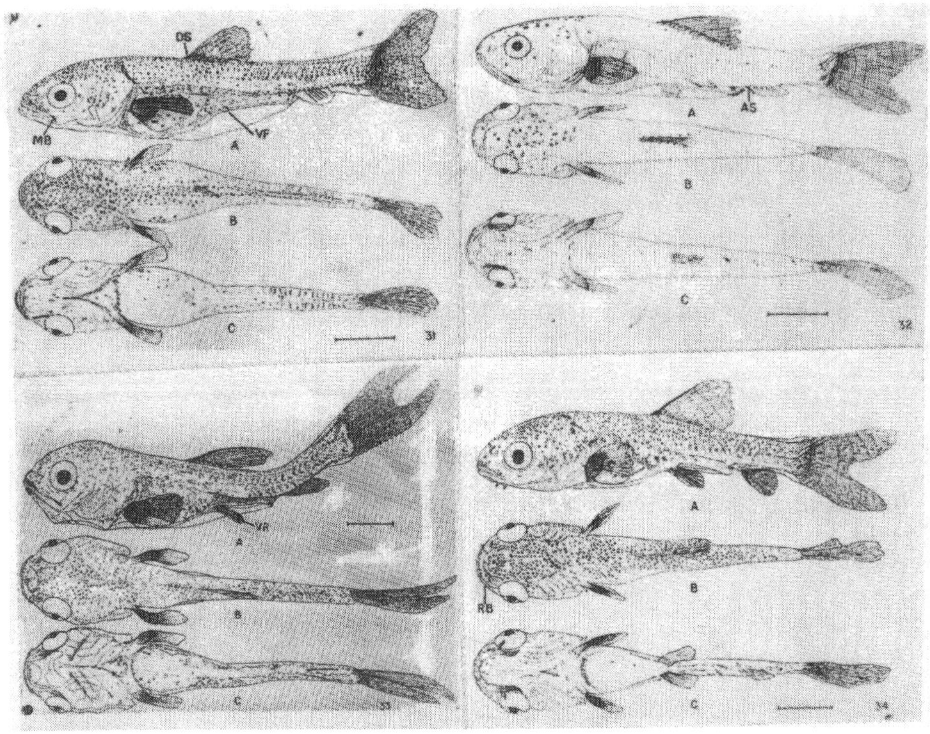
DS
MB
VF
A
B
C
31
A
AS
B
C
32
VR
A
B
C
A
B
RB
C

Legend to Figures

Plate-V

Figs. 35-39 : Post fry and fingerling stages of *Tor tor.*

35-840 : hours post hatching (A-Lateral view, B-Dorsal view, C-Ventral view)

36-960 : hours post hatching (A-Lateral view, B-Dorsal view, C-Ventral view).

37-1104 : hours post hatching (A-Lateral view, B-Dorsal view, C-Ventral view).

38 -1248 : hours post hatching (A-Lateral view, B-Dorsal view, C-Ventral view)

39-1440 : hours post hatching-Fingerling.

(A-Lateral view, B-Dorsal view, C-Ventral view).

(Scale = 215 mm.)

(CS-Caudal spot: NSAM-Arc of melanophores around and nasal sacs; SC-Scales.)

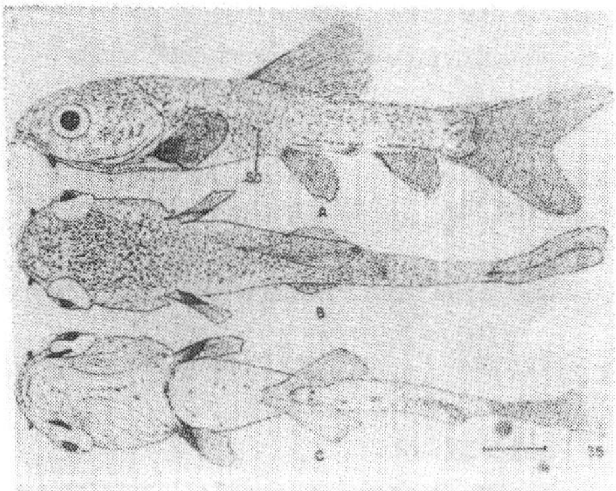
SC
A
B
C
35

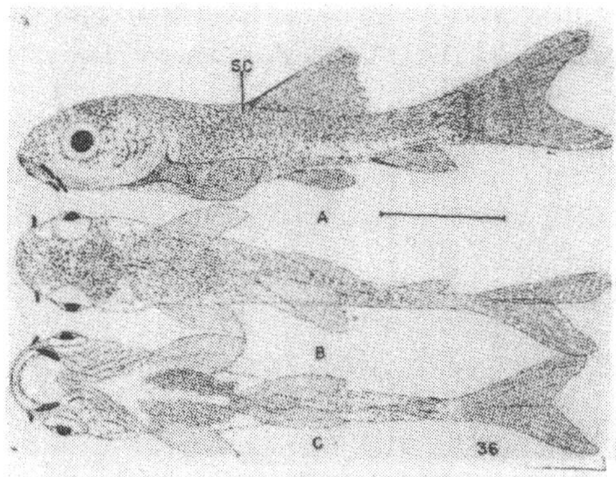
SC
A
B
C
36

CS
A
B
C
37

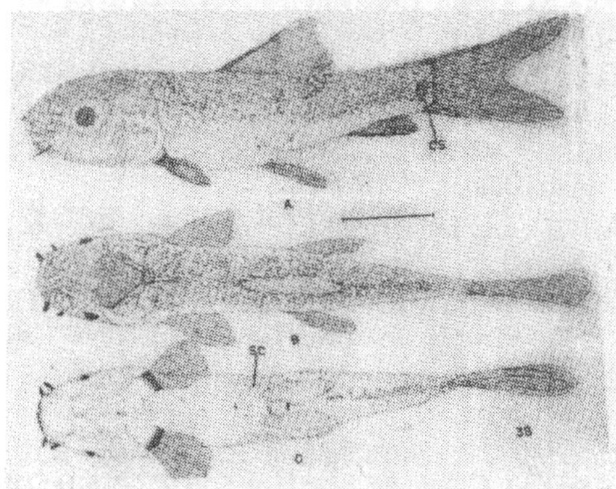
CS
A
B
SC
C

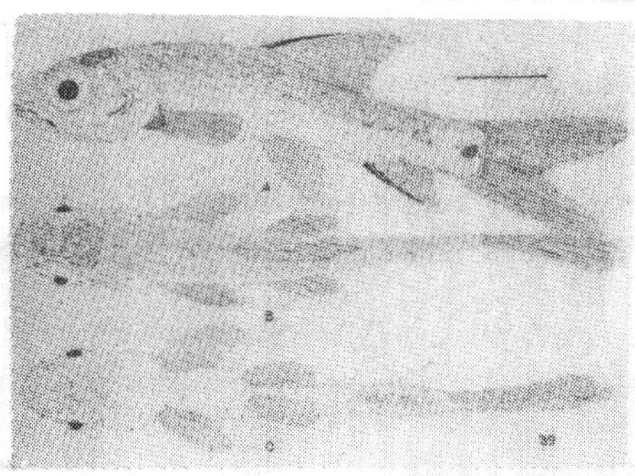

References

Ahmed Nazir (1948). On the spawning habits & early development of the copper Mahseer, *Barbus (Lissochilus) hexagonolepis* McClleland. Proc. Natn. Inst. Sci. India: 14, 21-28.

Balinsky, B.I. (1948). On the development of specific characters in cyprinid fishes. Proc. Zool. Soc. Lond. 118 (2) : 335-344.

Beavan, R. (1877). Handbook of freshwater fishes of India, London.

Codrington, K. de B. (1946). Notes on the Indian Mahseers. J. Bombay, Nat. Hist. Soc. XLVI, 2, 336-344.

David, A (1953). Notes on the bionomics & some early stages of Mahanadi Mahseer, J. Asiat. Soc. Calcutta. 19 (2) 197-209.

Desai, V.R. (1972). Notes on the early larval stages of *Tor putitora* (Hamilton) from Narmada River. J. Zool. Soc. India; 24 (1), 147-151.

—(1973). Studies on Fishery & Biology of *Tor tor* (Hamilton) from River Narmada. 2 Maturity, fecundity & larval development. Proc. Nat. Acad. Sci. India. Vol. 39, Part, B. No: 2, 228-248.

Dunsford, H.S. (1911). The report on Fish & Fisheries of Punjab Lahore.

Hora, S.L. (1940). The Game fishes of India (ix). The Mahseer or the large scaled Barbels of India (2) . The Tor Mahseer, *Barbus (Tor) tor* (Hamilton). J. Bombay Nat. Hist. Soc; 41 (3), 518-525.

Karamchandani, S.J., Desai, V.R; Pisolkar, M.D. & Bhatnagar, G.K. (1967). Biological investigations on the fish & fisheries of Narmada River (1958-1966). Bull-Cent. Inld. Fish Res-Inst. 10, 1-39.

Kasansky, V.I. (1915). Sur le development et al syste´matique de larves des carpiens. Trav. Lab. Ichthyol. Astrakhan, 3, No: 7, 1-23.

—(1925). Studies on the Morphology & Biology of Larvae of lower volga fishes. Trab. Lab. Ichthyol. Astrakhan, 5, No: 3, 1-109.

—(1928). To the Morphology & Systematisation of larva-stages of Cyprinidae Fishes of the type vobla *(Rutilus rutilus caspicus)*. Trav. Lab. Ichthyol Astrakhan, 6, No. 3, 1-27.

Khan, H. (1939). Study of the sex organs of Mahseer (*Barbus Tor* H.B). J. Bombay Nat. Hist. Soc, 41, 653-656.

Kulkarni, C.V. (1971). Spawning habits, eggs & early development of Deccan mahseer, *Tor Khudree* (sykes). J. Bombay. Nat. Hist. Soc; 67 (3), 510-521.

Mc Donald, A. St. J. (1948). Circumventing the mahseer & other sporting fish in India & Burma (Bombay).

Malhotra, Y.R., M.K. Jyoti & Subash C. Gupta (1980). Periodization of larvae of *Aspidoparia morar* (Bleeker) with special reference to melanophore occurrence & their role in identification of larvae. Matsya. No. 6, 58-70.

Nevill, C.A. (1915). The Breeding Habits of the Mahseer *(Barbus tor)*. J. Bombay Nat. Hist. Soc. XXIV.

Schnakenbeck, W. (1936). Üntersuchungen über die Entwicklung von Süsswassen fischen-I. Z. Fischen, 34, 647-81.

—(1914). Untersuchungen über die Entwicklung von Süsswassen fischen-II.Z. Fischen. 38. 269-321.

Spence, R and Prater, S.H. (1933). Game fishes of Bombay, the Deccan and the neighbouring Districts of the Bombay Presidency. J. Bomb Nat. Hist. Soc; Vol. 35; 29-66.

Thomas, H.S. (1897). The Rod in India, London.

Chapter 2

Seasonal Variations in Biochemical Composition in Some Freshwater Fishes, Part - III, *Channa Punctatus* (Bloch)

*B.N. Pandey, Anupa Sharan, Rumana Perween and M. Kumar**

Abstract

Estimates were made on the proximate body composition of *Channa punctatus* (Bloch) (based on 120 fishes, examined 10 every month for 12 months). The mean values of water, protein, lipid, carbohydrate and ash contents were 73.99%; 19.68%; 2.68%; 2.02%; and 1.61%; respectively. Marked seasonal variations in above mentioned constituents were observed. Water content was minimum in Post-spawning months (Sept./Oct.) and maximum in spawning months (July/Aug.). The per cent of protein varied from 18.96% in March to 20.14% in June, a secondary peak in protein content was also observed in January. Maximum (3.38%) and minimum (1.23 to 1.25%) levels of body lipid were recorded in Dec. and July/Aug. respectively. High values of carbohydrate were recorded from Nov. to May, while minimum values were recorded in June. Ash content showed small values during post-spawning and high values in pre-spawning month. The water content followed an inverse pattern to that of lipid, while protein showed a direct correlation with moisture content, significant statistical relationship was obtained between the changes in ambient water temperature and seasonal variation in different biochemical composition in this fish. The best fit was obtained by showing the relationship by two separate regression lines, the first between 18.5 to 29.0°C and the second between 29.0 - 33.4°. The significance of above studies has been discussed in this chapter.

Introduction

Biochemical studies of fish tissues are of considerable interest for their specificity in relation to the food values and for evaluating their physiological needs at different period of life. Increase in size may cause changes in the chemistry of the fish, but phenomena seen in larger specimens, sometimes result from the onset of sexual maturity and it is not always possible to distinguish between the two factors (Love, 1970). The metabolic processes increase considerably during pre-monsoon periods, when they prepare for spawning (Tandon & Joshi, 1974). Since the water, protein, fat and carbohydrate contents in the body of most of the fishes follow a definite seasonal cycle, the investigation on inter-relationship between above parameters and their correlation with ambient water temperature in different months of the year becomes imperative in the present study. This makes it possible to have a better understanding of their ecobiochemical adaptation. The criterion of protein deposit may be preferred over that of fat and carbohydrate as these two constituents

* P.G. Department of Zoology, Magadh University, Bodh Gaya - 824234.

are considered to be temporary stores of energy undergoing rapid changes. The total protein content seem to increase with initial growth of fish and also during different months of the year. Protein occupy the most important place in human nutrition hence it is essential to work out conditions when the fishes have highest amount of protein in the body of *Channa punctatus* and to correlate it with the increase in ambient water temperature.

Materials and Methods

Ten live specimens of *Channa punctatus* were collected from local ponds in the 2nd week of every months (From June 90 - May 91) and maintained in glass aquaria with a continuous flow of water for seven days for acclimatization.

After recording the weight, fishes were sacrificed and kept in hot air over (70°C) till constant weight of the fishes were obtained. The difference between the wet from dry weight gave an estimation of water content. Protein estimation was carried out by Duma's method (Bernard, 1954) by the following formula—Total $N_2 \times 6.25$ = Total protein content. Lipid was extracted by Soxhlet method using petroleum ether as solvent. To obtain Ash, the powder of body tissue was kept in muffle furnace between 500-600°C for about 3-4 hour. The difference between the dry weight and sum of total lipid, protein and ash was assumed to be carbohydrate content. Temperature of pond were recorded three times in a day every months for a period between June 85 —May 86.

The regression lines between water temperature and the above biochemical parameters were statistically calculated by least—square regression method.

Results

The data showing water temperature, moisture content, lipid, protein, carbohydrate and ash in different months of the year in *C. punctatus* are summerised in Table 1. The statistical relationship among various parameters are presented in Table 2.

Table-1: Data showing water temperature and biochemical composition in *Channa punctatus* (Bloch)

Months	*Water Temp. (C)*	*Water Content (%)*	*Lipid (%)*	*Protein (%)*	*Carbohydrate (%)*	*Ash (%)*
June	33.4	74.07	2.97	20.14	1.82	1.00
July	29.1	75.42	1.25	19.87	1.98	1.48
August	29.5	75.08	1.23	19.65	1.91	2.23
Sept.	29.25	72.88	3.21	19.54	1.92	2.45
October	28.10	72.82	2.71	19.36	1.98	3.13
November	24.0	73.21	2.21	19.83	2.05	2.70
December	19.40	73.97	3.38	19.83	2.06	0.84
Jan.	18.55	73.43	3.17	20.12	2.08	1.2
February	23.8	73.69	3.24	19.09	2.10	1.88
March	26.06	73.92	3.18	18.96	2.11	1.93
April	30.01	74.51	3.01	19.89	2.13	0.46
May	31.5	74.86	2.58	19.99	2.16	0.41
		73.99%	2.68%	19.68%	2.02%	1.64%

Table 2 : Equation showing the relationship between seasonal changes in ambient water temperature and variation in Body composition in *Channa punctatus* (Bloch)

S. No.	*Parameters*	*Equation* $Y = a + b.x$ *Range 18.5 – 29.0°C*	*Correlation Co-Efficient*	*Equation* $Y = a + b.x$ *Range 29.0 – 33.4°C*	*Correlation Co-Efficient*
1.	Moisture	$Y = 71.86 + 0.079. \times$	0.960	$Y = 67.71 + 0.223. \times$	0.965
2.	Lipid	$Y = 5.756 - 0.123. \times$	−0.875	$Y = 5.116 - 0.106. \times$	−0.975
3.	Protein	$Y = 21.52 - 0.084. \times$	−0.638	$Y = 16.47 + 0.111. \times$	0.946
4.	Carbohydrate	$Y = 2.28 - 0.009. \times$	−0.420	$Y = 2.906 - 0.029. \times$	−0.352
5.	Ash	$Y = 1.332 + 0.134. \times$	0.774	$Y = 7.770 - 0.217. \times$	−0. 280

The average value for different biochemical constituents were found to be as follows :

Moisture content	73.94%	(Range = 72.82 – 75.42%)
Lipid	2.68%	(Range = 1.23 – 3.38%)
Protein	19.68%	(Range = 18.96 – 20.14%)
Carbohydrate	2.02%	(Range = 1.23 – 3.38%)
Ash	1.64%	(Range = 0.41 – 3.13%)

The above constituents varied through out the year. Moisture content was minimum in October (72.86%) and maximum in June (75.42). It shows an inverse relationship with lipid. Lipid was minimum in summer months July/August (1.23/1.25%) and maximum in December (3.38%).

The protein content in fish body fluctuated through out the year from 18.96 to 20.14%. It shows two peak one in June (20.14%) and second one in the month of January (20.12%). It was recorded minimum in March (18.96%) and maximum in June (20.14%).

The lowest percentage of carbohydrate was observed in the month of June (1.82%) and highest in May (2.16%).

The Ash content was lowest in May (0.14%) after wards there was gradual increase, reaching maximum value in October (3.13%). Moisture, lipid and protein contents show good correlation with ambient water temperature.

Discussion

The study of biochemical composition of fish body is an important aspect because they constitute a rich source of high nutrients and calorific value.

The literature on biochemical composition in fishes in relation to water tempera-ture is relatively limited.

Atwater (1898) and Pearse (1925) determined the biochemical composition of fish body. Drummond (1918) observed the nutritive value of fish protein.

Parker and Venstone (1966) observed that the body composition of young Pink Salmon (*Oncorhynchus gorbuscha*, Walbaum) vary with temperature, size, age, diet and stage of development.

Brett et. al. (1969) found that body composition of fingerling sockeye salmon (*Oncorhynchus nerka*, Walbaum) was greatly affected by ration size and temperature.

Elliot (1975) studied the body composition of Brown trout (*Salmo trutta*) in relation to temperature and ration size and reported that water content increases and both protein and fat content decreases with the increasing temperature.

Perween (1988) studied the biochemical composition of *catla catla*.

Kamra (1988) worked on biochemical composition of *H. familes* and reported that with every 10°C rise in temperature the value of moisture, protein and ash increases, while value of lipid and carbohydrate decreases.

In the present study it indicates that breeding period of this fish seems to be associated with the hydration of tissues, while the dehydration of tissue starts after breeding (Post-Spawning). The water content shows an inverse relationship with lipid.

Temperature plays a great role in the variation of lipid and this finding is consistent with Prasad *et. al.* (1982) in *M. aculeatum*, Perween (1988) in *C. catla* and Kamara (1988) in *H. fossilis*.

Lipid increase in winter but decreases in summar (during breeding periods).

It has been observed that in *C. punctatun* two heads of protein was found one in June & other in January. The highest percentage of protein in June is marked with peak period in maturation; showing depletion in protein at the end of spawning. The present finding is consistent with the finding of Jafri Khawaja (1968), Savant & Bal (1969), Wagh (1986) in *Chela* & Perween (1988) in *Catla catla*. These investigator have reported that during spawning season fish flesh contain maximum amount of protein and minimum were also observed the spent and immature stages. The high values of protein during pre-spawning season may be due to the fundamental requirement of nitrogen for maturation. The low values of protein in post spawning season can be attributed to transfer of muscle protein towards energy required for spawning.

Fish contain small amount of glycogen in their body. In the present investigation the percentage of carbohydrate in *C. punctatus* was found maximum in Feb. to May i.e. during Prespawning period, it shows sharp decline during June to Sept. (Spawning & Post spawning period). This finding is consistent with Kamra (1988) & Perween (1988).

The ash content vary from 0.41% to 3.13%.The maximum per cent of ash content was recorded in the month of October and minimum in May.

Acknowledgements

The authors are thankful to Prof. D.P. Sinha Head, Dept of Zoology, Magadh University, Bodh-Gaya, 824234, for laboratory facilities and encouragement. First author (B.N.P.) is thankful to I.C.A.R. New Delhi, for financial help (Project No. 4 (II) 83-As (F)).

References

Atwater, W.O. (1898). The chemical composition and nutritive values of food-fishes and aquatic invertebrates. *Rep. U.S. Commiss. Fish* for 1888. pt. 16. (679-868).

Brett, J.R., Shelbourn J.E. and Shoop. C.T. (1969). Growth rate and body composition of fingerling sockeye Salman, *Oncorhychus nerka* in relation to temperature and ration size *J. Fish. Res. Ed. Canada. 26* : 2363-2394.

Drummond, J.C. (1918). The nutritive value of certain fish. *J. Physiol. 52*, 95-109.

Elliot, J.M. (1977) : Seasonal changes in the abundance and distribution of planktonic rotifers in Grasmese (English lake district). *Fresh. Wat. Bid. 7:* 147-166.

Jafri, A.K. and Khawaja, D.K. (1968). *Hydrobiologia*, 32. 245-256.

Kamra, A. (1988). The influence of formulated diets on the growth pattern in *H. fossilis* (Bloch), *Ph. D. Thesis*, Magadh University, Bodh Gaya.

Love, R.M. (1970). The chemical biology of fishes. A.P. London and New York, 547 C.

Parker, R.R. and Vanstone W.E. (1966). Changes in chemical composition of Central British Columbia Pink Salmon during early sea life. *J. Fish. Res. Ed. Can.* 23 1353-84, 88, 226.

Pearse, A.S. (1925). The chemical composition of certain freshwater Fishes. Ecology. 6(1). 7-16.

Perween, R. (1988). Studies on some aspects of dietary requirements and other factors effecting growth in *Catla catla* (Ham). *Ph.D. Thesis*. Magadh University.

Prasad, S. Pandey, B.N. and Sinha, D.P. (1982). Seasonal variations in fat and water content of the Indian fresh water mud eel *Macrognathus aculeatum*. *Proc. J. of Animal Nute* and *Phy*. pp. 117-126. Vol. 49. (3). pp. 117-172.

Savant, V.V. and Bal, D.V. (1969). *Proc. Ind. Acad. Sci. 708* (1969), 66.

Tandon, R.S. and Joshi, B.D. (1974). : Seasonal variations in blood glucose and serum calcium levels of fresh water fishes, I *Clarias batrachus, Zeit*, Tierphysiol, *Tierena, Futtermitt,* 33(2) : 108-112.

Wagh, S.B. (1986). Seasonal Variation in biochemical composition of fresh water cyprinid fish, *Chela clupoides* (Hamilton, Buchanan) Part - I. Muscle. Ind. *J. of Comp. Physiol. Vol. 4. No : 1*, 49-54.

Chapter 3

Some Relationships Between Size Structure and Fertility of Rotifer Populations

*S.S.S. Sarma**

Abstract

Size structure of six rotifer species viz., *Brachionus calyciflorus, B. diversicornis, B. rubens, Kellicottia longispina, Keratella quadrata* and *Pompholyx sulcata* was studied in relation to the fertility based on field collections. The smallest ovigerous female of a species was chosen as the basis for classifying a proportion of the population as the adults and the individuals smaller than this size were treated as non-adults.

The frequency distribution of individual sizes for most of the observed species followed normal distribution pattern. The ratio of minimum to maximum body size of the rotifer species when plotted against adult frequency, yielded a significant negative relation. When the egg ratio (eggs per adults) was plotted against the frequency of adult population, an inverse relation was observed. Using the size structure of living population of *Pompholyx sulcata*, dead individuals (empty loricas) were size-classified. The results indicated that heavy mortality in *P. sulcata* was experienced by adults rather than neonates.

Introduction

Body size measurements in rotifers reflect the patterns of development in individuals. Studies on the size distribution of individuals are important to understand the population structure of a species. It is generally known that rapidly growing populations have higher numbers of juveniles while those of senecent stage have predominantly adults (Colinvaux, 1973). Though population studies on rotifers have been well documented under both field and laboratory conditions, not much information is available on the size structure of rotifer populations (Lindstrom, 1983; Yufera *et. al.*, 1993).

Planktonic rotifers carry eggs which usually reflect the food availability in the ambient environment and the phase of the population growth (Edmondson, 1965). The individuals of population bearing eggs can be considered as adults, but for the rest, it is difficult to regard the non-ovigerous females as adults or young since at any time all size classes may be simultaneously present (Sarma & Rao, 1990). Consequently, in egg-ratio estimations the total number of eggs are expressed for the total females in the population (Urabe, 1992). It is possible that such egg ratios are more useful as a measure of birth rate and to predict oscillations in the populations

* Laboratory of Animal Ecology University of Ghent, K. L. Ledeganckstraat 35, B-9000, Ghent Belgium.

(Hofmann & Hofle, 1993), rather than true measure of fertility. In the present study. I report the size structure of selected rotifer species in relation to their fertility.

Material and Methods

Zooplankton samples were collected from local waterbodies (at 18–20°C) using plankton net (20 µm mesh) and immediately preserved in 5 per cent formalin. They were analyzed in the laboratory under stereo microscope. Rotifers were identified to species level using Koste (1978).

For quantitative study on the age structure of rotifer species, frequency of occurrence of different-sized individuals was determined. The samples were transferred to transparent shallow containers and separated exactly 100 first encountered females for each species. Lorica lengths were then measured (spines excluded) using a compound microscope under 15×40 magnification. For this, first each of the specimens was drawn using a camera lucida and the figures were calibrated using a stage micrometer. This helped in measuring the specimens more leisurely and in keeping a record of the observations for later use. Thus the measured rotifers included specimens of different sizes (and hence ages) including ovigerous ones. They were then clubbed into different size groups based on body length at intervals of 5 to 20 µm depending on species. Whenever ovigerous females were encountered, the size as well as the number of eggs were recorded. In all 6 species viz., *Brachionus calyciflorus, B. rubens, B. diversicornis, Kellicottia longispina, Keratella quadrata and Pompholyx sulcata* were analyzed.

Rotifers were recognized as mature by the presence of attached eggs. The smallest ovigerous female of a given species was considered as the basis for classifying a proportion of the population as the mature and the sizes of individuals smaller than this were treated as non-adults. Using the total number of eggs and adults, the egg ratio was calculated.

Results

The frequency distribution of individual sizes in a given rotifer population followed typical normal distribution pattern. *Brachionus calyciflorus* had the body size range of 180-350 µm. About 60 per cent of them were adults occurring within 300-350 µm size (*Fig. 1*). The size range of *B. diversicornis* was between 140-290 µm and about 70 per cent of them were adults with range of 240 to 280 µm lorica length (*Fig. 2*). In *B. rubens* frequency of 200-270 µm body-sized individuals was about 50 per cent of rest of other sizes. An individual in this size range was an adult (*Fig. 3*). In *Kellicottia longispina,* about 85 per cent of the constituting population had 115-154 µm lorica length (*Fig. 4*). This was the size range of ovigerous females. In *Keratella quadrata* ovigerous females with 140-169 µm body size constituted around 90 per cent of the population (*Fig. 5*). Similarly, in *Pompholyxsulcata* 80 per cent of the population had an adult body size range of 85-110 µm (*Fig. 6*).

When the egg ratio (total number of eggs divided by the total number of adults) was plotted against the frequency of adult population across different rotifer species, a statistically significant negative relation was obtained (Correlation

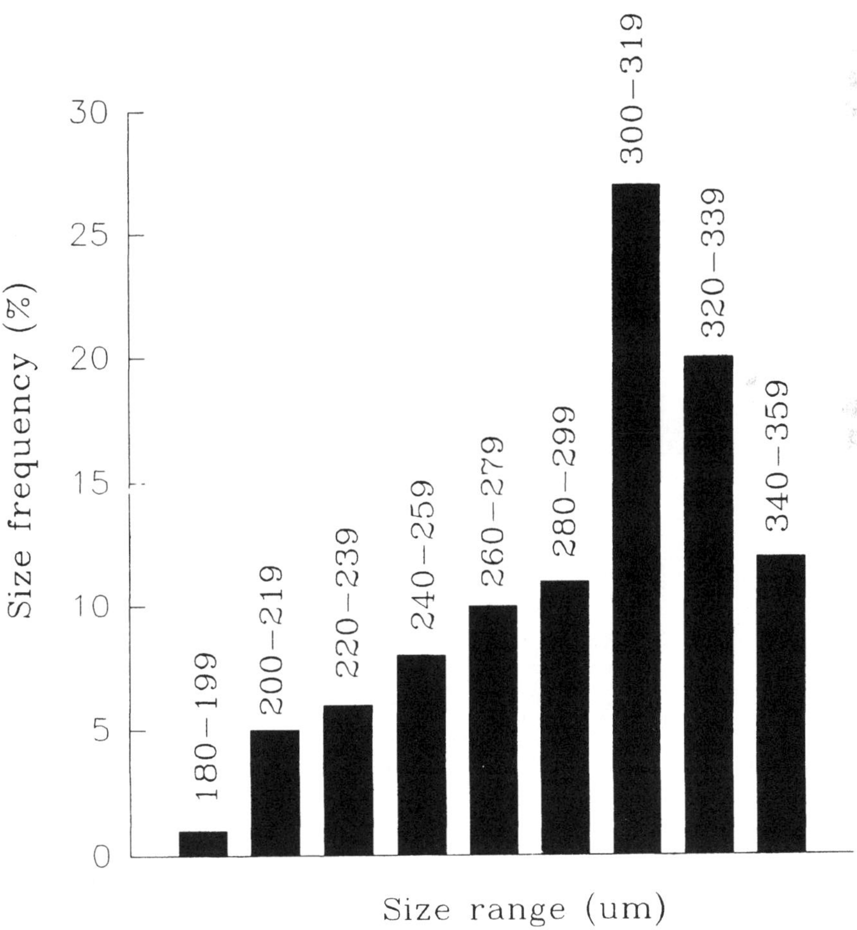

Fig. 1: Size Structure (frequency %) in the population of *Brachionus calyciflorus*

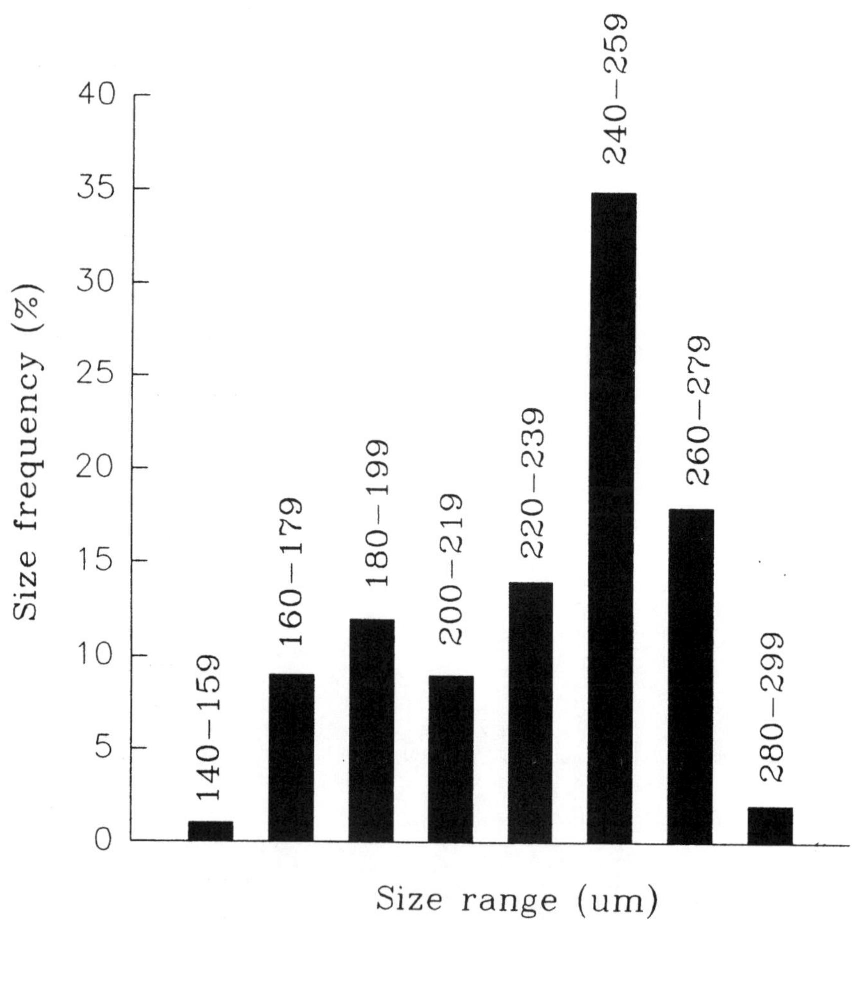

Fig. 2: Size structure (frequency %) in the population of *B. diversicornis*

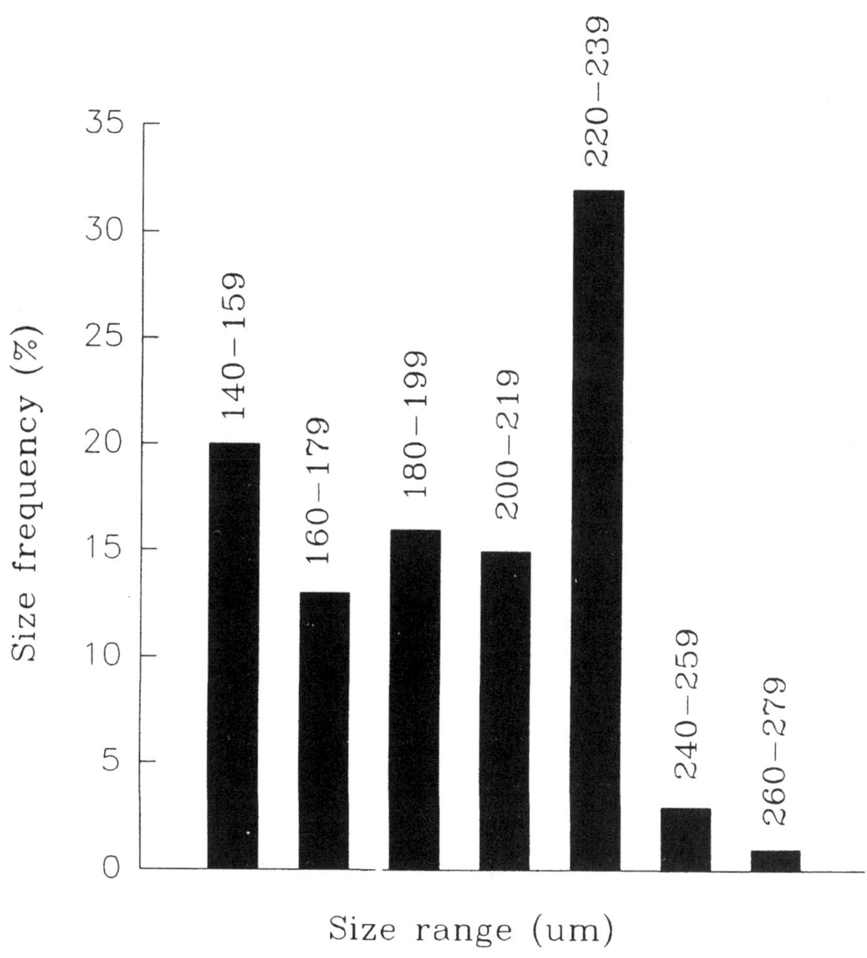

Fig. 3: Size structure (frequency %) in the population of *B. rubens*

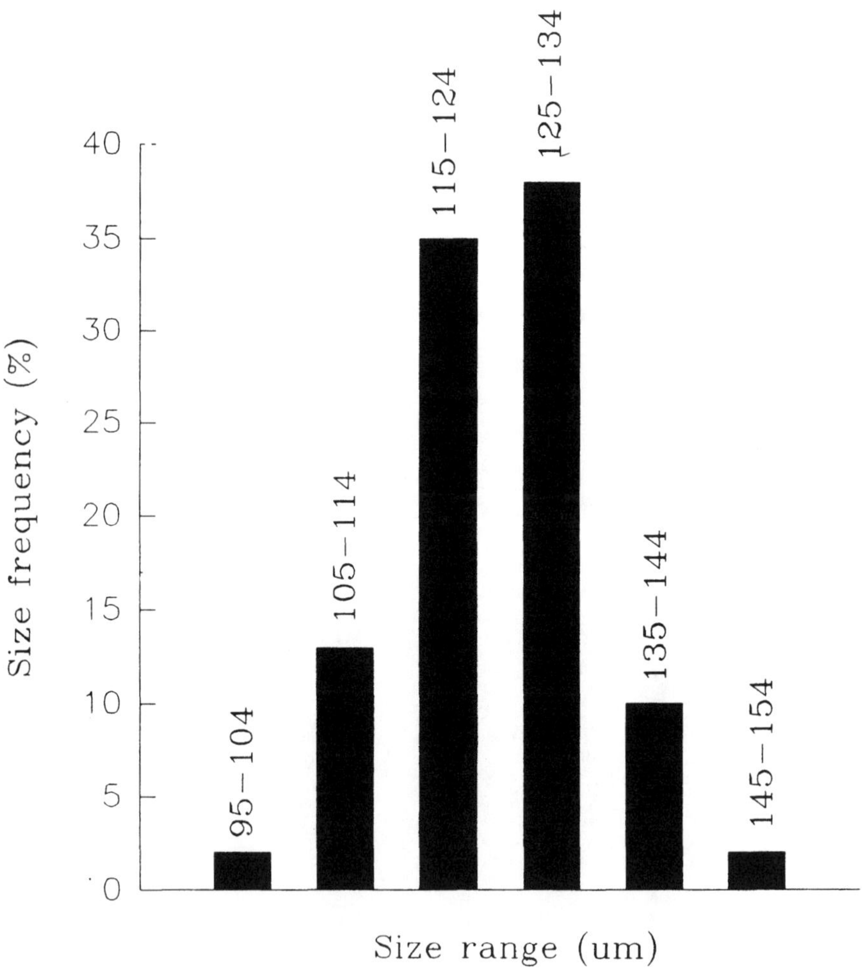

Fig. 4: Size structure (frequency %) in the population of *Kellicottia longispina*

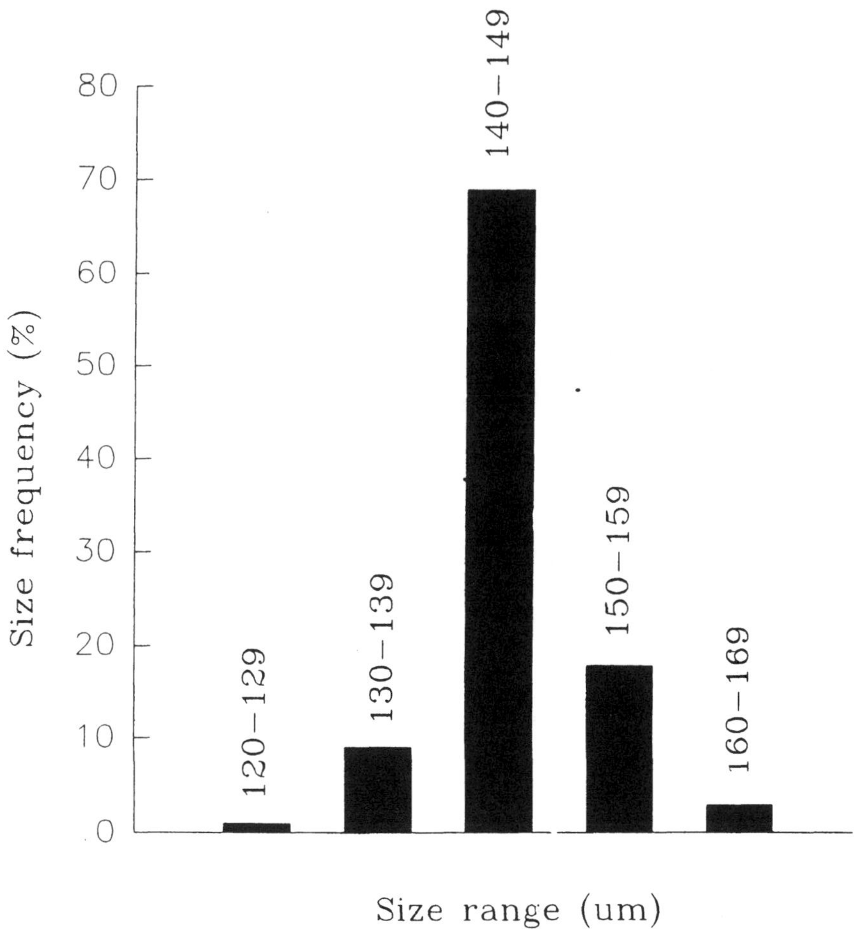

Fig. 5: Size structure (frequency %) in the population of *Keratella quadrata*

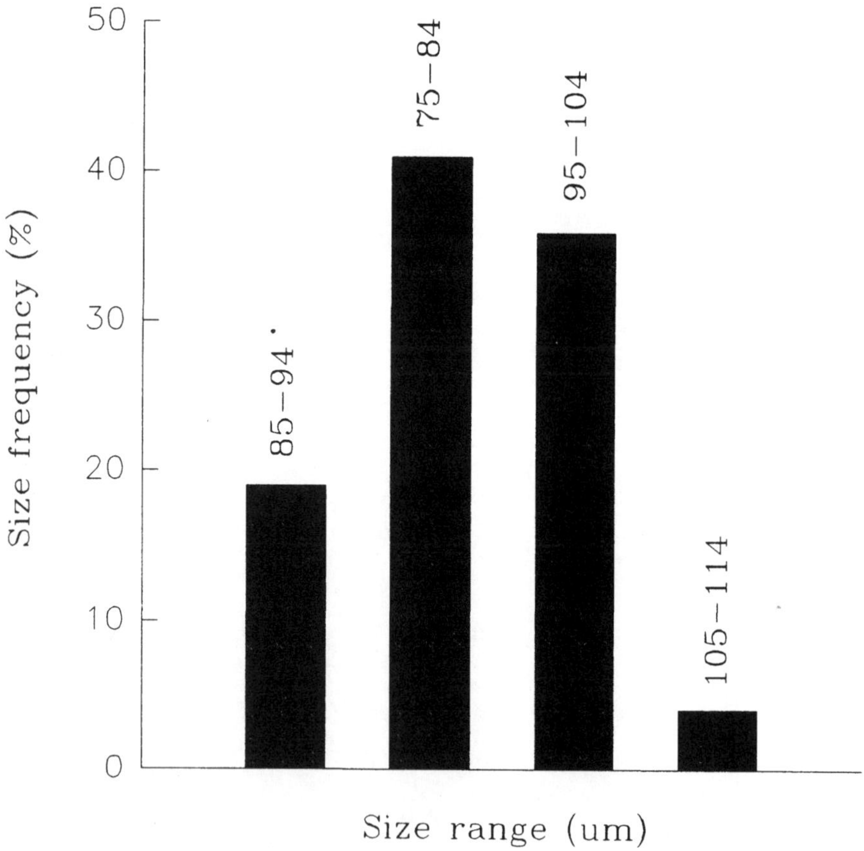

Fig. 6: Size structure (frequency %) in the population of *Pompholyx sulcata*

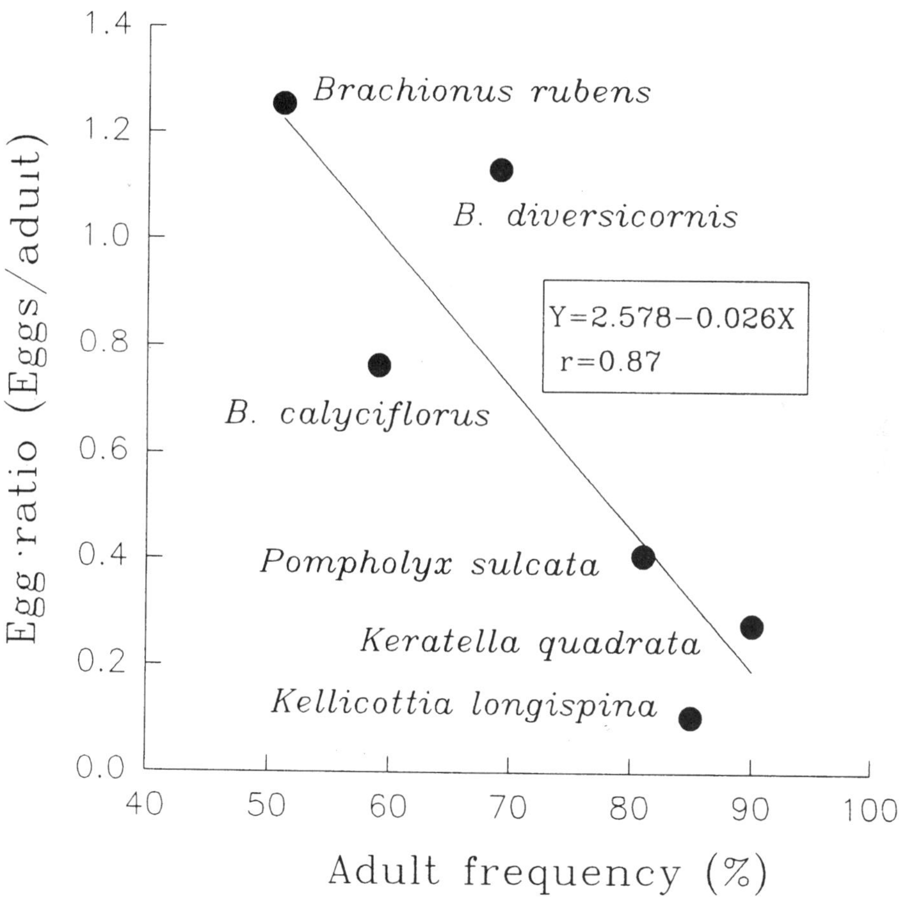

Fig. 7: Relation between adult frequency and fertility (eggs per adult) in the observed rotifer species

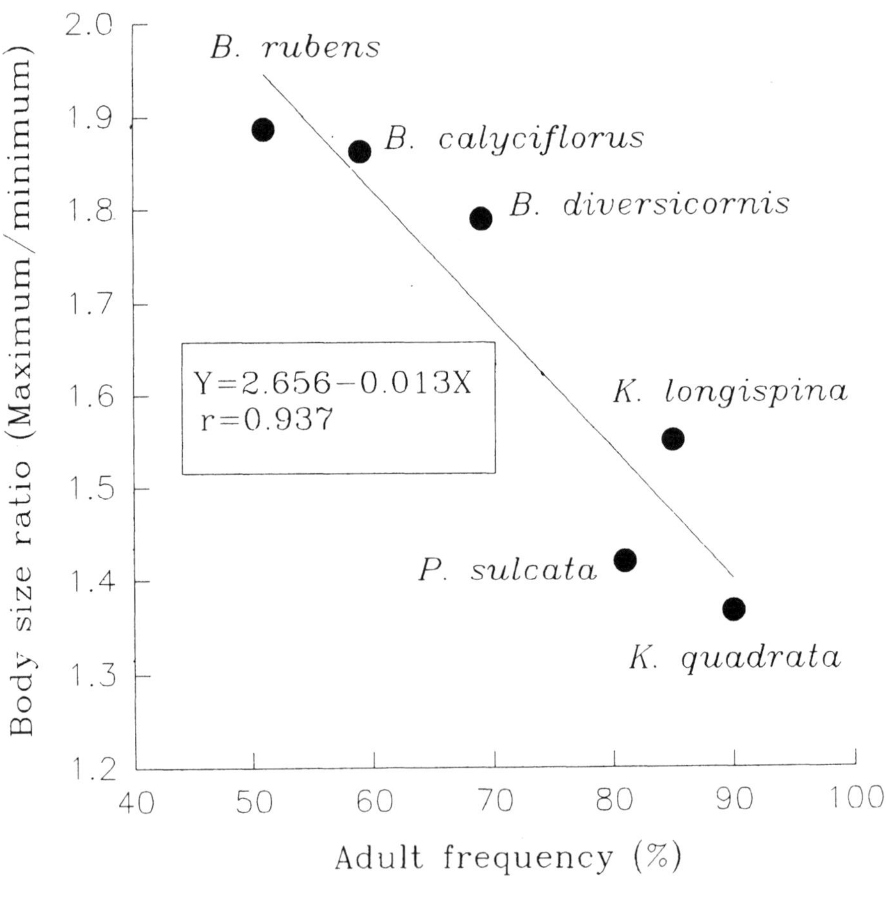

Fig. 8: Relation between adult frequency and body size ratio (length smallest/largest) in the rotifers

coefficient r = 0.87, $P < 0.05$, Fig. 7). Similarly, the ratio of minimum to maximum body size of the 6 rotifer species when plotted against adult frequency, a significant negative relation emerged (r = 0.94, $P < 0.05$, Fig. 8).

Discussion

Population structure of rotifers is usually expressed in terms of total number of rotifers living and the total number of eggs they carry. Only in rare cases, the size distribution of individuals in the populations was recorded (Lindstrom, 1983). Classifying adult rotifers in terms of body size (e.g. length) is more advantageous than biomass. For example, in *Euchlanis dilatata*, the neonates matured and attained constant length in less than 2 days but the biomass varied greatly throughout the lifespan (Ejsmont-Karabin et al., 1993). Rotifers having the largest proportion of egg bearing females usually reflect the growing phase of the population (Amren, 1964). The large number of eggs can be expected to result in the large numbers of young ones. Consequently, when the proportion of eggs to the adults is plotted as function of adult numbers, we can expect an inverse relation. This implies that when a population is in the senecent phase, its members are in no position to reproduce actively and hence there is a fall in the egg ratio. Alternatively, when a population is in the juvenile phase (log phase), its members reproduce at a faster rate resulting in the increased egg ratio (Edmondson, 1965). With the availability of large numbers of eggs, one may expect large number of neonates and pre-adults and hence the relative proportion of adults in the population falls. This trend is visible in Fig. 7. This has also another consequence: with the presence of large numbers of eggs, at any time in the population one may get just hatched or very young neonates with the same probability as one would get an adult rotifer. Thus, considering that the smallest sized individual from a population reflects the youngest one and the largest one representing the adult, their body size ratio would be an indicator of the population growth phase. This is evident in the Fig. 8 where an inverse relation existed among six different rotifer species. However, in a few cases, deviations from this pattern may be encountered. For example, larger neonates might be due to hatching of resting eggs rather than amictic eggs (R. Pourriot, personal communication). Similarly, populations with high proportion of mictic females may shift the size frequency, because of differences in the body sizes of amictic and mictic adults (Gilbert & Williamson, 1983).

Table 1 shows the fertility of the rotifer species as number of eggs per adults. As is evident from the *Table 1*, that *Brachionus* species had higher egg ratios than the other genera. Since the size of the amictic eggs of the observed rotifer species differs widely, it may be interesting to compare the body size - egg size relations (Pourriot & Rougier, 1991; Walz *et al.*, 1994). It may be noted that the egg ratio expressed here is different from that available in literature on rotifers. In the present study, the egg ratio was not the ratio of total eggs to total rotifers, but eggs to adults. This was possible here, because on the basis of body size a given individual was classified as adult or non-adult. This approach seems more reasonable than treating all females of a population as egg bearers. Thus, by eliminating neonates and pre-adults, the egg ratio recorded here may be closer to the reality.

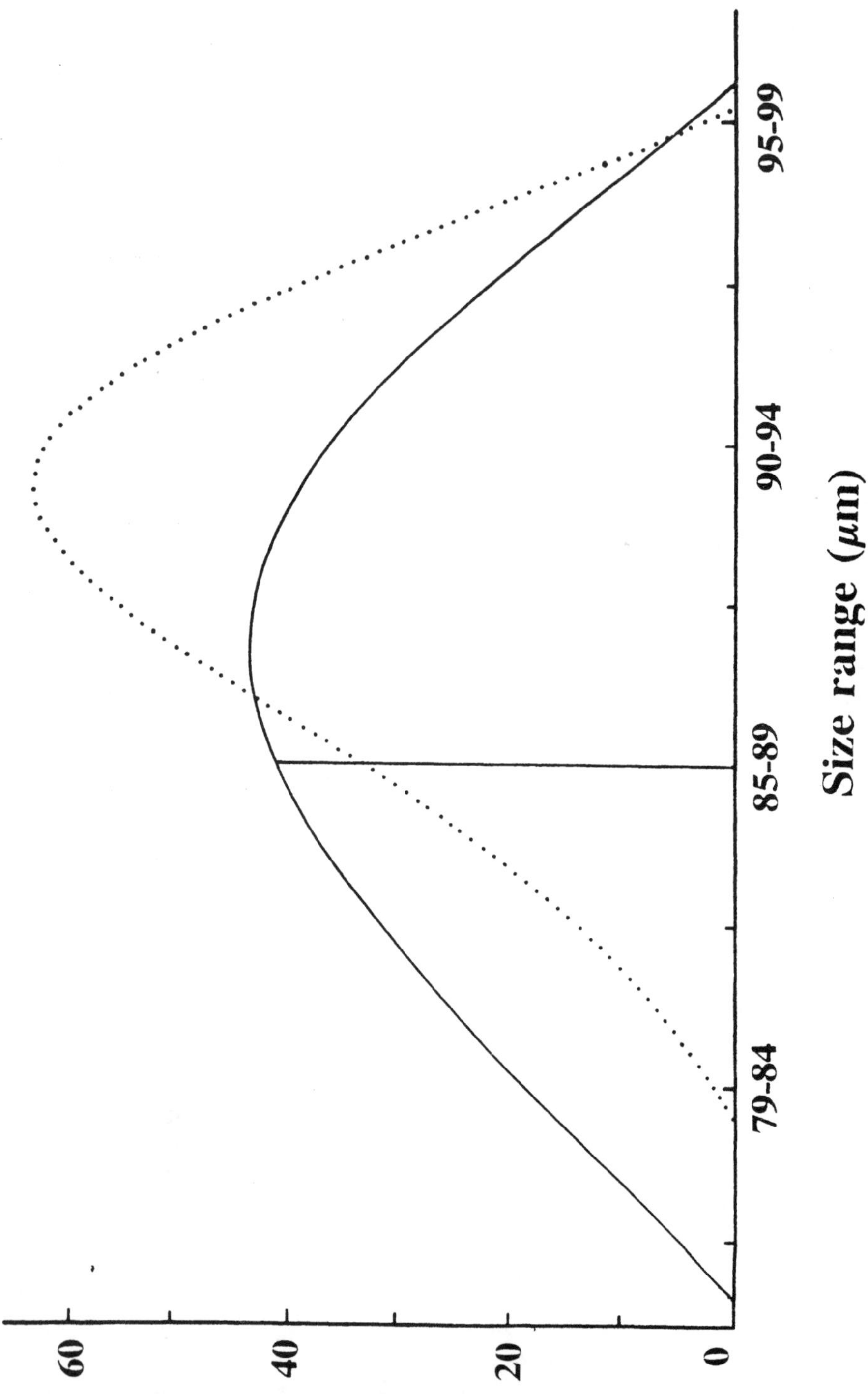

Fig. 9: Size structure of *Pompholyx sulcata* based on living (solid line) and empty loricas (dotted line). Vertical line represents the minimum body size range for ovigerous females

Table 1: Frequency of adults and the amictic eggs in six different rotifer species

Species	*Frequency of adults (for 100 females) (N_{ad})*	*Frequency of eggs (for 100 adults) (N_{eg})*	*Egg ratio (N_{eg}/N_{ad})*
Brachionus calyciflorus	59	45	0.763
B. diversicornis	69	78	1.130
B. rubens	51	64	1.255
Kellicottia longispina	85	9	0.106
Keratella quadrata	90	25	0.278
Pompholyx sulcata	81	33	0.407

The assumption based on the fact that rotifers have short juvenile and long reproductive phases and hence the population mainly consists of adults (see Urabe, 1992) is not always true. If this is considered, then much shorter hatching times of rotifer eggs (Herzig, 1983) should result in almost exclusively of neonates in rapidly reproducing populations. It is more applicable under conditions of ceased or suppressed egg production wherein the existing neonates mature and constitute the population wholly with adults. Alternatively, one may assume from the fact that planktonic rotifers continuously reproduce (and thus have eggs at any time) and therefore, all non-ovigerous individuals may be treated as non-adults. This too does not appear reasonable because we have substantial evidence to support that adult females in food-limited environments show decreased egg production (Hofmann & Hofle, 1993; Pozuelo & Lubian, 1993).

The size dependent population structure shown in this study has another implication: one may use this criterion for the dead rotifers too. In nature, mortality of strongly loricate species can be estimated based on the presence of empty loricas (Threlkeld, 1983). A quantification of such loricas can only give mortality as a single unit but not differentially. In other words, if the mortality was heavy in the adults (or in the young ones), it cannot be established. In the present study, an attempt was made to apply the size-dependent population structure to the patterns of mortality using empty loricas in *Pompholyx sulcata*. Thus when the frequencies of different-sized empty loricas were plotted against those of living ones, a clear shift of mortality towards larger sized individuals was recognized (Fig. 9). This implies that heavy mortality in *P. sulcata* was experienced by the adults rather than neonates.

Acknowledgements

I am grateful to Prof. H. J. Dumont (Belgium) for facilities and ICSC-World Laboratory (Switzerland) for financial support. Thanks are also due to Prof. R. Pourriot (France) and Dr. N. Walz (Germany) for comments on the manuscript.

References

Amren, H., 1964. Ecological studies of zooplankton populations in some ponds on Spitsbergen. Zool. Bidr. Upps. 36 : 161-191.

Colinvaux, P.A., 1973. Introduction to Ecology. John Wiley & Sons, New York, 621 pp.

Edmondson, W.T., 1965. Reproductive rate of planktonic rotifers as related to food and temperature in nature. Ecol. Monogr. 35 : 61-111.

Ejsmont-Karabin, J. K., Siewertesen & R.D. Gulati, 1993. Changes in size, biomass and production of *Euchlanis dilatata lucksiana* Hauer during its lifespan. Hydrobiologia 255/256 : 77-80.

Gilbert, J.J. & C.E. Williamson, 1993. Changes in growth and size of *Keratella cochlearis* (Gosse) in relation to some environmental factors in cultures. Hydrobiologia 104 : 325-328.

Herzig, A., 1983. Comparative studies on the relationship between temperature and duration of embryonic development of rotifers. Hydrobiologia 104 : 237-246.

Hofmann, W. & M. G. Hofle, 1993. Rotifer population dynamics in response to increased bacterial biomass and nutrients : a mesocosm experiment. Hydrobiologia 255/256 : 171-175.

Koste, W., 1978. Rotatoria. Gebruder Borntraeger, Stuttgart, 2 Vols.

Lindstrom, K., 1983. Changes in growth and size of *Keratella cochlearis* (Gosse) in relation to some environmental factors in cultures. Hydrobiologia 104 : 325-328.

Pourriot, R. & C. Rougier, 1991. Importance volumetrique des oeufs chez les rotiferes planctoniques. Ann. Limnol. 27: 15-24.

Pozuelo, M. & L.M. Lubian, 1993. Asexual and sexual reproduction in the rotifer *Brachionus plicatilis* cultured at different salinities. Hydrobiolgia 255/256 : 139-143.

Sarma, S.S.S. & T.R. Rao, 1990. Population dynamics of *Brachionus patulus* Muller (Rotifera) in relation to food and temperature. Proc. Indian Acad. Sci. B99: 335-343.

Threlkeld, S.T. 1983. Empty loricas and the dynamics of *Kellicottia longistpina* in a subalpine, oligotrophic lake. Hydrobiolgoia 104 : 367-372.

Urabe, J., 1992. Midsummer succession of rotifer plankton in a shallow eutrophic pond. J. Plank. Res. 14 : 851-866.

Walz, N., S.S.S. Sarma & U. Benker 1994. Egg size in relation to body size in rotifers : an indication of reproductive strategy? Paper presented at VII International Rotifer Symposium, June 6-11, Mikolajki (Poland). Full paper communicated to Hydrobiologia.

Yufera, M., E. Pascual & J. Guinea, 1993. Factors influencing the biomas of the rotifer *Brachionus plicatilis* in culture. Hydrobiologia 255/256: 159-164.

Chapter 4

Effect of Accessory Pneumectomy on Some Haematological Values of Air Breathing Fish *Clarias batrachus**

*B.D. Joshi***

Abstract

Changes in some haematological parameters for a period of 15 days were studied following accessory pneumectomy in the fish *Clarias batrachus*. The TEC, Hb, PCV and MCH values showed an initial fall post one day of accessory pneumectomy, attaining lowest values of various days and then tended to recover by 15th day. The TLC value showed a sudden rise, maximum being by 85.5% on 4th day. The changes in sham operated fishes were of lower volume than in the accessory pneumectomised fish, when compared to the control value. The DLC exhibited sharp rise in haemoblast, monocytes and basophils, a moderate rise in thrombocytes and eosinophils and a fall in lymphocytes. The number of erythroblast cells increased in circulation during the same period.

Introduction

The field of fish haematology is comparatively of recent interests to, bio-physiologists and the ichthyologists. As a result various aspects of experimental haematology, with a view to understand the role of various organ of fishes like, spleen, pituitary, liver, gills are being worked out in myriad ways. (Fange and Sjobeck 1975, Hickey, 1976, Hughes, Kikuchi and Wateri, 1981, Yamamoto, Itazawa and Kobayashi, 1984; and Kita and Itazawa 1989).

Few of the tropical freshwater fish species are equipped with a pair of accessory respiratory organ *Clasrias batrachus*, a siluroid cat fish is one among such species. This species is very hardy with high TEC and Hb contents in its blood and can survive for many hours, out side the water, in a damp environment. In three earlier publications author and his coworkers have already reported the effect of hypophysectomy, (Joshi and Sharma, 1981 & 1982) and partial hepatectomy (Joshi *et. al.* 1982), besides effects of asphyxiation (Joshi, 1978) on various blood parameters. And since, one of the major role of blood is with regard to the respiratory-physiology, among all vertebrates, therefore it attracted authors attention, as to what may happen to important haematological values of such fish, if its accessory respiratory

---

* Paper presented in 32nd I.U.P.C., held in Glasgow University Glasgow, Britain, in August 1993.

** Deptt. of Zoology, Gurukula Kangri University, Hardwar, 249404 (U.P.) India.

organ is removed. The paper presents results of one such experimental study on the fish *clarias batrachus.*

Materials & Methods

Stock of fish *clarias batrachus*, fully acclimatized to laboratory conditions was used in this experimental study. The fish were anaesthetized using 2 per cent paraldehyde solution. Each fish was measured and weighed at the beginning of experiment. To remove accessory respiratory organ. The bony part of head, just above the pectoral spine was cut partly on three sides, so as to expose the suprabranchial chamber. Then with the help of scissor, the attachment of accessory respiratory organ to the dorsolateral wall of suprabranchial chamber was cut and organ completely removed, using a forcep. Much bleeding occurred for about 1-2 minutes, which was cleaned using sterilized cotton swabs soaked with 0.2% fresh $Kmno_4$ solution. The suprabranchial chamber was, then cleaned and given cibazol puffing, over the scar. Then the incision was stitched from three sides and fish put back in an aquarium. The aquarium contained 250 litres of chlorinated tap water. The fish gained full consciousness within next 8-12 minutes post operation. During a 4 hrs period 30 specimens were thus operated on first day and 30 specimens on second day. Thus a total of 60 specimens were subjected to accessory pneumectomy (AP). Simultaneously 20 fishes were given a sham operation, where only the bone was cut to make a window to peep into suprabranchial chamber. It bled and wiped dry using a cotton swab and puffing of cibazol powder. And then without removing the accessory respiratory organ, the cut end was stitched given cibazol puffing and fish released into another aquarium. Twenty specimens of clarias were subjected to sham operation. Out of sixty pneumectomized specimens 26 specimens succumbed to pneumectomy while only four specimens succumbed to the sham operation. Observations were made on 1st, 2nd, 4th, 8th, 12th and 15th day post operations (AP). On each day of observation 6 specimens from AP group and 2 specimens from sham group were sacrificed. Estimations were made for TEC, TLC, Hb, PCV and MCH values, besides for differential blood cell counts comprising DLC and DEC (Differential erythrocyte count). Simultaneously a batch of 15 specimens of *clarias batrachus* was maintained as control group under laboratory conditions out of this control group fifteen specimens in batches of five were sacrificed on 1st, 4th and 15th day of observation. Values for each fish to each group were taken individually and then pooled for respective group, for the respective day. These studies were made in the month of October. During the whole period the fish groups were maintained under laboratory conditions as usual (vide Tandon and Joshi, 1973) and provided with minced goat liver on alternate days. The water of aquaria were given regular aeration with the help of electric aerators and chlorinated tapwater changed every day. The std. body length and body weight for all the fishes used were 19.7 ± 1.2 cm and 156.750 ± 9.5 gm, respectively. The methods of making haematological studies were the same as followed earlier by Tandon and Joshi (1973) vide Gradwohl (1943).

Table 1 : Changes in Some Haematological Values of *Clarias batrachus* following Accessory Pneumectomy Values are Mean ± S.E.

Status, Time & No. of observation	*TEC ($\times 10^6$/cmm)*	*TLC (/cmm)*	*Hb (gm%)*	*PCV (%)*	*MCH (pg)*
Control (10)	3.02 ± 0.87	12780.0 ± 642	13.2 ± 0.22	40.4 ± 0.06	43.7 ± 3.8
Sham (12)	2.86 ± 0.26	14705.0 ± 976	12.3 ± 1.07	38.8 ± 0.12	43.0 ± 2.9
Post Pneumectomy					
1st Day (5)	2.30 ± 0.07	17850.0 ± 540	9.6 ± 0.30	34.7 ± 0.22	41.7 ± 2.4
2nd Day (5)	2.12 ± 0.34	23708.0 ± 436	9.8 ± 0.32	32.4 ± 1.2	46.2 ± 2.7
4th Day (5)	1.66 ± 0.72	23708.0 ± 805	9.6 ± 0.28	28.8 ± 0.90	57.8 ± 5.1
8th Day (5)	1.98 ± 0.34	18735.0 ± 314	9.2 ± 0.60	32.6 ± 0.80	46.4 ± 2.9
12th Day (5)	2.46 ± 0.07	19615.0 ± 505	10.4 ± 0.30	35.8 ± 1.10	42.2 ± 2.0
15th Day (5)	3.0 ± 0.62	16650.0 ± 706	11.8 ± 0.40	39.8 ± 0.88	39.3 ± 1.9

Results and Observations

The TEC value fell by 23.8% as compared to control value on first day. This trend of fall continued till 4th day, when the value in pneumectomised fish was 1.66×10^6/ cmm (Table 1). However, 4th day onward, the TEC showed a continuous rise (Fig. 1) from the lowest value on 4th day, which was 4.5% less as compared to control value, to the level of 3.0×10^6/.cmm on 15th day, nearly equal to the control value. Now as a corallary when it is compared against the value for sham group, the pneumectomised fish shows that the fall was higher in it against a fall of 5.23% only in the sham group as compared to the control value. The TLC value showed varying degree of fluctuations during 15 day period of the study. The value rose sharply by 39.6%, immediately after post one day of pneumectomy. This trend of rise continued till 4th day, when it was 23,708 ± 805/cmm. as compared to control value of 12,780 ± 642/cmm (Fig.1). This rise was 85.5% higher, compared to control value. During next eleven days the TLC tended to deplete, except on when it rose little from 18,735 on 8th day to 19,675 on 12th day, to decline again to 16,650 ± 706/.cmm on 15th day when it was still 30.2% higher than the control value (Table 1). On the other hand the rise in sham group was 1.5%, when compared to control value.

The haemoglobin contents of pneumectomised fish showed almost concomitant fall and recovery with the TEC value, though the degree of difference varied much. On the first day of post pneumectomy the Hb contents depleted by 27.2% with a maximum fall of 30.3% occurring by 8th day, as compared to control value (vide Table 1). The Hb content showed almost a stable condition during first four days. The value showed a good recovery when the Hb value was 11.8±0.40 gm% 15th day, being 10.6% less, than the control value. Similarly when fluctuation in Hb value of the sham fishes is compared with the control group then it becomes apparent that it depleted by 6.8% only. On the other hand the PCV value of the pneumectomised *Clarias batrachus* fell by 14.10% on first day, post operation. This trend of fall continued till 4th day, almost parallel to TEC, but not equal to it (Fig. 1). The maximum fall in PCV was 28.7% on 4th day, against the control value. It then gradually recovered being 39.8 ± 0.88% on 15th day in the pneumectomised fish (Table 1). And when value of sham fish was compared to the control group, the PCV had fallen by 3.9% only, almost concomitant to TEC value in the same group. The MCH value showed a marginal initial fall of 4.5%, on first day, post pneumectomy and then soon registered a rise of 5.7% by 2nd day. This trend of rise continued till 4th day, with a peak rise of 32.2% (Table 1), when the MCH value was 57.8 ± 5.2 pg in the pneumectomised fish, against the control value of 43.7 ± 3.8 pg and then it started falling and was even lower by 3.4% on 12th day and by 10% on 15th day, as compared to control. During the same period of study the MCH value of sham group has fallen merely by 1.6% against the control.

The Differential Leucocyte Count (DLC) revealed interesting fluctuations (Table 2). There has been a high rise in TLC. The DLC showed that the initial reaction in the fish was that the large haemoblasts, small haemoblasts, monocytes, basophils, eosinophils and thrombocytes rose during post pneumectomy period,

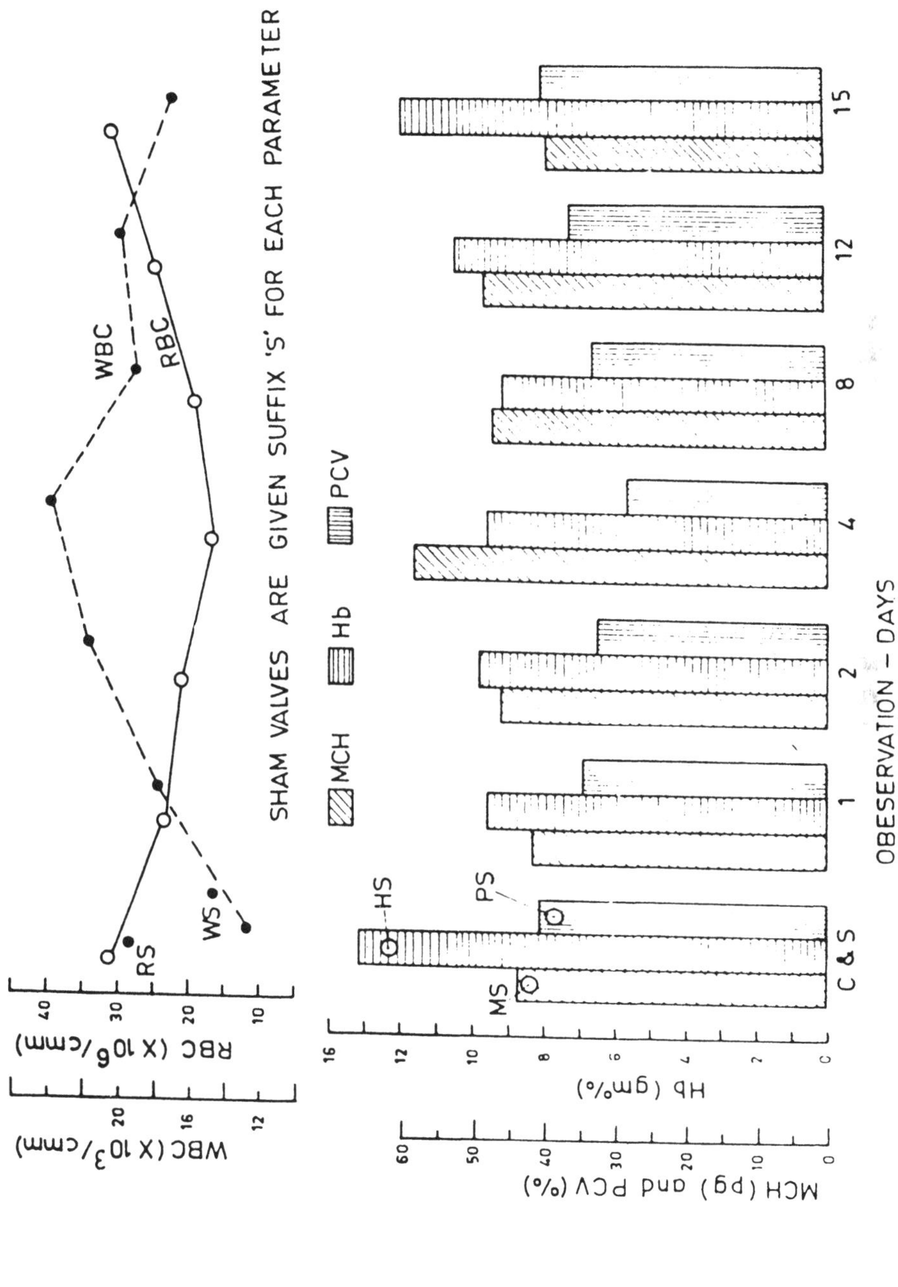

Fig. 1: Effects of Accessory Pneumectomy on some Blood Values of *Clarias batrachus*

Table 2 : Changes in Differential Blood Cell Counts (%) of *Clarias batrachus* Following Pneumectomy. All Values are Mean ± S.E.

Status, Time & No. of observations	Differential Leucocyte Counts: Haem – Large	Haem – Small	Mon	Lymph – Large	Lymph – Small	Nevt.
Control (10)	2.0 ± 0.6	0.8 ± 1.1	0.9 ± 0.7	6.0 ± 3.2	38.4 ± 3.2	17.4 ± 2.1
Sham (12)	3.9 ± 1.1	4.7 ± 1.0	1.1 ± 0.8	4.2 ± 0.2	30.5 ± 1.7	14.6 ± 2.1
Post Pneumectomy						
1st Day	4.2 ± 0.9	8.0 ± 0.9	1.2 ± 0.9	2.1 ± 0.2	23.2 ± 0.9	15.9 ± 1.7
2nd Day	3.9 ± 1.1	9.2 ± 0.9	1.2 ± 0.3	1.9 ± 0.7	20.7 ± 2.2	23.6 ± 3.4
4th Day	7.6 ± 1.2	14.5 ± 2.2	0.7 ± 1.2	—	24.2 ± 1.7	12.4 ± 2.9
8th Day	4.1 ± 0.6	11.6 ± 3.0	0.5 ± 0.1	—	14.6 ± 3.2	18.7 ± 1.7
12th Day	10.9 ± 0.8	10.2 ± 1.7	3.0 ± 0.1	2.8 ± 0.2	27.5 ± 2.4	12.6 ± 1.8
15th Day	4.8 ± 0.9	7.8 ± 2.0	1.2 ± 0.4	0.7 ± 1.2	32.8 ± 3.4	10.8 ± 2.7

	Differential Leucocyte Counts: Bas.	Eos.	Thromb.	Differential Erythrocyte Counts: E.B.	E.C.
Control (10)	0.9 ± 1.6	0.2 ± 0.2	33.4 ± 2.4	2.3 ± 0.7	97.7 ± 0.7
Sham (12)	—	6.4 ± 1.4	34.6 ± 3.9	2.9 ± 1.3	97.1 ± 1.3
Post Pneumectomy					
1st Day	2.2 ± 0.8	2.8 ± 1.0	41.3 ± 3.5	4.0 ± 0.9	96.0 ± 0.9
2nd Day	3.0 ± 0.9	1.6 ± 0.3	34.9 ± 2.7	4.2 ± 1.4	95.8 ± 1.4
4th Day	4.0 ± 2.1	—	36.4 ± 4.8	4.4 ± 0.8	95.6 ± 0.8
8th Day	3.7 ± 1.7	—	36.8 ± 5.3	4.5 ± 1.7	95.5 ± 1.7
12th Day	2.8 ± 2.2	—	30.2 ± 2.9	4.3 ± 1.8	95.7 ± 1.8
15th Day	1.8 ± 2.0	0.8 ± 0.2	39.5 ± 3.7	4.0 ± 0.9	96.0 ± 0.9

ABBREVIATIONS: Haem = Haemoblast, Mon = Monocyte, Lymph = Lymphocyte, Nevt = Neutrophil, Bas = Basophil, Eos = Eosinophil, Thromb = Thrombocyte, E. B. = Erythroblast and E. C. = Erythrocyte.

while the large and small lymphocytes and neutrophils declined. However, on 2nd day onward the individual leucocyto type exhibited varying mode of fluctuations. The overall picture reveal that during 15 days period of experimentation the lymphocytes and neutrophils declined. The eosinophils and large lymphocytes were altogether absent on certain days of observation (Table 2), in the pneumectomised fish. The thrombocytes, except on 12th day, showed higher counts, than the control level. In the sham operated fish too, all types of leucocytes have increased, except the lymphocytes (Table 2).

The erythrocyte and erythroblast ratio showed concomitant fluctuations with those of the TEC value. In general the erythroblast i.e. immature red blood cells at various stages of maturity increased manifold showing that erythrocytes have depleted in their population, therefore, more erythroblast were released in circulation. The maximum number of erythroblast was found on 15th day. When the TEC value has almost fully recovered during these 15 days of observations and all haematological parameters, had nearly reached to the control level.

Discussion

The foregoing results, following the accessory pneumectomy in *Clarias batrachus,* offered quite interesting changes during these 15 days of observations. Depletion occurred in all the red cell related blood values, viz. TEC, Hb, PCV and MCH. The initial depletion can be attributed to the excessive blood loss in the fishes subjected to pneumectomy. This is partly reflected from the blood values of fishes of sham group. However, the operational loss and trauma would be of higher magnitude in the pneumectomised fish, than in the sham operated fish. Taking RBC as the essential basic components of most of these red cell related values, the initial depletion has to be attributed to the excessive loss only. But it is almost inexplainable that during next four days what leads to the further depletion in TEC. Hb and PCV values. However, the fall in Hb and PCV values was not totally parallel to TEC value i.e. while the TEC value dropped by 23.8%, 29.8% and 45.3% during first four days of post operation, the PCV value fell by 14.1%, 19.8% and 28.7% only. This means that during these four days greater number of erythroblasts have been released in circulation, besides that greater amount of Hb has been incorporated to the erythrocytes or erythroblasts.

This is evidenced from the value of Hb and ultimately the MCH. The MCH values reveal that there is a clear rise in the amount of Hb per cell. Now the differential blood cell counts further support that erythropoieses as well as leucopoieses have been enhanced as the population of erythroblasts have considerably increased by 73.9%, 82.2% and 91.3% in circulation by 1st, 2nd & 4th day. This further confirms that, there has occurred much loss of blood during operation and consequently erythropoiesis has increased to make up the loss. However, the increase in TEC, is mainly due to release of greater number of immature blood cells, is apparent, (Table 2), which also contained higher amount of haemoglobin during 2-8 days as compared to the erythrocyte in fish of control group, is clear from MCH value (Table 1).

The present results find some support, regarding the basic behavioural response of various blood values as exhibited by *Clarias batrachus* following accessory pneumectomy from the earlier works of Soivo *et al.,* (1974), who pointed out that haematocrit values of several salmonids increased under reduced oxygen tension in vitro. It has been attributed to the swelling of the erythrocytes due to increase of lactic acid and, further that oxygenated erythrocytes gave a reduced haematocrit and increased MCHC values of *Salmo gairdnerii.*

Among fishes various tissues are known to play part in the process of haemopoieses, viz. kidneys, epigonal organs, G I tract, liver and spleen. The results of Fange and Sjobeck (1975) following splenectomy in the dogfish *Scyliorhinus canicula* (Elasmobranchii) showed almost similar trend of changes in haematocrit, haemoglobin and MCHC values during a course of three weeks. This is to substantiate the role of spleen in haemopoiesis. In the present study the influence of accessory respiratory organs depicts the fact that the active habits of the fish require higher amount of red blood cells and haemoglobin in this fish of amphibious mode of respiration.

This also permit us to infer that during such stressful conditions. Where much blood loss has occurred, the fish *Clarias batrachus,* is capable of coping the situation and haemopoiesis can be increased almost 100% within a short span of 4 days is evidenced from the increase in erythroblasts (91.3%) on 4th day, whereas the Hb has increased by 32.2% in the RBC. The TLC continued to rise till 4th day, (23,708.0/cmm,) about 85.5% higher than in the control fish and about 70% more than in sham fish. The rise in TLC is on account of two reasons, first the haemopoiesis in general has been enhanced, due to operational stress, to cope with the blood loss. And secondly the defence portfolio of immunological system would have demanded greater number of certain types of cells.

In the present case maximum rise has occurred in the population of haemoblast cells (i.e. immature leucocyts). This is obviously due to stress of blood loss. The eosinophils, basophils and thrombocytes also rose conspicuously by first day, possibly on account of their role in checking blood loss (Wintrobe, 1976), by way of blood coagulation and for immunological reasons. But what cause high rise in basophils population is not well understood. The tendency of internal physiological milieu to response under such circumstances, is generally to cope with the facing situation. The earlier work on the effect of partial hepatectomy on some haematological parameters of the same species *Clarias batrachus* showed that the TEC, TLC and Hb values rose during first 8 days post hepatectomy, while in the sham fish the TEC and Hb fell significantly, but the TLC and MCH rose (Joshi *et. al.* 1982). But how the accessory pneumectomy among fishes affects blood values has not been hitherto worked out by any earlier author.

But still, it is apparent that the pneumectomised fish tries to compensate for blood loss, by accelerating erythropoiesis and incorporating more haemoglobin within erythrocytes or erythroblasts. On the other hand the increased TLC suggest that fish has enhanced its leucocyte population and thrombocyte population to check further loss of blood due to haemorrhage and defend body against foreign invasion. This is the reason of increase in immature white blood corpuscle in circulation. The rise in erythroblasts and MCH value has a direct relationship with the loss of blood on one hand and also a considerable loss in the available area of gaseous exchange through accessory respiratory organs, on the other. A high mortality among pneumectomised fish, post operation also show that it is indeed very hard and traumatic for a fish to survive, following much blood loss and loss of accessory respiratory organ. This is evidenced from the fact that only four fishes (2.5%) died post sham-operation while twenty six (43.5%) died post accessory pneumectomy. Therefore it can also be suggested that the pneumectomised fish faces a sort of hypoxia, where the role of accessory respiratory organs become more important which is a step towards aerial mode of respiration at the threshold of evolution of vertebrate life.

References

Fange, R. and Sjobeck-Johanson, M.L. (1975). The effect of splenectomy on the haematology and on the activity of 6-aminolevulinic acid dehydratase (Ala-D) in hemopoietic tissues of the dogfish *Scyliorhinus canicula* (Elasmobranchii). Comp. Biochem. Physiol., 52(A) : 577-580.

Gradwohl, R.B.H. (1943). In : Clinical Laboratory methods and diagnosis Vol. I. 3rd ed. St. Louis. The C.V. Mosby Co.

Hickey, C.R. (Jr.), (1976). Fish haematology, its uses and significances *N.Y. Fish Game J., 23* : 170–175.

Hughes, G.M., Kikuchi, Y. and Watari, H. (1982). A study of deformability of red blood cells of a teleost fish, the yellow tail *(Seriola quinqueradiata)* and a comparison with human erythrocyte. *J. exp. Biol., 96:* 209-220.

Joshi, B.D. (1978). Effect of asphyxiation on some haematological values of fish *Clarias batrachus. Proceed. All India Sem. Ichthyol.* (1977) : 5-8.

Joshi, B.D. and Sharma, T. (1981). Effect of hypophysectomy and pituitary extract treatment on blood and liver metabolites of a freshwater catfish *Clarias batrachus. Jap. J. Ichthyol.* 28(3) : 283-286.

Joshi, B.D. and Sharma, T. (1982). Effect of hypophysectomy and pituitary extract treatment on haematological and biochemical constituents of blood and liver of a freshwater fish *Heteropneustes fossilis. J. Anim. Morphol. Physiol.*, 29 : 42-54.

Joshi, B.D., Dabral, R. and Sharma, T. (1982), Haematological response of a freshwater teleost *C. batrachus*, following partial hepatectomy. *J. Anin. Morphol. Physiol., 29:* 149-156.

Kita, J. and Itazawa, Y. (1989). Release of erythrocytes from the spleen during exercise and splenic constriction by adranaline infusion in the rainbow trout. *Jap. J. Ichthyol.,* 36 (1) : 48-52.

Soivo, A., Westman, K. and Nyholm, K. (1974). The influence of changes in oxygen tension on the haematocrit value of blood samples from asphyxic rainbow trout *(Salmo* gairdnerii), *Aquaculture,* 3 : 395-401.

Tondon, R.S. and Joshi, B.D. (1973). Studies on the physiopathology of blood of freshwater fishes, infected with two new forms of trypanosomes. *Z. Wiss. Zool.* Leipzig, *185* : 207-221.

Yamamoto, K., Itazawa, Y. and Kobayashi, H. (1985). Direct observation of fish spleen by an abdominal window method and its application to exercised and hypoxic yellow tail. *Jap. J. Ichthyol.*, 31(4) : 427-433.

Wintroboe, M.M. (1976). In : Clinical haematology. 7th ed. Lea and Febier. Philadelph.

Chapter 5

Food and Feeding Habits of *Heteropneustes Fossilis* (Bloch) Inhabiting Gadigarh Stream, Jammu

S.P.S. Dutta

Abstract

Heteropneustes fossilis is euryphagus, molluscicarnivorous and bottom feeder. Its annual mean gut analysis has shown the presence of diatoms (2.83%), green algae (6.75%), macrovegetation (3.25%), hemipterans (0.33%), ephemeropterans (0.83%), dipterans (2.08%), odonates (2.58%), coleopterans (5.0%), parts of insects (4.17%), decapod crustceans (1.0%), *Gyraulus convexisulculus* (26.58%), *Viviparous bengalensis* (7.58%), *Melanoides tuberculatus* (5.0%), *Lymnaea* spp. (1.0%), bivalves (5.33%), broken shells (22.25%), fish (0.42%) and sand and mud (3.0%). Gastrosomatic index indicates low feeding during fish spawning (June), monsoon (August) and Winter (January) periods.

Introduction

Heteropneustes fossilis, locally named as "Lallai", is widely distributed in springs, ponds, streams, rivers etc. in Jammu. Due to the presence of accessory respiratory organs, it has also been netted from oxygen deficient waters. In the present communication, food and feeding of *H. fossilis* is described and compared with the earlier findings of Mookerjee *et. al.* (1946), Das and Moitra (1955), Menon and Chacko (1958), Bhat (1968), Fernando and Indrasena (1969), Srivastava (1979), Johal (1981), Panday (1983), Sharma (1989) and Shrivastava *et. al.* (1992).

Table 1 : Monthly percentage of different food items and quantitative feeding variations in *Heteropneustes fossilis* (Bloch).

Food items	*J*	*F*	*M*	*A*	*M*	*J*	*J*	*A*	*S*	*O*	*N*	*D*	*Mean*
Green algae	-	15	-	15	33	5	10	-	-	3	-	-	6.75
Diatoms	-	29	-	-	-	-	-	-	-	5	-	-	2.83
Macrovegetation	-	-	-	-	9	5	10	9	6	-	-	-	3.25
Decapod Crustaceans	3	6	-	-	-	-	-	-	-	-	3	-	1.0
Hemiptera	-	1	3	-	-	-	-	-	-	-	-	-	.33
Ephemeroptera	-	-	10	-	-	-	-	-	-	-	-	-	.83
Diptera	-	10	-	-	-	-	15	-	-	-	-	-	2.08
Odonata	-	1	30	-	-	-	-	-	-	-	-	-	2.58

(Contd)

* Department of Bio-Sciences, University of Jammu, Jammu - 180004.

Food items	*J*	*F*	*M*	*A*	*M*	*J*	*J*	*A*	*S*	*O*	*N*	*D*	*Mean*
Coleoptera	-	-	12	-	2	5	10	23	-	8	-	-	5.0
Parts of insects	9	-	-	-	20	-	-	-	11	10	-	-	4.17
Lymnaea spp.	-	12	-	-	-	-	-	-	-	-	-	-	1.00
Melanoides tubeculatus	-	6	-	-	2	-	30	-	-	-	22	-	5.00
Viviparous bengalensis	23	4	-	-	24	10	-	-	-	-	30	-	7.58
Bivalves	8	11	20	-	-	-	-	-	-	-	25	-	5.33
Broken Mollusca shells	19	-	-	77	-	-	52	-	-	-	19	100	22.25
Fish	-	5	-	-	-	-	-	-	-	-	-	-	.42
Sand & Mud	-	-	-	8	1	5	-	16	5	-	1	-	3.0
Gastrosomatic index (G.S.I.)	1.18	2.95	3.07	1.00	3.54	2.49	2.79	1.36	3.70	3.89	2.89	3.26	

Material and Methods

Study material was collected from Gadigarh stream (74°.50′E longitude, 30.40′N latitude at an elevation of 304 meters mean sea level), Jammu. In the laboratory, fishes were blotted and their length and weight recorded. Next the stomach was dissected out and weighed, with and without contents. For quantitative study, gastrosomatic index was calculated by applying the formula:-

$$\text{Gastro Somatic Index (G. S. I.)} = \frac{\text{Wt. of food}}{\text{Wt. of fish}} \times 100$$

The percentage of different food items in stomach (sometimes) from other parts of alimentary canal) was recorded by eye estimation (Pearse, 1915; Job, 1940 and Pillay, 1952).

Results and Discussion

The monthly average percentage of gut contents of *H. fossilis* is given in Table 1. Molluscs formed its main food item accounting for 67.74%, followed by insects, 14.99%; plant matter, 12.83%; sand and mud, 3.0%; decapod crustaceans, 1.0% and fish 0.33%.

Among molluscs, *Gyraulus convexisulculus* (26.58%) formed the most dominant group, followed by broken shells (22.25%), *Viviparous benglanensis* (7.58%), bivalves (5.33%), *Melanoides tuberculatus* (5.0%) and *Lymnaea* spp. (1.0%) Molluscs remained perennial and pre-dominant each month (Table 1).

Insects, represented in fish diet by Hemiptera (0.33%), Ephemeroptera (0.83%), Diptera (2.08%), Odonata (2.58%), Coleoptera (5.0%) and broken parts of insects (9.17%), ranked second in the fish diet (Table 1). Except during April, November and December, insects were consumed by fish throughout the year. Annual highest percentage of insect food was seen in March (55%) and lowest in May (2%).

Inspite of perennial presence and abundance in stream, decapod crustaceans viz *Macrobrachium dayanium, M. lamarrei* and *Caridina* sp. were seen only thrice in fish food.

Fish did not share much importance in the diet of *H. fossilis* and was seen only once in the month of February (Table 1).

Plant matter in fish food viz. green algae, diatoms and macrovegetation shared annual mean percentage of 6.75%, 2.83% and 3.25%, respectively. Considerable quantities of plant matter were consumed by fish in February and May (Table 1). Annual mean low percentage of plant diet (12.83%) suggests its incidental entry along with molluscs, where green algae and diatoms show periphytonic growth. A detailed study on the food composition of *H. fossilis* has revealed that fish is mollusci-carnivorous and bottom feeder.

Earlier, Dass and Moitra (1955) designated it as a bottom feeder, on decomposed aquatic plants, bryozoans and their statoblasts, molluscs, fish scales and large quantities of sand and mud in Lucknow. According to Srivastava (1957), bottom feeder. *H. fossilis* feeds on larval stages of insects, insects and their parts, copepods, cladocerans, water mites, oligochaetes, fish fins and scales, detritus and debris in La Martinier and Kathauta lakes, Lucknow. Bhatt (1968) recovered molluscs, *Cypris*, daphnids, rotifers, dipterans, coleopterans, odonates, ephemeropterans, hemipterans, eggs of invertebrates, planarians, weeds, debris and algae. Johal (1981) recorded algae (4%), Plant matter (2%), higher plants (4%), rotifers (3%), crustaceans (20%), insects and their larvae (11%), nematodes (4%) oligochaetes (28%) and fishand fish scales (24%), and documented it as a carnivorous and bottom feeder. Pandey (1983) described it as bottom feeder. According to Sharma (1989), carnivorous and bottom feeder, *H. fossilis* feeds on eggs (13.9%), protozoans (4.9%), round worms (1.8%), insects and their larvae (45.7%), crustaceans (15.2%), molluscs (2.5%); vegetable matter, algae and green matter (2.5%), materials in advanced stage of digestion (12.0%) and sand and mud (1.5%) in Jammu waters. Shrivastava (1992) worked out the dominance of insect larvae and molluscs throughout the year. The other food items recorded include algae, eggs of invertebrates, cladocerans copepods, fish and organic matter.

Menon and Chacko (1958) noticed protozones (5%) crustaceans (16%), insects (22%), molluscs (15%) and fish (42%) and documented it as piscivorous.

Present carnivorous feeding of *H. fossilis* is different from the omnivorous (algae, 10%, higher plants, 17% protozoans, 18% worms, 10%; crustaceans, 20%; insects, 15% and mud and sand, 10%) findings of Mookerjee *et. al.* (1946) in Madras and Fernando and Indrasena (1969) in Ceylon.

Quantitative feeding in *H. fossilis* has indicated two periods of intensive feeding. Annual February-May, except in April, intensive feeding has a correlation with spring and summer biotic richness in water and prespawning building up of the gonads to the fish spawning in June, when feeding recorded a trough. Feeding also remained high during September-December and coincided with late and postmonsoon rise in biotic-diversity in stream. Monsoon low autogenetic production and retention of flora and fauna may account for July and August low records of quantitative feeding. Reduced metabolism may explain winter (January) trough in gastrosomatic record in *H. fossilis* and is in agreement with the findings of Malhotra and Gupta (1990), Dutta (1991, 1993 a, b) and Shekhar *et. al*. Contrary to the present observations, Sharma (1989) documented January maximum and May minimum record of quantitative feeding in *H. fossilis*. Bhatt (1968) noticed intensive feeding during postmonsoon and winter months, with a peak in September. During summer, it goes on declining till it records its lowest observation in June. Feeding is also quite low in July and August.

From the foregoing discussion, it is clear that fish feeding pattern varies not only in different areas but even in different waters of the same area. This variability may be due to ecological factors including availability and abundance of preferred fish food.

References

Bhatt, V.S. (1968). Studies on the biology of some freshwater fishes. Part VII. *Heteropneustes fossilis* (Bloch). Ind. J. Fish *15*, (1-2) : 99-115.

Das, S.M. and Moitra, S.K. (1955). Feeding habits of freshwater fishes of Uttar Pradesh, Curr. Sci., *24* (12): 417-418.

Dutta, S.P.S. (1991). Food and feeding habits of *Rasbora rasbora* (Ham.) inhabiting Gadigarh stream, Jammu. Geobios new Reports, *10* (2) : 135-137.

—(1993 a). Food and feeding habits of *Danio rerio* (Ham.) inhabiting Gadigarh stream, Jammu. J. Freshwater Biol. Assoc., *5* (2) : 165-168.

—(1993 b). Food and feeding habits of *Lepidocephalichthyes guntea* (Ham. Buch) and *Noemachilus biota* (Ham. Buch) inhabiting Gadigarh stream, Jammu. Him J. Environ. Zoology, *7* : 43-46.

Fernando, C.H. and Indrasena, H.H.A. (1969). The freshwater fisheries of Ceylon. Bull. Fish Res. Stni, Ceylon, *20* (2) : 101-134.

Job, T.J. (1940). An investigation on the nutrition of the perches of the Madras coast. Rec. Ind. Mus., *42*: 289-364.

Johal, M.S. (1981). Food and feeding habits of some fishes of Punjab. Vest. cs. Spotec. Zool. 45 : 87-93.

Malhotra, Y.R. and Gupta, A. (1990). Seasonal fluctuations in food and feeding of *Puntius conchonius* inhabiting lake Mansar, Jammu. J. Freshwater Biol. Assoc., *2* (2) : 147-151.

Pandey, K.C. (1983). Concepts of Indian fisheries. K. Nath and Company, Meerut, India.

*Pearse, A.S. (1915). On the food of the small shore fishes in the waters near Madison. Wisconsin. Bull. Wis. Nat. Hist. Soc., *13* : 7-22.

Pillay, T.V.R. (1952). A critique of methods of study of food of fishes. J. Zool. Soc. India: 4 : 185-200.

Shrivastava, S., Rao, K.S. and Dubey, R. (1992). Diet feeding patterns of some freshwater catfish from Kshipra River, India. Pak. J. Zool. *24* (1) : 27-29.

Sharma, V.K. (1989). Studies on the stomach contents of some freshwater fishes of Jammu. M. Phil. Dissertation Univ. of Jammu.

Shekhar, C., Malhotra, Y.R. and Dutta, S.P.S. (1993). Food and feeding habits of *Schizothorax richardsonii* (Gray and Hard) inhabiting Neeru nullah, Bhaderwah, Jammu. J. Indian Inst. Sci., *73*: 247-251.

Srivastava, V.K. (1979). The correlation between bottom fauna and the food of fishes. Proc. Nat. Acad. Sci., India, 29 : 217-223.

* Original not seen

Chapter 6

Observations on the Use of Ovaprim for Induced Spawning of Indian Major Carps

P.K. Roy

Abstract

The paper presents a resume on experiments carried out on breeding of Indian major carps, *Catla catla, Labeo rohita,* and *Cirrhina mrigala* in a cement cistern through administration of 'Ovaprim'—a new spawning hormone. The Experiments proves that ovaprim can be used for breeding of the fish in simple enclosures, with facilities of some input and output of water and a dose of .4 ml and 0.5 ml per kg. body weight, for males and females respectively are required for the purpose.

Introduction

In aquaculture industry, fish seed is a very important component for fish culture. To increase seed production considerable attempts have been made by several workers viz. Alikunhi et. al. 1963, 1964 and 1965, Chaudhury et. al. 1966, Singh et. al. 1970 for successful breeding of carps by pituitary hormone injection. With the increase in demand for pituitaries in recent years and difficulties in their mass procurement Russian Workers felt the need of a substitute which could be procured commercially. Choriogin (a trade name of CG) was first used in 1946 on Loach with great success. Among other possible substitute several preparations such as 'Szhk', prepared from pregnant mare serum, 'Estrovest' and 'Hypophysine Forte, were tried by the Bulgarian fish culturists on carp and trouts (as cited by Gerbilskii, 1965). The author concluded that these preparations reduced the duration of spawning process. Tang (1965) states that treatment with fish pituitaries in combination with chorionic gonadotropin increases the effectiveness and better success is achieved. The chorionic gonadotropins (CG) behave primarily as lutenizing hormone (Burrows 1949). Since LH is responsible for the growth and maturation of Gonads of fish. Ramaswamy and Lakshman (1959) injected a pond catfish with 250 I.U. of chorionic gonadotropin (Physex) and obtained ripening of eggs.

Recently Agrivet Farm care Glaxo India Ltd. has marketed ovaprim in glass vial which is used as substitute of carp pituitary gland. During July, 1992 experiments were conducted on Indian Major carps, Catla, Rohu and Mrigal by using ovaprim in Barasagar Dighi fish farm at Malda of West Bengal.The experiments were successfully conducted in a cement cistern for mass breeding of the fish.

Table 1: Details of Experiments conducted on the use of Ovaprim for induced spawning of Indian major carps

Expt. No.	Date	Species	Female		Male		Dose and time of injection	Water temp. °C	No. of spawners responding	Qty/No. of egg	Percentage of fertilization	No. of larvae
			No.	Wt. (kg)	No.	Wt. (kg)						
1.	23.7.92	*Catla catla*	4	28	4	30	♀ = .5 ml/kg. B. Wt ♂ = .4ml / kg.B. Wt 1730 hrs.	32	3 (20 kg)	200 lit. / 30 lakhs	91	21 lakhs
2.	26.7.92	*Labeo rohita*	20	30	24	26	♀ = .5 ml/kg. B. Wt ♂ = .4ml / kg.B. Wt 1800 hrs.	33	18 (27 kg)	324 lit. / 45 lakhs	94	33 lakhs
3.	29.7.92	*Cirrhina mrigala*	16	22	18	26	♀ = .4 ml/kg. B. Wt. ♂ = .4ml / kg.B. Wt 1730 hrs.	31	14 (20 kg.)	200 lit. / 32 lakhs	90	23 lakhs

Materials and Methods

Ovaprim is readily available in market. For injection to females and males, a doze of .5ml and .4ml per km body weight was used respectively.

A simple rectangular cement cistern of 14.2 × 6 m and 1.25 m of depth provided with an overflow outlet at one end and a guarded bottom outlet was used for the experiments. For introduction of water in the cistern, a 5 H.P. diesel pump was used. After injection, both females and males were released in the cistern and a water flow was maintained for 2 hours, after about 1,1/2 hours of introduction.

Results and Discussion

In all the experiments, conducted with use of ovaprim, 75 to 90 per cent of fishes bred and 90 to 94 per cent of fertilization was observed (Table 1). In all experiments, dose to the females were kept .4 ml to .5 ml per kg body weight respectively and in case of males, only .4 ml per kg body weight was given in all the species. The water flow was started after 1 hour to 1,1/2 hours of injection and continued for 2 hours, chasing of the females by the males were observed after 5 to 8 hours of the injection. From the 3 experiments with ovaprim, altogether 46 nos males (total weight 82 kg) and 40 nos. Females (total weight 80 kg) were utilised to produce 1.07 crores eggs with very high percentage of fertilization. This proves that Ovaprim can be commercially used for seed production of Indian major carps.

Acknowledgements

I express my grateful thanks to Prof. Y. Sreekrishna, Director, CIFE, Bombay and Prof. Kuldip Singh, Principal, IFTC, Barrackpore for their keen interest and encouragement during the study.

References

Chaudhuri, H. Singh, S.B. and Sukumaran, K.K.1966. Experiments on large scale production of fish seed of the chinese grass carp, *Ctenopharyngodon idellus* (C & V) and Silver carp, *Hypophthalmichthys molitrix* (C & V) by induced breeding in ponds in India. Proc. Indian Acad. Sci. (8) 6(2) : 80-95.

Ramaswamy, L.S. and Lakshman, A.B. (1959). Action of Mammalian hormones on the spawning of cat fish J. Scient. Ind. Res. 18C : 185-191.

Tang, Y.A.1965. Progress in the hormone spawning of pond fishes in Taiwan. Proc.Indo-Pacif. Fish, Coun, 11(2) : 122-127.

Chapter 7

Aquatic Odonata and Hemiptera of Jammu and their Role in Aquaculture

*Baldev Sharma, Neeru Dhalla & Renu Salaria**

Insects representing the culmination of evolutionary development of terrestrial arthropods are also one of the most successful animals which have mastered almost all the types of habitats which are available upon the earth. As such we find the insects which are adapted to the most diverse types of environmental conditions. Amongst the best habitat adaptations, the adaptation to aquatic mode of life is one of them.

The developmental stages otherwise of many terrestrial groups which inhabit aquatic environment required considerable and special types of modifications to withstand the stresses and strains of such a dramatic change in the environment and have added functional plasticity to their physiological triggers as well.

Aquatic insects abound almost all the types of water bodies, perennial and annual ones with no exception and thus they form an important component in various aquatic ecosystems operating in various types of water bodies.

The importance and position which aquatic insects enjoy in Fisheries, Limnology and general aquaculture does not need to be over emphasized. They serve as food for fish, frogs, birds and man. Some are of medical importance as they help in the biological control of mosquitoes. A number of aquatic insects are used as fish bait and some as pollution indicating and controlling agents. On the other hand we find that unlike Odonates which are totally harmless to aquatic biota and human beings as they don't bite or sting, there are some aquatic insects such as aquatic bugs which are harmful to man as they not only cause painful bites at occasions but also are responsible for killing a large number of small fish, their fries and fingerlings during sucking out body fluids from these hosts for food.

The various aspects of the role which are played by insects in aquaculture can be divided into following categories.

I. As Food for Fish

Being one of the major animal group amongst the aquatic communities insects serve as food for so many insectivorous and omnivorous fishes. Few earlier records

* Deptt. of Biosciences, University of Jammu, Jammu-180004

on the role of aquatic insects as fish food has been investigated by Wilson (1923); Chowdhary (1980); Jhingran (1982) has reported the following fishes of Indian plains which feed on aquatic insects in one or more stages of their life cycle as shown in Table A.

Table A

S. No.	*Name of the fish*	*Insectivorous stage*
(1)	***Catla catla***	Adult
(2)	***Labeo kontius***	Adult
(3)	***Barbus carnatleus***	Fingerlings & adult
(4)	***Anabas testudincus***	-do-
(5)	***Channa striatus***	Fry and fingerlings
(6)	***Channa marultius***	-do-
(7)	***Heterophenstes fossilis***	Fingerlings & adult

It is on record that about 85% of 20,000 fish species are carnivorous. It is also on record that exposed to *Chironomous* larvae, *Cyprinus carpio* consumes 10 larvae (30.9 mg of live weight) in about 16 minutes as against 299 *Daphnia* (28.3 mg) in 85 minutes. This suggests that the aquatic insect larvae which are rich in nitrogen content may be very much useful as fish prey organisms. Moreover, the efforts needed on the part of the fish to complete its ration requirement would be far less when feeding on the insect larvae as compared to daphnids. (Salaria, 1992).

Among all aquatic insects, the coleopterans, odonates, hemipterans and dipterans constitute the major components of the dietery of the fish species. Various species of fish have been recorded to utilize fairly good percentages of odonate larvae as well as coleopteran insects in Jammu region of J & K state (Table B).

Table B

S. No.	*Name of the fish*	*% age composition in food*	
		Odonata	*Coleoptera*
1.	***Oxygaster bacalia***	10.5	20.33
2.	***Osteobrama cotio***	9.0	3.33
3.	***Ompok bimaculatus***	13.57	
4.	***Mystus vittatus***	2.0	15.63
5.	***Mastacembelus pancalus***	10.0	5.0

(c.f. Dutta & Malhotra, 1991)

That himepteran, *Hespercorxia* sp. forms the food of common carp fingerlings and adults has been recorded by Chowdhary (1980).

II. As Biological Control Agents

Both odonates as well as hemipterans have great role to play in the biological control of harmful insects as they voraciously feed on the abnoxious insects like mosquito larvae, may fly naids, gnats, midges and other smaller insects.

Among Odonata *Telephlebia godeffroyi* feeds deliciously on a large number of mosquitoes at once and ***Aeschna brevistyla*** take gnats and mosquitoes in dozens while on the wing. Odonate larvae relish almost any thing they can catch but more particularly smaller specimens of their own or other species of Odonata. When young, they feed readily on Protozoa. A little later they are expert at catching water fleas and other smaller crustacea. Still later, they prefer the larvae of May-flies or mosquito to any thing else.

In an experiment, a larva of ***Synthemis macrostigma*** kept without food for a week, when was offered mosquito larvae, it used its mask with such a vigour that no less than sixty larvae were swallowed within ten minutes by it. (Tillyard, 1917).

Among Hemipterans, ***Laccotrephes*** and larvae of ***Ranatra*** sp. feed on mosquito larvae and Corrixids feed mostly on small midges.

III. Destruction Caused in Nursery Ponds

The nursery ponds designed to rear spawn of major carps are invariably populated with a large number of aquatic insects. Their destructive role in carp nurseries has been described by Khan and Hussain (1947), Pakrasi (1954), Alikunhi (1952), Chawdhuri (1960), Gorai and Ray Chaudhuri (1962) and Julka (1965).

Amongst Hemipterans, species belonging to the families *Notonectidae* (Back swimmers) ***Belostomatidae*** (Gaint water bugs) and *Nepidae* (Water scorpions) cause a heavy loss to the carp spawn. Notonectids are known to attack carp spawn, small fish, tadpoles, and even small frogs (Distant, 1906). Gorai and Ray Chaudhuri (1962) observed ***Anisops bouvieri*** feeding on fish larvae in addition to plankton, mosquitoes, chironomid larvae and mayfly naids in juvenile stages.

Giant water bugs (Family Belostomatidae) are highly predacious and can feed on bigger aquatic animals like frogs and fishes. They are known to secrete a toxic salivary substance which kills the prey in a very short period of time.

Water scorpion, ***Laccotrephes*** spp. ***(ruber*** and ***maculatus)*** attack advanced carp fry and feed on them. The water stick insect ***Ranatra elongata*** and ***R. filiformis*** both are also predators of carp spawn and fry. Tripathi and Sharaf (1974) observed that ***Anisops*** larvae measuring 14 to 15 mm and ***Ranatra*** larvae measuring 45 to 48 mm can consume 182 and 162 carp eggs within 24 hours.

IV. For Pollution Control

This aspect is wholly-soly dependent on the detritus-feeding insects. The hemiperans and other insects inhabiting the bottoms of streams, lakes and ponds mostly feed on the detritus and other organic matter, debris and humus thus constituting important pathways of energy flow and are thus very beneficial for the productivity of aquatic environment. By ingesting a large mass of detritus and other organic matter bottom dwelling insects decrease the rapidity with which the fresh water systems are turned shallow following eutrophication. Strategy of suspension feeding is most fascinating among insects. Some of them e.g. larvae of may-fly/ ***Isonchia,*** caddisflies ***Brachycentrus*** and black flies ***Simulium*** sp. have developed filtering devices like nets and mucus and passively collect the particles which are swayed by running waters.

V. As Food for Other Animals

Two north American species *Arctocorixa mercenaria* (Say) and *A. abdominalis* (Say) are found in such huge numbers in Mexico, that they can furnish eggs for food for the natives and their adults are captured in such large numbers, that they can be dried and exported in tons for the consumption of the domestic poultry, song birds and fish. Many Notonectides are also used for the same purpose. In certain South American and Asiatic countries, the Giant water bugs (Belostomatids) are eaten by the natives besides manufacturing medicines out of them (Essig, 1941). Even their eggs are collected from the water bodies and are used as food in certain South East Asian countries.

VI. Insects in Limnology and Aquaculture

As primary convertors of plant material, the aquatic insect communities have a very important role to play in aquaculture. The corixids and other hemipterans suck all sorts of organic ooze, including both animal and plant matter, into their mouths. They also break the cell walls and extract the contents of *Spirogyra* and other water plants thus because of their herbivorous nature the hemipterans can be placed at the lowest trophic level in the animal food chain as they scoop up and ingest small benthic, planktonic as well as periophytic organisms. *Notonecta* can even feed on animals which are much larger than their own size. Even vertebrates such as frog and their tadpoles, small fish in addition to insects are also not spared by the notonectids. This very habit of notonectids compelled Mellanby (1963) to recommend that notonectids should never be kept in an aquarium along with other animals. Even human bathers also become victims of their painful bites.

Similarly members of family Belostomatidae and Napidae are also predators preying on other aquatic insects, snails, fish tadpoles, salamenders etc; the later group also looks for eggs of various aquatic animals and they lie in wait for their prey. The members of the family Gerridae are also predacious in habits and prey upon such small insects which fall into the water or feed or rest upon dead insects and other animals available to them. On the other hand, the Odonates are voracious feeders. They are extremely predatory camivores voraciously feeding on small fishes, certain other smaller aquatic insects and some are even cannibals, eating the young ones of their own species.

Thus with all these observations the role of aquatic insects in the aquaculture, fisheries and limnology cannot be underestimated.

The present authors have surveyed the various water bodies of Jammu region for the Odonate (Dhalla, 1992) and Hemipteran (Salaria, 1992) fauna and in the following pages an inventory of various species of these two groups have been recorded along with their brief diagnostic features, habitat and figures.

ODONATA

Sub-order-ZYGOPTERA

Sub-family - PSEUDAGRIONINAE

Pseudagrion rubriceps Selys (Fig. 1)

Larvae yellowish brown. Head rectangular with eyes occupying dorsolateral position. Antennae filiform; Labium triangular. Median lobe of prementum without median clift. Premental setae 1 & 1 and palpal satae 4 + 4. Movable hook long about two-third of the palpus. Mandibles dentate. Abdomen long terminating into caudal lamellae. Male gonapophyses paired, triangular and mesal on 9th segment while in female, elongated structures arising from beyond the 9th segment extending beyond the 10th segment present. Caudal lamellae flat, deplex with well developed tracheation.

Habitat—Larvae are very active swimmers and can be collected from muddy waters with plenty of vegetation around.

Pseudagrion laidlawi Fraser (Fig. 2)

Similar to ***P. rubiceps*** except that these larvae are greenish brown. Movable hook, slender and about half the length of the palpus. Female gonapophyses elongated and arising from the anterior of the 9th segment extending up-to the middle of the 10th segment. Caudal lamellae with rounded apex.

Habitat—Actively swimming larvae inhabiting bottom, along the sides of the lake.

Sub-family-ISCHNURINAE

Ischnura delicata (Hagen) (Fig. 3)

Greenish larvae with yellowish marking. Head rectangular with anterolateral position of eyes. Median lobe of prementum without median cleft. Premental setae 4 & 4, palpal setae also 4 + 4. Caudal lamella foliaceous, transparent with greenish tracheation, broader in the middle and tapering towards both ends.

Habitat—Large populations of larvae inhabitating the littoral zones of large water bodies like lakes. They usually remain attached to the submerged vegetation at rest but were found very active while swimming.

Ischnura senegalensis (Rambur) (Fig. 4)

Similar ***I. delicata*** except that the larvae are brownish to black. Premantal satae 3 & 3, Palpal setae 5 + 5. Movable hook medium sized, about half the palpus. Caudal lamellae foliaceous, tapering at both ends.

Habitat—Inhabit the bottom of the water body along with the growing vegetation.

Sub-order-ANISOPTERA

Sub-family-ICTININAE

Ictinogomphus rapax (Rambur) (Fig. 5)

Head somewhat squarish; eyes large capping head dorsolaterally. Antennae 3-segmented thickly beset with setae; labium flat. Rudiments of meso & meta thoracic wings represented by pleural ridges upto the first abdominal segment. Abdomen somewhat oval, hairy, broadest in the middle and beset with lateral spines. Anal appendages short, beset with setae.

Habitat—They burrow in shallow sand. Can be collected from the sides of the lakes near the surface of the sediment.

Sub-family-ANACTINAE

Anax parthenope parthenope (Selys) (Fig. 6)

Head widest over vertex across the eyes which are anterolateral in position. Antennae 7 - segmented filiform; Labium long, broader distally and narrowing proximally. Distal margin of prementum with a deep median cleft; movable hook, longer than the palpus; rudimentary wing buds present; abdomen ten segmented. Lateral abdominal spines present on segments 7th to 9th.

Habitat—These larvae inhabit littoral zones with plenty of growing & decaying vegetation.

Sub-family-LIBELLULINAE

Orthetrum sabina (Drury) (Fig. 7)

Head hairy prostero-laterally. Antennae medium-sized; Labium spoon-shaped; Distal margin of prementum convex while distal margin of palpus formed into crenations. Abdomen elongated, tapering and hairy posteriorly; Mid - dorsal abdominal protuberances on segments 4th to 7th.

Habitat—Larvae sluggish and bottom dwellers in lakes.

Potamarcha obscura (Rambur) (Fig. 8)

Similar to *Orthetrum sabina* except that premental setae are 11 & 11 and palpal setae 8 + 8 Abdomen oval, broadest in the middle. Small lateral abdominal spines on segments 8th and 9th.

Habitat—Sluggish forms and habitating lake bottoms covered with sand and mud. They camouflage very well with the surroundings.

Sub-family-SYMPETRINAE

Crocothemis servilia servilia (Drury) (Fig. 9)

Resembles *Potamarcha obscura* but for the premental setae 14 & 14 and palpal setae 10 + 10 abdomen elongated but abdominal spines altogether absent.

Habitat—Very active swimmers. Found in association with the vegetation growing along with certain sedentary objects in the water body.

Acisoma panorpoides panorpoides Rambur (Fig. 10)

Larvae hairy, small and compact in appearance. Premental setae 11 & 11; palpal setae 8 + 8. Pair of meso-and meta-thoracic wing buds present extending upto the 7th abdominal segment. Characteristic feature is the presence of tuft of long mid-dorsal abdominal retae on segments 2nd to 10th.

Habitat—Inhabiting the aquatic vegetation along the sides of the lake.

Brachythemis contaminata (Fabricius) (Fig. 11)

Larvae yellowish and appears shining due to the deposition of sand and mud. Premental setae 14 & 14; palpal setae 7 + 7. Mid-dorsal abdominal spines present on segments 4th to 9th.

Habitat—These insects inhabit shallow water of the lake. They are very active while moving and when at rest, usually, remain embedded in sand or mud.

Sympetrum commixtum (Selys) (Fig. 12)

Resembling ***Brachythemis contaminata*** in general appearance except that the labium is spatula-like; premental setae 18 & 18; palpal setae 11 + 11. Distal margin of prementum is produced into a blunt hump. Abdomen oval with crescent-shaped brown markings dorsally. Mid-dorsal abdominal spines on segments 3rd to 8th. Lateral abdominal spines on segment 8th and 9th extending upto the end of the anal appendages.

Habitat—Larvae are very active and inhabit the sides of the lake along with the rotting leaves and vegetation.

Sub-family-TRITHEMISTRINAE

Trithemis aurora (Burmeister) (Fig. 13)

Head somewhat oval with antero-lateral eyes. Premental setae 10 & 10; paipal setae 7 + 7. Mid-dorsal abdominal spines present on segments 3rd to 9th. Lateral spines present on segments 8th and 9th.

Habitat—Larvae sluggish and found among stable objects in water.

Sub-family-TRAMEINAE

Tramea virginia (Rambur) (Fig. 14)

Labium spoon-shaped, large and when held close, covers the anterior half of the head almost up to the eyes. Premental setae 13 & 13; Palpal satae 10 & 10. Lateral abdominal spines present on segments 8th and 9th.

Habitat—Very active swimmers and at rest, dwell amidst weeds on the sides of the lake. Very rare in occurrence and were not commonly found like the other odonate larvae.

HEMIPTERA

Family-GERRIDAE

Gerris sp. (Fabr.) (Fig. 15)

Medium sized somewhat flattened, macropterous long legged insects with head pointed in front and widest at the eyes. Antennae four segmented. Eyes large, ocelli often greatly reduced. Claws inserted preapically on tarsus. Face between eyes yellowish with a black V-shaped spot in middle; two longitudinal black fascia starting from base of head to anterior margin of black eyes.

Habitat—Fast flowing streams

Eotrechus sp. Kirk (Fig. 16)

Resembling *Gerris* in general appearance except that claws are inserted at apex and it has a round yellowish brown spot in the middle of anterior lobe of pronotum.

Halobates sp. Eschsch. (Fig. 17)

Presence of a fringe of long hair on the intermediate tibia and tarsi is characteristic.

Habitat—Though Distant (1904) records it to be marine but it has been reported from fresh water as well.

Gerastratus foveatus Distant (Fig. 18)

Basal joint of protarsus tougher than apical segment. Body pale with black spots on different parts dorsally.

Habitat—Fast flowing streams and pools.

Metrocoris compar Buch. (Fig. 19)

Basal joint of protarsus shorter than apical one. Pronotum with U-shaped pale spots laterally on anterior lobes and two big U-shaped spots, two semilunar spots laterally and an inverted V-shaped pale spot medially near posterior border.

Habitat—Ponds as well as flowing waters.

Family - MESOVELLIDAE

Mesovelia mulsanti (Buch.) (Fig. 20)

Head triangular with prominent eyes. Antennae four segmented; Rostrum extending beyond the middle of fore coxae. Hemielytra not developed fully. A pair of moderately developed claws present at the apex of each leg. Body pale beneath with greyish pubescence.

Habitat—Small ditches and ponds.

Family-VELLIDAE

Microvelia signalensis Kirk. (Fig. 21)

Head small triangular with prominent eyes. Ocelli absent. Pronotum with a longitudinal central carina. Hemi-elytra well developed. Head and pronotum piceous, body piceous beneath with silvery-grey pubescence.

Habitat—Ponds as well as small ditches.

Family - HYDROMETRIDAE

Hydrometera vittata Stal. (Fig. 22)

Head elongated, beyond one and half times longer than pronotum with a median carina extending through out the length of head and pronotum. Rostrum shorter than head. Hind legs longer than middle one. Hind tibia longer than femur and middle tibia.

Habitat—Found on the surface of shallow waters close to the shore, amongst all sorts of stranded plant parts and vegetation.

Family - PLEIDAE

Plea buenoi Kirk. (Fig. 23)

Head broader than long. Vertex with a short elongated median dark spot and with a transverse black fascia below it. Rostrum short. Pronotum wider than long. Fore wings cover the abdomen completely, membrane not clear, veins not visible. Hind legs longer than middle ones. Claws well developed. Body yellowish with black blotches at certain areas dorsally with a silver shine.

Habitat—Amongst the vegetation in still waters.

Family - CORIXIDAE

Corixa hieroglyphica Duf. (Fig. 24)

Head peculiar somewhat crescent in outline with convex dorsum overlapping pronotum. Antenna four segmented. Rostrum short. Fore legs raptorial. Middle legs long used for walking and clinging. Hind legs also elongate and ornamented with stiff swimming hair. Pronotum yellowish with transverse black lines.

Habitat—Found in ponds.

Micronecta striata Kirk. (Fig. 25)

It is characterised by the presence of 3-antennal segments and uncovered scutellum. Head pale yellow with a small triangular brownish spot at the base.

Habitat—Preferring ponds and shallow and calm back waters.

Family -NOTONECTIDAE

Enithares sp. Spin. (Fig. 26)

Medium sized, head inserted into pothorax. Eyes large ocelli absent. Antennae four segmented. Rostrum long curved. Pronotum short and transverse. Hemielytra well developed. Hind legs long, natatory. Claw single sharp and pointed. Head ochraceous; scutellum black with pale U-shaped marking in the centre.

Habitat—Found in ditches.

Anisops sardea Herr. Schaff. (Fig. 27)

It differs from ***Enithares*** sp. by possessing moderately elongated transverse pronotum pale in colour with three black spots.

Habitat—Surfaces of freshwater ditches.

Family–BELOSTOMATIDAE

Belostoma indicum Lep. & Sev. (Fig. 28)

Flat bugs with leathery exoskelton which is usually brownish in colour. Head broad and conically extended between eyes. Eyes large occupying much of the head. Thorax broad. Membrane distinct. Posterior tibia and tarsi prominently ampliately compressed. Pronotum with two median pale lines coverging towards apex. Total body length 63 mm.

Habitat—Amongst vegetation inside the ponds.

Sphaerodema rusticum (Fabr.) (Fig. 29)

Body depressed. Head subtriangular but not conically produced. Membrane somewhat small. Hemielytra without any spot. Anterior tarsal claws short. Body thoroughly punctate, ochraceous, pronotum centrally dark brownish. Legs pale with black markings. It is smaller than *Belostoma*; the body length average 23 mm.

Habitat—Ponds as well as running water.

Sphaerodema molestum Duf. (Fig. 30)

Similar to *S. rusticum* is general appearance. But its hemielytra possesses an oval spot and anterior tarsal claws are moderately long. Body length averages about 17 mm. Force femur and tibia with three transverse fascis on each segment.

Habitat—Ponds and lakes.

Family–NEPIDAE

Ranatra elongata Fabr. (Fig. 31)

Body subcylindrical, elongate with paired posterior respiratory siphons which are longer than body length. Antennae three segmented located in depressions beneath and between eyes. Middle and hind legs long and slender and greatly removed from fore legs which are prehensile. Eyes brownish black, head yellowish brown dorsally & pale ventrally. Prosternum pale with a broad longitudinal black fascia running from apex to base. Body length averages 41.55 mm in male and 42.5 mm in female.

Habitat—Found in ponds.

Ranatra filiformis Fabr. (Fig. 32)

Smaller as compared to *R. elongata.* Abdominal appendages shorter than body length and male averages 25 mm in length of abdominal appendages is 22 mm. Female nearly 25 mm long but abdominal appendages average about 18 mm.

Habitat—Found in ponds.

Laccotrephes ruber (Linn.) (Fig. 33)

Body oblong. Pronotum as long as broad, sinuetely narrowed anteriorly. Scutellum slightly longer than broad. Hemielytral membrane complete, reticularely veined. Para-sternites distinct not concealed by the ventro-leteral tergites as in case of genus *Ranatra*. Abdominal appendages longer than body. Male Body length nearly 28 mm and length of abdominal appendages is 31 mm in average. Female body length averages 27.5 mm and abdominal appendages nearly 32 mm in length.

Habitat—Rice fields and small ditches.

Laccotrephes maculatus (Fabr.) (Fig. 34)

Similar to that of *L. ruber* except that body is small and abdominal appendages are also smaller than body length. Male body length averages 18 mm and length of abdominal appendages is 14 mm where as in female body length averages 19 mm and abdominal appendages are 14 mm long.

Habitat—Collected from ponds as well as lakes.

Family–NAUCORIDAE

Heleocoris sp. Stal. (Fig. 35)

Only Nymphal stage has been collected. Head large broad at base. Eyes small convergent. Antennae short. Rostrum three segmented. Thorax wide. Fore legs raptrorial with one segmented tarsus. Middle tibia shorter than femur, hind tibia slender and a bit longer than femur. Tarsi two segmented with a pair of claws. Abdomen six segmented.

Habitat—Amongst vegetation on banks of small streams.

Acknowledgements

Authors are thankful to the Head, Department of Bio-sciences, University of Jammu for providing facilities and respected Prof. Y.R. Malhotra, presently Vice-Chancellor, University of Jammu for suggestions and his keen interest in these studies.

I. EXPLANATION OF THE FIGURES

Fig. 1: *Pseudagrion rubriceps* Selys
Fig. 2: *Pseudagrion laidlawi* Fraser
Fig. 3: *Ischnura delicata* (Hagen)
Fig. 4: *Ischnura senegalensis* (Rambur)
Fig. 5: *Ictinogomphus rapax* (Rambur)
Fig. 6: *Anax p. parthenope* (Selys)
Fig. 7: *Orthetrum sabina* (Drury)
Fig. 8: *Potamarcha obscura* (Rambur)
Fig. 9: *Crocothemis s. servilia* (Drury)
Fig. 10: *Acisoma p. panorpoides* (Rambur)
Fig. 11: *Brachythemis contaminata* (Fabr.)
Fig. 12: *Sympetrum commixtum* (Selys)
Fig. 13: *Trithemis aurora* (Burmeister)
Fig. 14: *Tramea virginia* (Rambur)
Fig. 15: *Gerris* sp. (Fabr.)
Fig. 16: *Eotrechus* sp. Kirk.
Fig. 17: *Halobates* sp. Eschsch.
Fig. 18: *Gerastratus foveatus* Distant
Fig. 19: *Metrocoris compar* Buch.
Fig. 20: *Mesovelia mulsanti* (Buch.)
Fig. 21: *Microvelia signalensis* Kirk.
Fig. 22: *Hydrometera vittata* Stal.
Fig. 23: *Plea buenoi* Kirk.
Fig. 24: *Corixa hieroglyphica* Duf.
Fig. 25: *Micronecta striata* Kirk.
Fig. 26: *Enithares* sp. Spin.
Fig. 27: *Anisops sardea* Herr. Schaff.
Fig. 28: *Belostoma Indicum* Lep. & Sev.
Fig. 29: *Sphaerodema rusticum* (Fabr.)
Fig. 30: *Sphaerodema molestum* Duf.
Fig. 31: *Ranatra elongata* Fabr.
Fig. 32: *Ranatra filiformis* Fabr.
Fig. 33: *Laccotrephes ruber* (Linn.)
Fig. 34: *Laccotrephes maculatus* (Fabr.)
Fig. 35: *Heleocoris* sp. Stal.

II. ABBREVIATIONS USED

AAP - Anal appendages
AB - Abdomen
AN - Antenna
CL - Claw
EP - Epiproct
EY - Eye
FL - Fore leg
FWS - Fore wing sheath
H - Head
HL - Hind leg
HWS - Hind wing sheath
LS - Lateral spines
MDS - Mid-dorsal spines
ML - Median leg
PR - Paraproct
PT - Prothorax
TH - Thorax

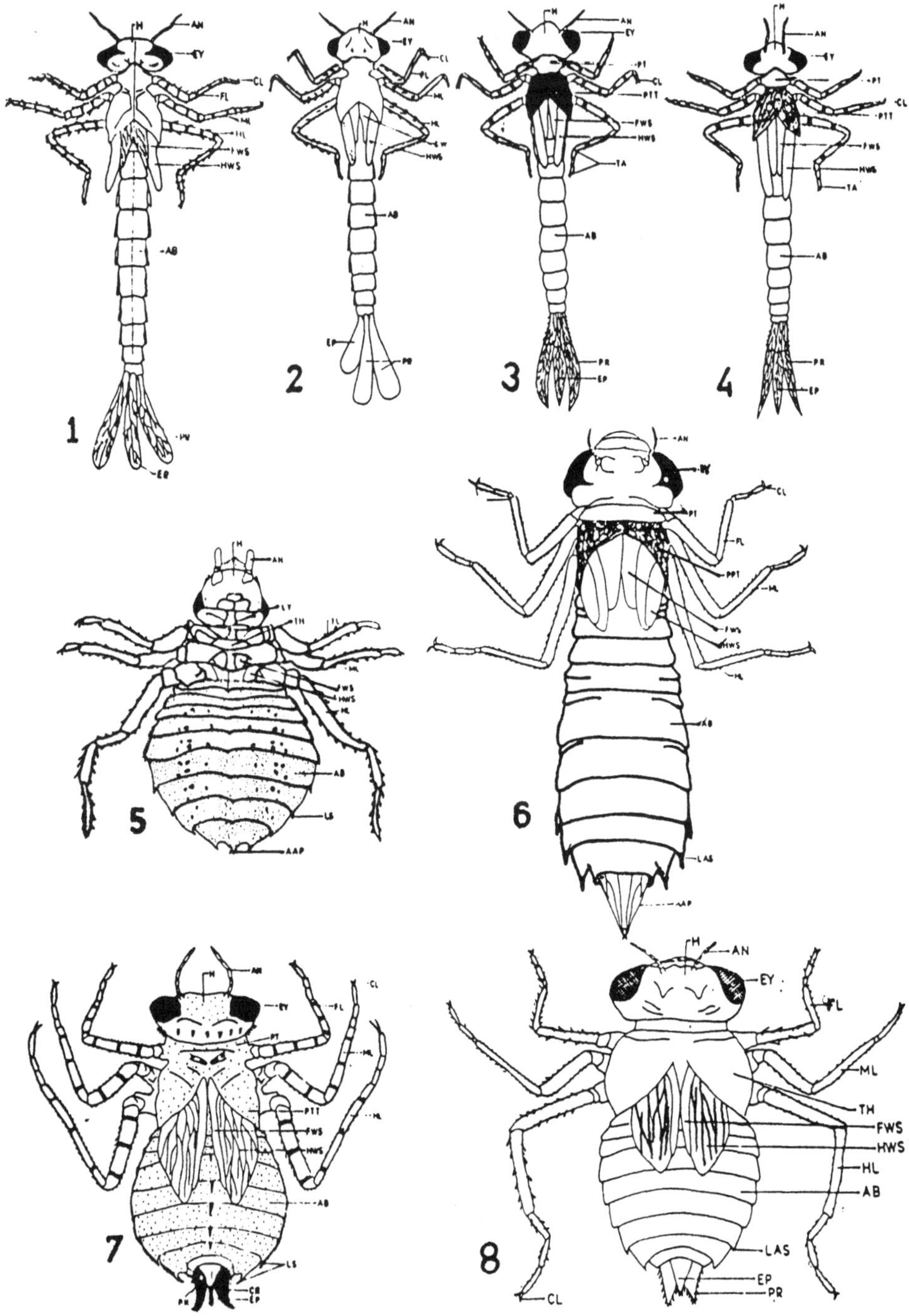
1
2
3
4
5
6
7
8
H
AN
EY
FL
ML
TH
FWS
HWS
HL
AB
LAS
EP
PR
CL

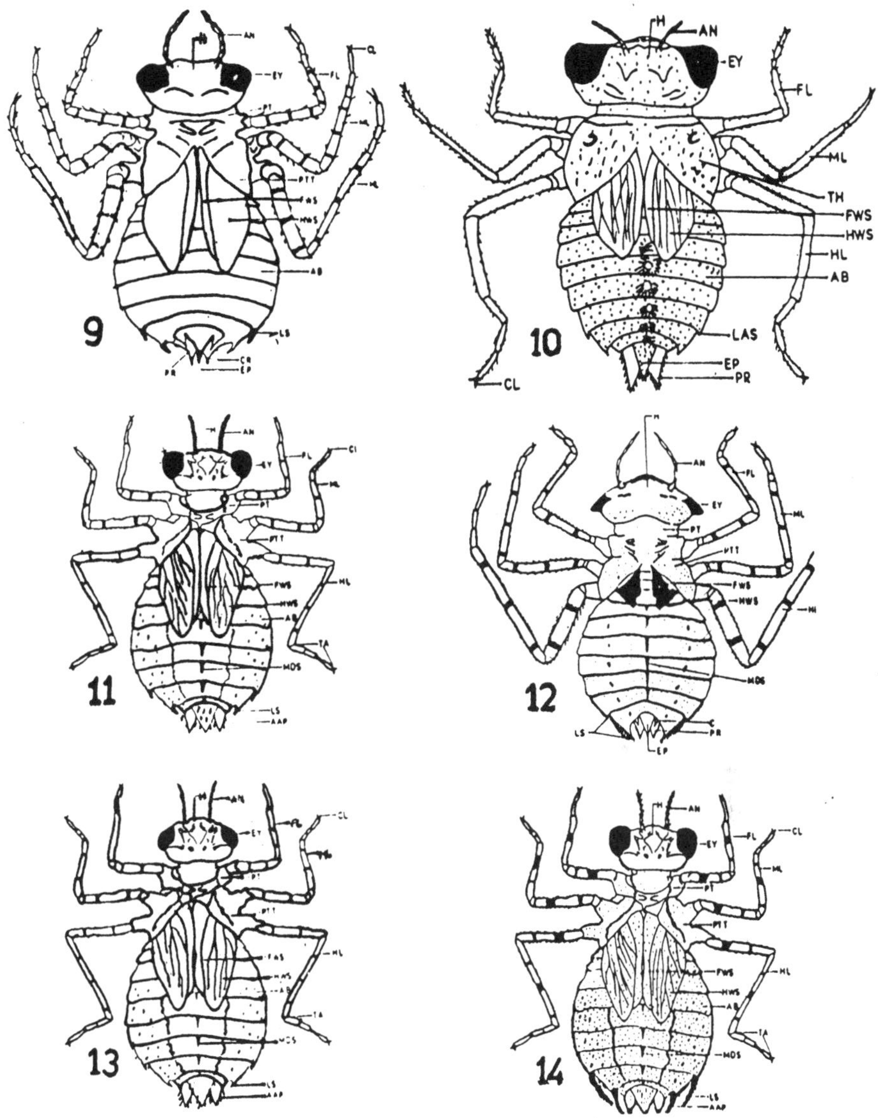
9
10
H
AN
EY
FL
ML
TH
FWS
HWS
HL
AB
LAS
EP
PR
CL
11
12
13
14

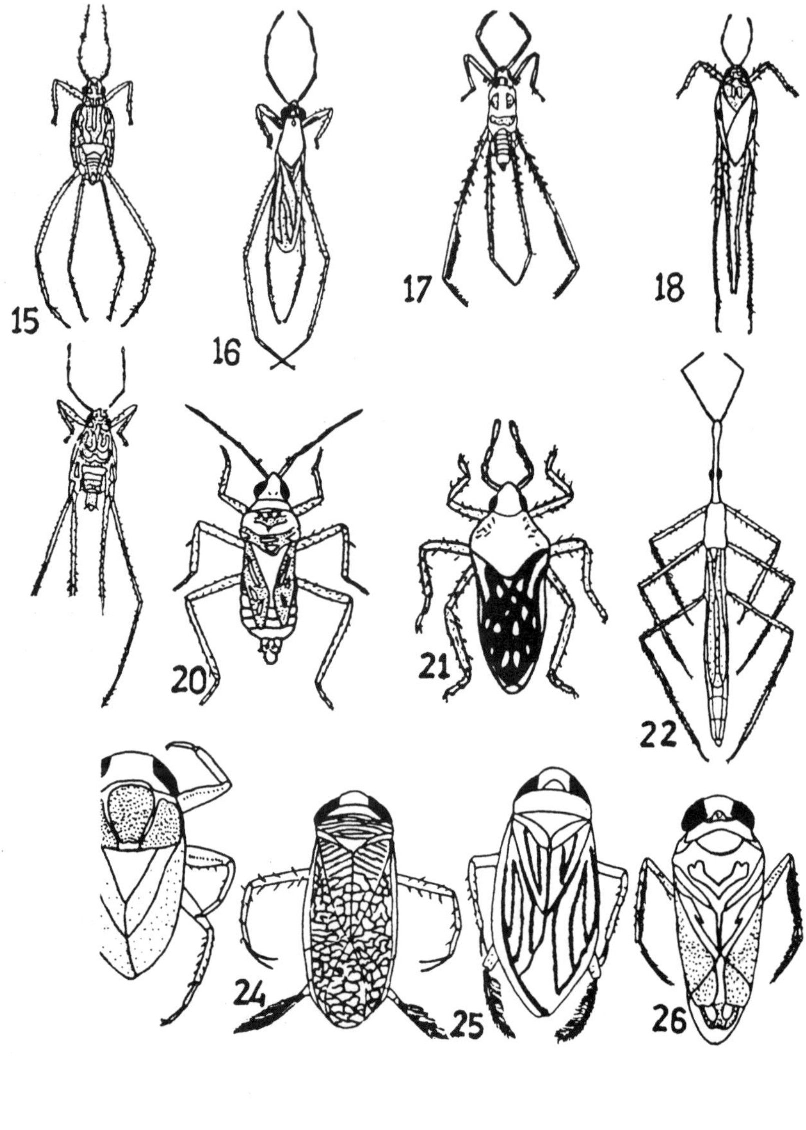
15
16
17
18
20
21
22
24
25
26

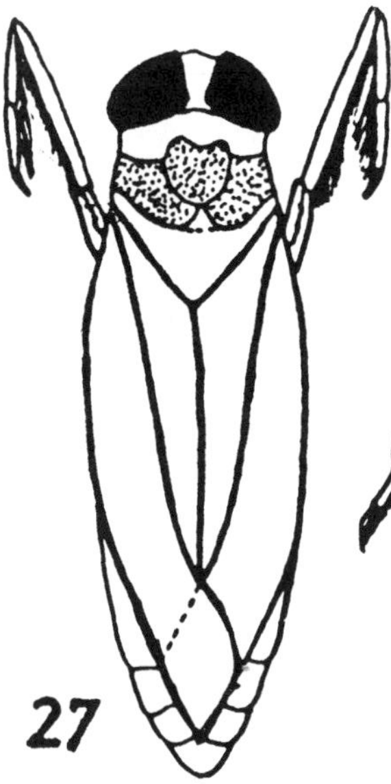

27

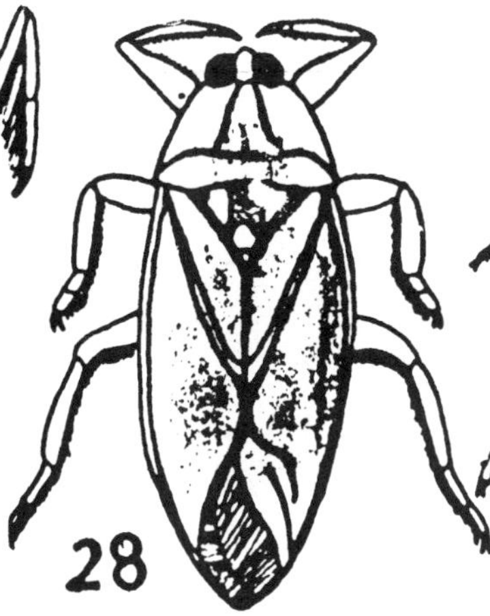

28

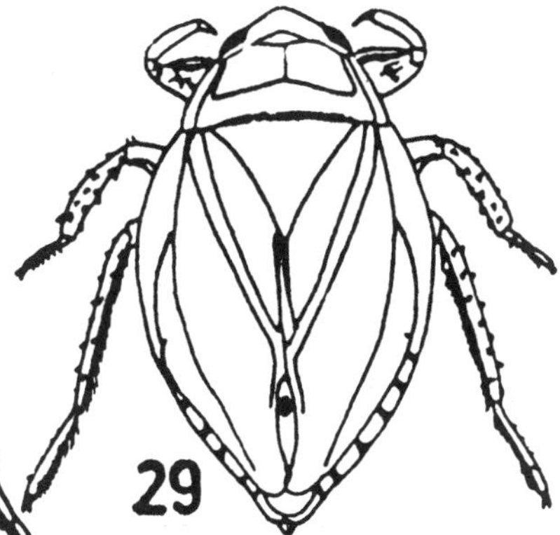

29

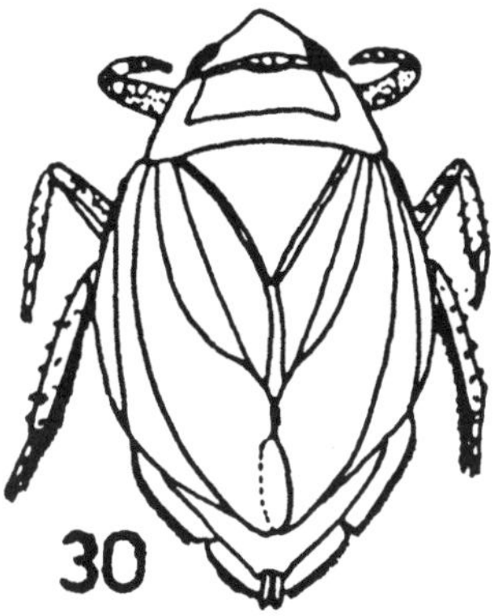

30

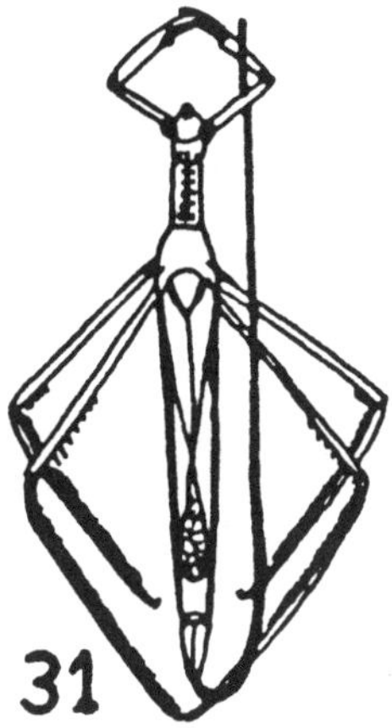

31

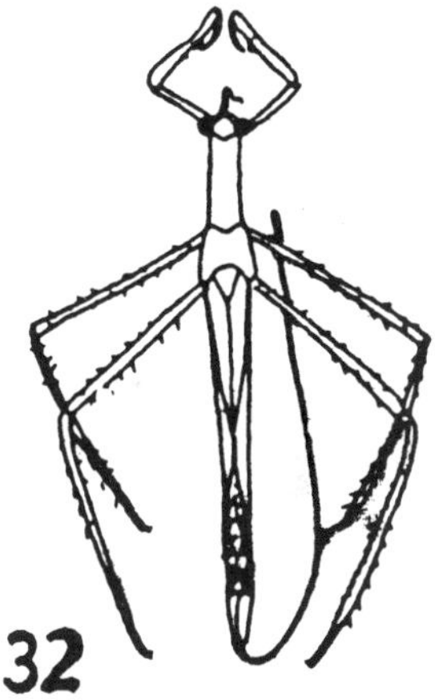

32

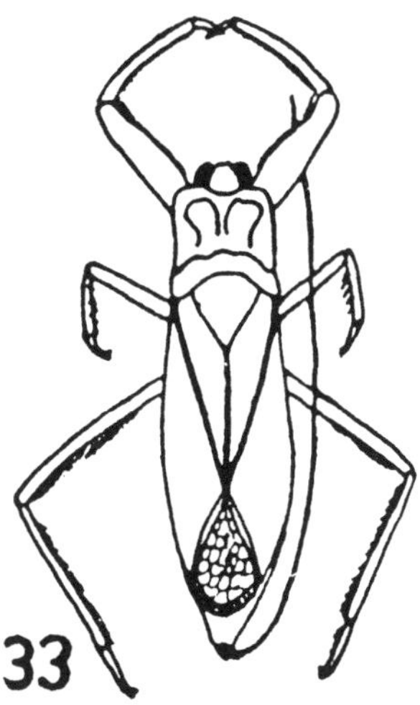

33

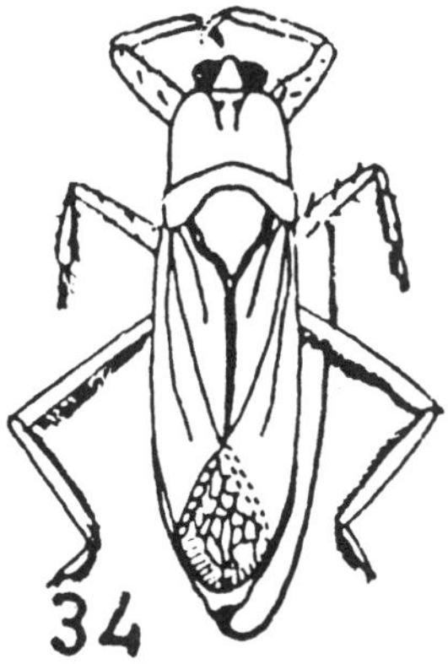

34

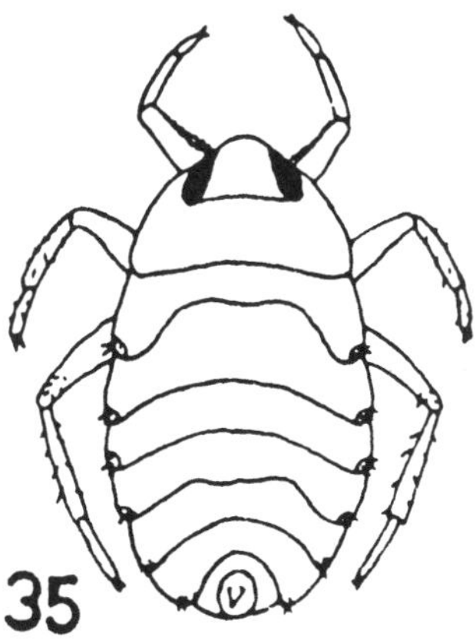

35

References

Alikhunhi, K.H. (1952). On food of young carp fry. J. Zool. Soc. India 4 : 77-84.

Chawdhuri, H. (1960). Contribution to the techniques of pond fish culture in India. Unpublished D. Phil. thesis. University of Calcutta, Calcutta.

Chowdhary (1980). Studies on the bioecology and distribution of aquatic insects of J & K. Ph. D. thesis Kumaun University, Nainital.

Dhalla, N. (1992). Survey of Odonate larvae with observations on their ecology in Lake Mansar, Jammu. M. Phil. dissertation. University of Jammu, Jammu.

Distant, W.L. (1906). Fauna British India, Rhynchota (Heteroptera). 3 : 17-50.

Dutta, S.P.S. and Malhotra, Y.R. (1991). Food and feeding habits of some fishes of Gadigarh, Jammu Proc. Nat. Acad. Sci. India 61 (B) II.

Essig, E.D. (1941). College Entomology, Satish Book Enterprises, Moti Katra, Agra. 263-306.

Gorai, A.K. and Ray Chaudhuri, D.N. (1962). Food and feeding habit of ***Anisops bouvieri*** - Kirk. (Notonectidea). J. Asiat. Soc. 4 (314): 135-39.

Jhingran, V.G. (1982). Fish and fisheries of India (Revised & Enlarged IInd Edition. Hindustan Publishing Corporation (India), Delhi.

Julka, J.N. (1965). Observation on the biology of aquatic bugs injurious to pond fishes (*Anisops bouvieri Kirk*). Proc. Ind. Acad. Sci. 61 (B) 1 : 49-61.

Khan, H.A. and Hussain, A. (1947). A note on the destruction of carp fry by larvae of aquatic insects. Proc. Ind. Sci. Congr. 34(3) : 184.

Mellanby, H. (1963). Animal life in fresh water (London, Chapman & Hall). Redwood Burn Ltd. Trowbridge & Escher. 148-158.

Pakrasi, B. (1954). Preliminary observation on the contact of aquatic insect in nursery ponds. Proc. Ind. Acad. Sci. 38 B(5) : 211-13.

Salaria, R. (1992). Taxonomic survey of aquatic Hemipterans of Jammu region of J & K State. M. Phil. dissertation. University of Jammu, Jammu.

Tillyard, R.J. (1917). The biology of dragonflies. Cambridge Univ. Press. 8 : 327-332.

Tripathi, S.D. and Sharaf, R.K. (1974). Predatory role of weed fishes vis-a-vis aquatic insects, tadpoles and froglets. J.N.K.V.V. Res. J. B. (2) : 123-127.

Wilson, C.B. (1923). Life history of the scavenger water beetle ***Hydrous (Hydrophilus) triangularis*** and its economic relation to the fish breeding. Bull. Bur. Fish. 39: 9-39.

Chapter 8

A Comparative Study of the Renal Organs of the Freshwater Teleostean Fish Part I—Morphology

*B.L. Kaul**

Abstract

The present study of the kidneys of six species of freshwater teleosts from Kashmir valley and the rest of India has revealed that these can be classified into six distinct types viz; *Schizothoracine, Cyprinine, Botine, Silurine, Carpine* and *Notopterine* types.

The study has revealed that the kidney structure is related with phylogeny and evolution. The author believes that it is also closely related with ecology.

Introduction

Kidneys of freshwater teleostean fish play an important role not only as organs of excretion but also in osmoregulation. Despite their importance in the physiology of a freshwater fish much attention of workers has not gone into the study of the morphology of these important organs. The earliest works on these organs are those of Muller (1846) and Hyrtl (1855) who studied the structure of kidneys of *Lepidosteus*. This work was later reinvestigated by Balfour and Parker (1882) for confirmation. Accounts of kidney structure are also found in the classical books of Sedgwick (1909), Lankester (1909), Kingsley (1926), Goodrich (1930) and Parker and Haswell (1940). Ogawa (1961) made a very useful comparative study of the external shape of teleostean kidney and related it to phylogeny. In India serious study of the teleostean kidneys has been attempted by Kamalaveni (1961), Nadkarni (1962) and Kaul (1971).

Materials and Methods

The fishes used in this study are listed in Table 1. In all six species were studied and several fish were used in a freshly killed state or after preservation in 5% formalin. Most of these fishes were collected from river Jhelum, Dal lake, Wular lake and several hill streams of Kashmir valley.

* Department of Zoology, Govt. College for Women, Parade Ground, Jammu (India) Pin 180001.

For making observations on the morphology of the kidney and associated organs, the specimens were dissected and air bladder, alimentary canal, liver, pancreas, gonads were removed. In order to study the head kidney, the oesophagus, pericardium, heart and gills were also removed.

Table 1: Six species of freshwater Teleosteans studied

S. No.	*Species*	*Order*	*Family*	*No. of specimens dissected*
1.	*Schizothorax esocinus* (Heckel)	Cypriniformes	Cyprinidae	25
2.	*Oreinus plagiostomus* (Heckel)	"	"	21
3.	*Cyprinus carpio specularis Linn*	"	"	18
4.	*Crossocheilus latius diplocheilus* (Heckel)	"	"	22
5.	*Botia birdi* Chowdry	"	Cobitidae	27
6.	Glyptothorax *kashmirensis Hora*	"	Siluridae	16

Observations

Morphological study of the kidney of freshwater teleosts has revealed the presence of two distinct parts viz. the head kidney and the kidney proper. There are also the associated organs viz ureters and urinary bladder. A detailed account of the observations made are given hereunder :

Schizothorax esocinus (Heckel)

(i) Kidneys

The kidneys are two flat structures closely applied to the dorsal body wall. They lie throughout the length of the abdomen, extending from the cardiac region to almost the posterior end. In front of the two functional kidneys are present a pair of solid, thick and upturned structures which correspond to the head-kidneys of the other teleosts. The kidneys are dark red in colour. The anterior 1/3 part of each kidney is free and has a wavy margin; the rest 2/3 part is fused with its counterpart. The region immediately behind the point of fusion is the thickest part and has concavity in the middle. A thin transparent layer of peritoneum covers the two kidneys. Ventral to the kidneys lies the air bladder and the former cannot be seen without removal of the latter.

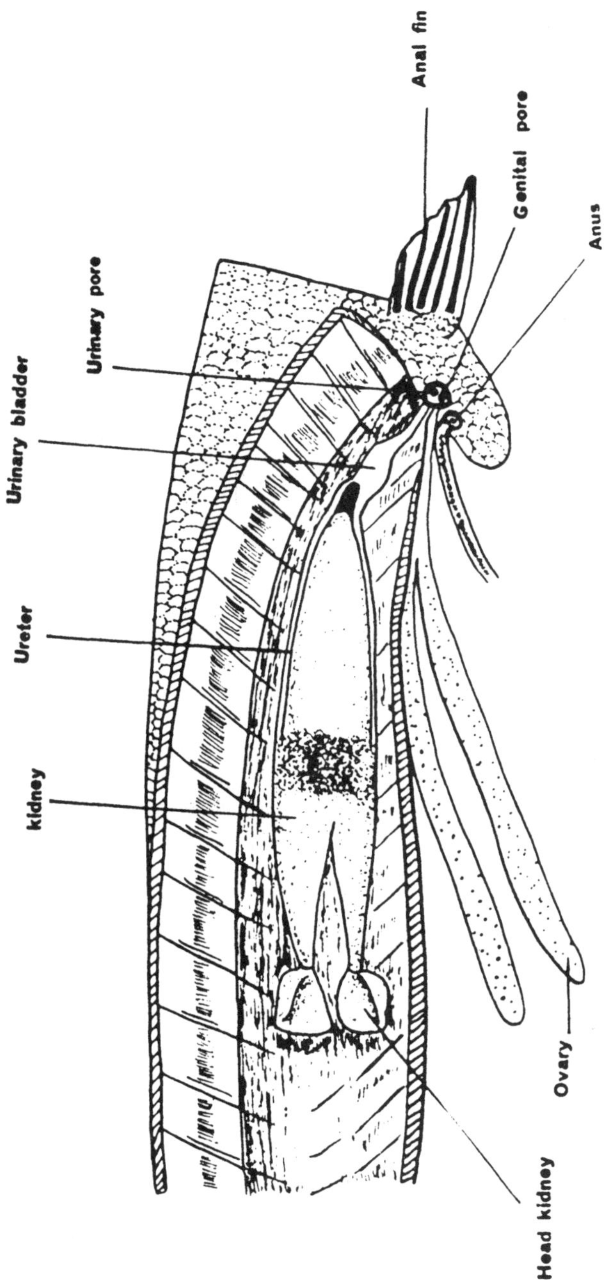
Anal fin
Genital pore
Anus
Urinary pore
Urinary bladder
Ureter
kidney
Ovary
Head kidney
1cm.

Fig. 1

A pair of spherical and white Corpuscles of Stannius were observed on the dorsal surface in the posterior part of kidneys.

(ii) Ureters

There are two ureters (mesonephric ducts) which can be seen as two white tubes running along the two borders of the fused kidneys. They start from about the middle part of the kidneys. Posterior to the kidneys the two ureters are fused and form a single tube. In some the ureters were observed to open separately into the urinary bladder.

(iii) Urinary Bladder

The common ureter becomes dilated to form the urinary bladder. It is narrow in front and behind but is swollen in the middle. It opens into a depression by a urinary pore in both the sexes.

Oreinus plagiostomus (Heckel)

(i) Kidneys

The kidneys are much elongated structures occupying almost the whole length of the abdominal cavity. They are fused along the middle and are quite massive structures. The anterior 1/5 part of each kidney is narrow and free. In front they are connected with the massive head kidneys projecting anterior to the heart. The 4/5 of the functional kidneys are fused, thick, and wide. The thickest part is the middle region. The central part is more or less concave. The fused kidneys are closely applied to the dorsal body wall and can be removed only with difficulty. They are dark red in colour and are masked ventrally by the air bladder. The kidneys gradually become narrow and tapering towards the posterior end. In preserved specimens the colour turns pink or yellow. Two assymmetrically placed Corpuscles of Stannius were observed as small white bodies in the posterior part of the kidneys on the dorsal surface.

(ii) Ureters

There are two long ureters which start out from the thick fused portion and pass along the outer margin of the kidneys upto the posterior end. They receive collecting tubules enroute. The two ureters open into the urinary bladder behind the kidneys.

(iii) Urinary Bladder

The ureters open into the urinary bladder which is narrow and elongated. The urinary pore is separate from the genital pore in both the sexes and lies behind it. There is a depression behind the anus into which the genital and the urinary pores open.

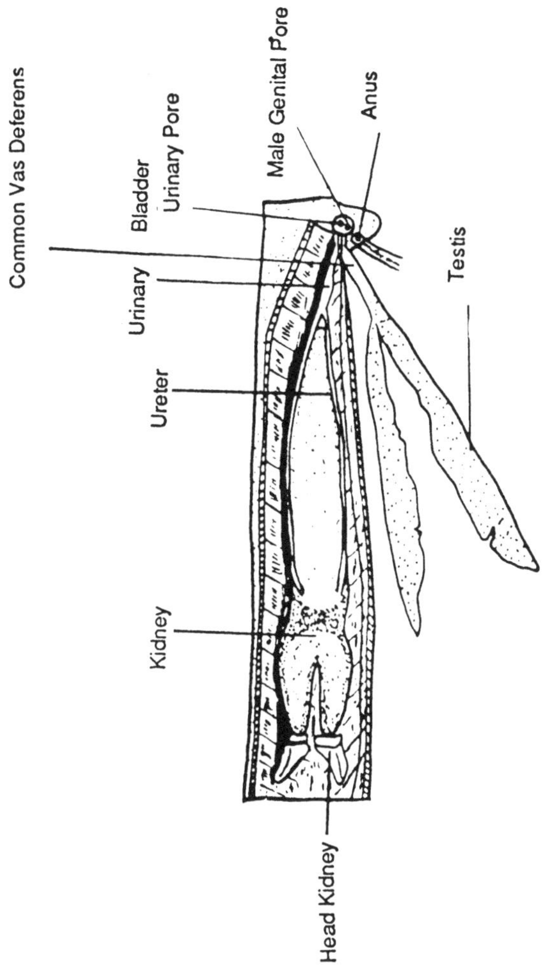
Common Vas Deferens
Bladder
Urinary Pore
Male Genital Pore
Anus
Urinary
Ureter
Testis
Kidney
Head Kidney

Fig. 2

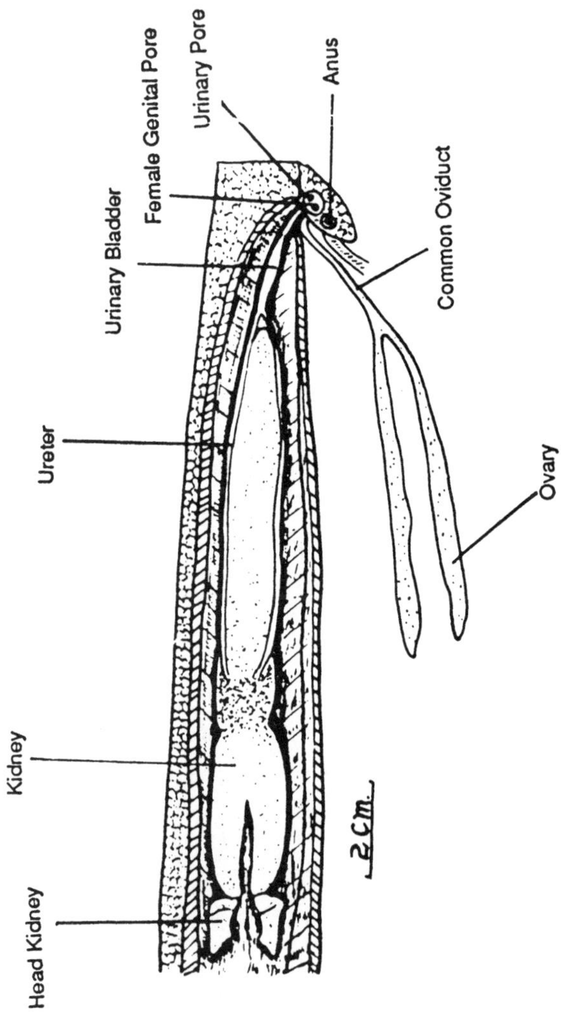
Head Kidney
Kidney
Ureter
Urinary Bladder
Female Genital Pore
Urinary Pore
Anus
Common Oviduct
Ovary
2 cm

Fig. 3

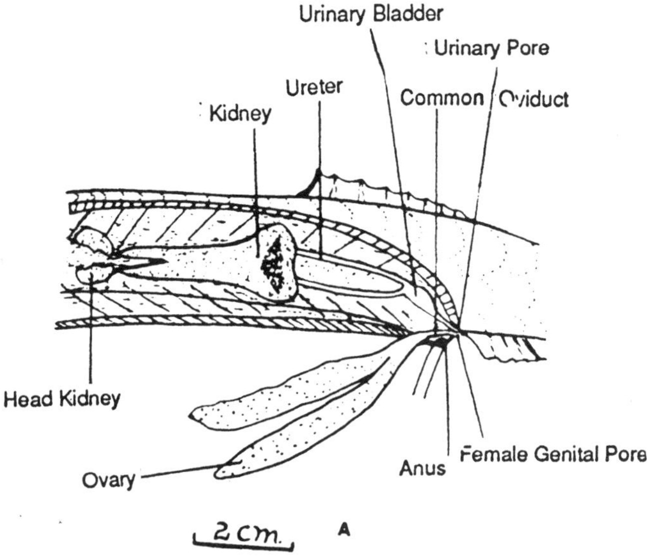
Urinary Bladder
Urinary Pore
Ureter
Common Oviduct
Kidney
Head Kidney
Ovary
Anus
Female Genital Pore
2 cm.

A

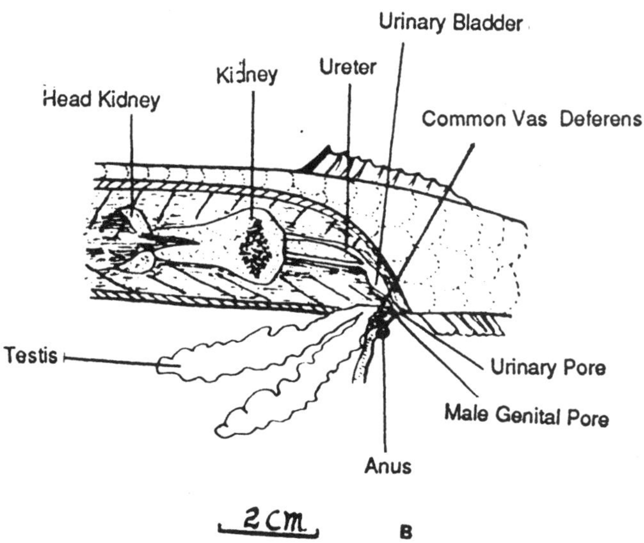
Urinary Bladder
Kidney
Ureter
Head Kidney
Common Vas Deferens
Testis
Urinary Pore
Male Genital Pore
Anus
2 cm

B

Fig. 4A, 4B

Cyprinus carpio specularis Linn

(i) Kidneys

There are two kidneys which are free in the anterior 1/4 part. The rest of each kidney is fused with its counterpart. In about the middle part the kidneys fan out transversely. They are the thickest in this region and are closely applied against the dorsal and dorso-lateral body wall. Since the transverse lobes are quite thick the middle bridge connecting them appears to be concave. Behind this region the kidneys gradually narrow towards the posterior end. The free anterior ends of the kidneys are connected in front with the large head-kidneys. The head-kidneys are transversely placed, are quite large and lie in front of the pericardium. The kidneys extend along the whole length of the body cavity and lie dorsal to the air bladder. They are deep red in colour in fresh specimens but their colour fades on preservation.

The corpuscles of Stannius are a pair of spherical white bodies in the posterior region of the kidneys on the dorsal surface.

(ii) Ureters

There are two ureters which start from the expanded part of the kidneys and run along the thin posterior part of the kidneys. They meet and open into the urinary bladder.

(iii) Urinary Bladder

The urinary bladder is wide anteriorly and in the middle. It narrows towards the posterior end. In one specimen it was observed that the common ureter opened into the side of the urinary bladder rather than at the anterior end so that the latter appeared like a diverticulum. The urinary pore opens into a depression behind the anus.

Crossocheilus latius diplocheilus (Heckel)

(i) Kidneys

There are two dorso ventrally flattened kidneys which are free in the anterior 1/4 part and fused for the 3/4 parts. The free anterior part appears like a V. They are elongated structures being widest and thickest in the region where they become fused with each other. The posterior part is narrow and tapering. The margin of the anterior free part is slightly wavy. There are two head-kidneys which are flat and thick. They project out antero-laterally and lie anterior to the pericardium. The kidneys are dark red in colour in fresh specimens and reddish-brown or light yellow in preserved specimens. The fused kidneys are closely applied to the vertebral column lying immediately above them. The air bladder covers them ventrally and unless it is removed they cannot be seen.

The Corpuscles of Stannius were not observed in this fish.

(ii) Ureters

The two ureters are visible as two white tubes coming out of the thick fused middle region. They run backwards and are closely applied to the sides of the kidneys. They meet on the ventral side of the fused kidneys in front of their posterior end and open into the urinary bladder.

(iii) Urinary Bladder

It is a wide chamber, being widest in its middle. It narrows posteriorly and opens by a urinary pore into a depression lying behind the anus in the ventral body wall. The genital pore opens in front of it.

Botia birdi Chowdhry

(i) Kidneys

There are two kidneys which are fused almost completely. The double nature of the kidneys can only be hinted at on account of a slight furrow at the anterior end. The kidneys are dorso-ventrally flattened, closely applied against the vertebral column and the dorsal body wall. They are narrow at the anterior end and become wider immediately behind. The wide part graduallynarrows towards the posterior end. There appear to be no head-kidneys in this fish.

A pair of Corpuscles of Stannius were observed as small structures in the posterior dorsal part of kidneys.

(ii) Ureters

There are two ureters which emerge from the wider segment of the fused kidneys. They pass backwards along the sides of the fused kidneys. Posteriorly they open into the urinary bladder.

(iii) Urinary Bladder

The urinary bladder is a wide sac which narrows posteriorly and opens into a chamber into which genital pores also opens.

Glyptothorax kashmirensis Hora

(i) Kidneys

The kidneys are elongated but differ from all other types in being Y shaped. They are almost completely fused except for the anterior end. They are thick and wide anteriorly and narrow down posteriorly. They are dark red structures and are covered by a thin layer of transparent peritoneal membrane. The kidneys are so closely applied to the vertebral column that it becomes difficult to remove them. Closely applied to the anterior most end are two transversely placed thick and flat head-kidneys.

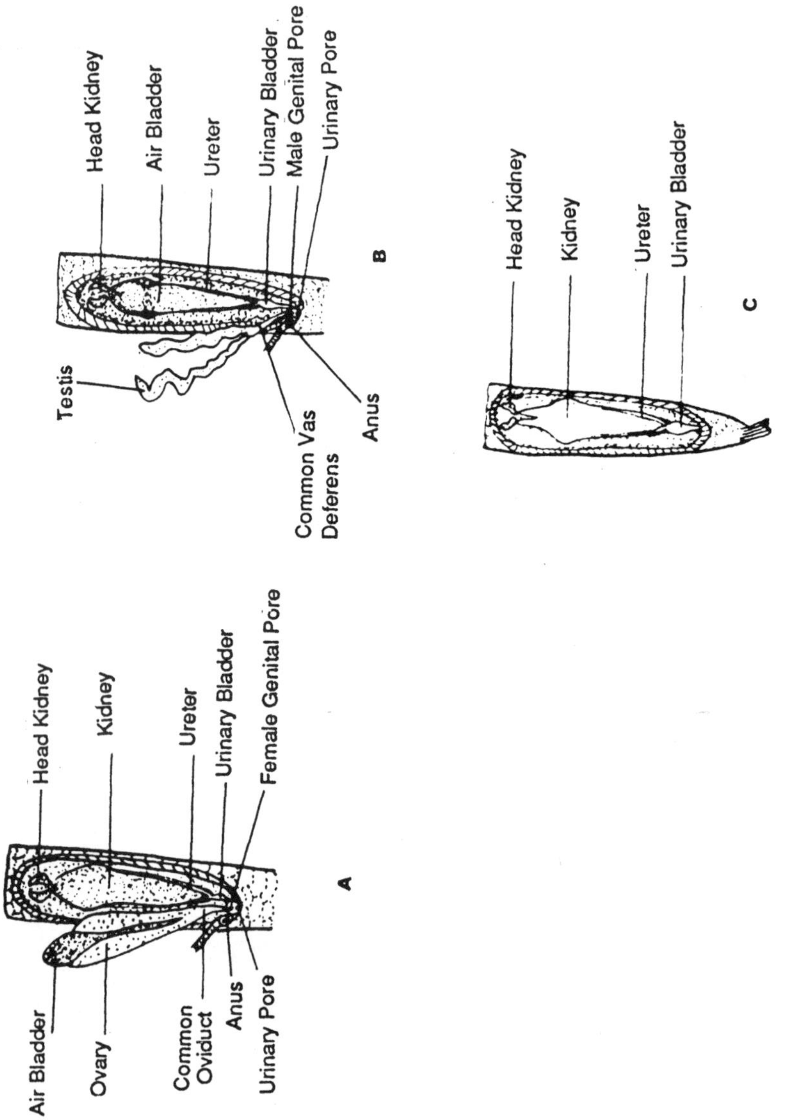
A
Air Bladder
Ovary
Common Oviduct
Anus
Urinary Pore
Head Kidney
Kidney
Ureter
Urinary Bladder
Female Genital Pore
B
Testis
Common Vas Deferens
Anus
Head Kidney
Air Bladder
Ureter
Urinary Bladder
Male Genital Pore
Urinary Pore
C
Head Kidney
Kidney
Ureter
Urinary Bladder
2cm

Fig. 5A, B and C

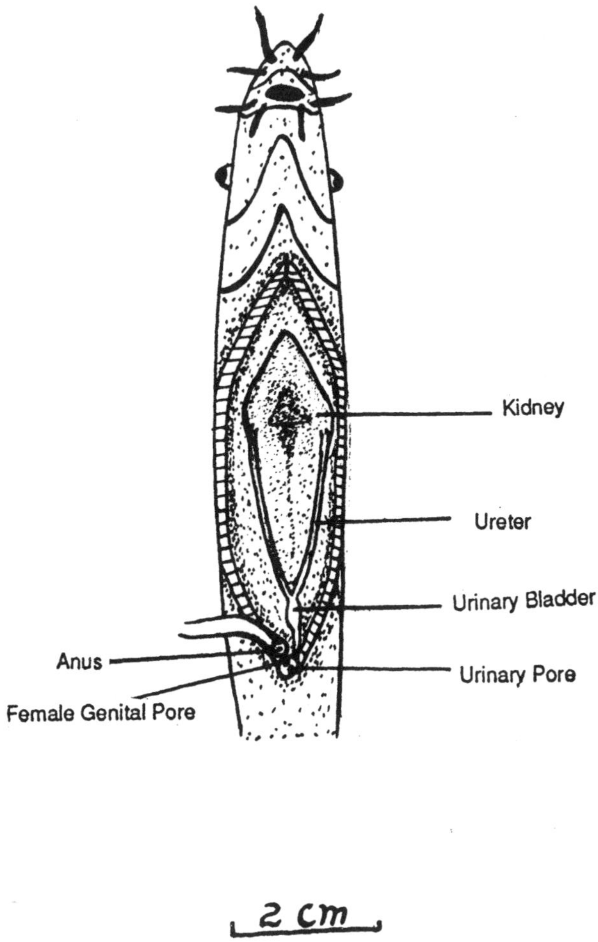
Kidney
Ureter
Urinary Bladder
Anus
Urinary Pore
Female Genital Pore
2 cm

Fig. 6

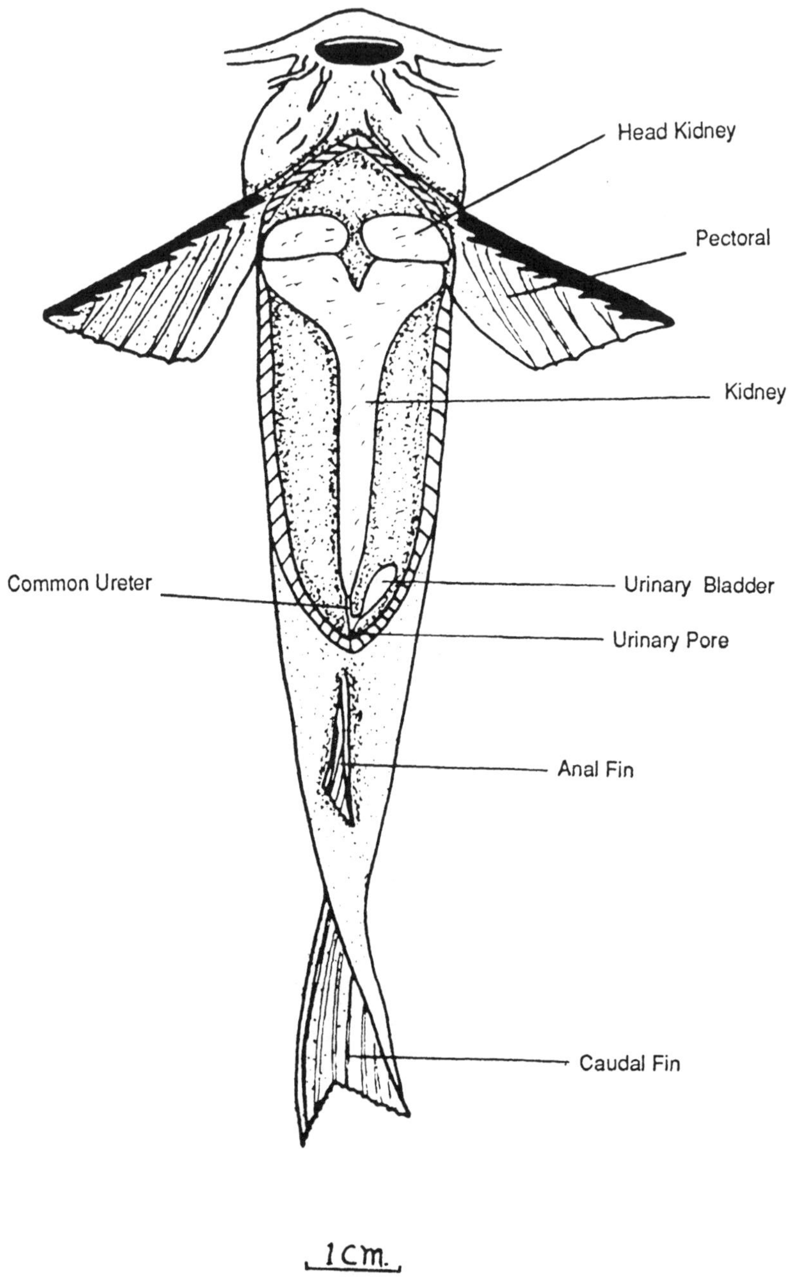
Head Kidney
Pectoral
Kidney
Common Ureter
Urinary Bladder
Urinary Pore
Anal Fin
Caudal Fin
1cm.

Fig. 7

The Corpuscles of Stannius are present as a pair of white spherical bodies in the posterior region of the fused kidneys. They are placed more or less laterally.

(ii) Ureter

There is a single common ureter which comes out of the posterior most end in the middle.

(iii) Urinary Bladder

It is a large sac-like diverticulum. In the male it sends a duct which joins the common vas deferens and opens with it on the urinogenital papilla by a common pore. In the female it sends out a small duct which opens by a urinary pore on a urinary papilla behind the genital pore.

Discussion

The major organs of excretion among fishes in general and teleosts in particular are kidneys. There is a great variation met with in the structure and form of kidneys among fishes.

The general structure and form of kidneys in *Schizothorax esocinus, Oreinus plagiostomus, Cyprinus carpio specularis, Crossocheilus latius diplocheilus, Botia birdi,* and *Glyptothorax kashmirensis,* conform to the basic teleostean plan. The kidneys are paired and their double nature is obvious from the incomplete fusion of the two in most cases (*Schizothorax, Oreinus, Crossocheilus* and *Glyptothorax*). Even in *Botia* its double nature can be identified by a median furrow and the two ureters lying on opposite sides of the fused kidneys. The double and separate nature of the teleostean kidneys was established by Balfour and Parker (1882), Goodrich (1930) and Bridge (1932). Kingsley (1926) pointed out the fusion of kidneys in fishes. The degree of fusion of the kidneys in teleosts is subject to considerable variation as reported by Ogawa (1961). However, an exceptional condition of a teleostean kidney was reported by Owen (1938) in *Cyclothone* in which the kidneys are represented by just a pair of tubules. The present observations on the nature of fusion of kidneys in *Schizothorax, Oreinus, Cyprinus, Crossocheilus, Botia* and *Glyptothorax,* also tend to indicate that the degree of fusion is considerably variable in teleosts studied.

The presence of head-kidneys in teleosts is a fairly constant feature. Except in *Botia birdi* the author has found that the head-kidneys are always present and separate, while in *Botia* the head-kidneys after fusion appear to have degenerated. They are small massive structures and lie near the pericardium. Head-kidneys have been described in teleosts by many authors including Balfour (1881), Sedgwick (1905), Audige (1910), Kamalaveni (1961), Ogawa (1961), Nadkarni (1962), Lal (1963), Sharma (1969) and Hickman and Trump (1969). There is general agreement about the observations of Audige (1910) that the head-kidneys in fishes are remnants of the larval pronephros. Audige (1910) found that head-kidneys in some young fish contain two large glomeruli derived from the pronephros.

The non-renal nature of the head-kidneys in teleosts is clear from their histology. In all the species studied they are composed of lymphoid tissue and contain no renal tubules. The observations of Balfour (1881), Ogawa (1962), Lal (1963), Sharma (1969) and Hickman and Trump (1969) have mentioned the presence of haemopoietic tissue in the head-kidneys of teleosts. The present author has also found presence of haemopoietic tissue in *Schizothorax.*

In *Schizothorax esocinus* the anterior thirds of the kidneys are free while the remainder is fused. A similar condition occurs in *Oreinus plagiostomus*. In *Cyprinus carpio* a somewhat different condition obtains. The anterior fourths of the kidneys are free, for the rest they are fused. The first fused part is quite thick and fans out transversely, posteriorly the fused kidneys narrow down gradually.

In *Crossocheilus latius* the kidneys have wavy margins anteriorly. They are free for their anterior fourths and for the rest they are fused together. A different condition is met with in *Botia birdi* . The two kidneys are fused almost completely with one another. The double nature of the kidneys can be identified on account of a slight median furrow at the anterior end.

In *Glyptothorax kashmirensis* the two kidneys are fused for the most part except anteriorly, where they are thick and wide. Apart from lending support to the generalisations made by Ogawa (1961) this variation however, is specific.

Thus on the basis of the nature of fusion between the kidneys definite morphological types in kidneys can be recognized.

Type I: Kidneys free and narrow anteriorly but fused posteriorly for the most part. This condition is observed in *Schizothorax, Oreinus* and *Crossocheilus* amongst Kashmir fishes and in *Tor putitora* (Lal 1963). This can be called the *Schizothoracine* type of kidneys.

Type II: Kidneys free and narrow anteriorly the thick and fused part in the middle fanning out transversely; the posterior part fused but narrow. This condition is observed in *Cyprinus carpio.* This can be called the *Cyprinine* type of kidneys. Nandkarni (1962) has described a slightly modified type in *Barbus sarana.*

Type III: Kidneys fused almost completely as observed in *Botia birdi* and *Lapidocephalichthyes thermalis* (Nadkarni 1962). This may be called *Botine* type of kidneys.

Type IV : Kidneys fused and thin for the most part posteriorly; anteriorly free and expanded. This may be called the *Silurine* type of kidneys and has been recorded by Kamalaveni (1961) without classification in *Clarias batrachus, Eutropichthyes vacha, Mystus aor, Wallago attu, Heteropneustes fossilis* and *Ophicephalus.*

Type V: In *Labeo rohita, Labeo boggut, Catla catla, Cirrhina mrigala, C. reba* and *Rasbora rasbora,* Kamalaveni (1961) and Nandkarni (1962) observed that the kidneys are bifid anteriorly, fused for a short length, separate again and then becoming reunited posteriorly. No such condition was observed by the author and as such, it is classified as a separate type V (Table 2) and may be called *Carpine type.*

Type VI: Further Kamalaveni (1961) found that in *Notopterus* and *Mastacembelus* the kidneys are bilobed and separated for the greater part of their length. A similar but slightly modified form was observed by Nadkarni (1962) in *Mystus bleekeri.* This is classified as type VI (Table 2) by the author and the name *Notopterine* type suggested for it.

These six morphological types of kidneys are found in Indian and Kashmir fresh water teleosts described so far (Table 2).

Table 2 : Classification of the Kidney Types of Some Indian Freshwater Teleosts

Type	*Condition of the kidneys*	*Fish in which found*	*Author*
Type-I	The kidneys free and narrow anteriorly but fused posteriorly for the most part	*Schizothorax esocinus; Oreinus plagiostomus; Crossocheilus latius;*	Kaul (1971)
		Tor putitora	Lal (1960)
Type-II	The kidneys free and narrow anteriorly, the middle fused thick part fans out transversely, the posterior part fused and narrow.	*Cyprinus carpio*	Kaul (1971)
		Barbus serana (slightly modified)	Nadkarni (1962)
Type –III	The kidneys fused almost completely	*Botia birdi; Lepidocephalichthyes thermalis*	Kaul (1971) Nadkarni (1962)
Type –IV	The kidneys fused for the most part posteriorly. The anterior ends free and quite thick.	*Glyptothorax kashmirensis; Clarias batrachus,*	Kaul (1971)
		Eutropichthyes vacha; Mystus aor; (Heteropneustes fossilis; Wallago attu; Ophicephalus.	Kamalaveni (1961)
Type –V	The kidneys bifid anteriorly, fused for a short length separate again, and then reunited posteriorly.	*Labeo rohita; Catla catla; Cirrhina mrigala; Rasbora rasbora*	Kamalaveni (1961)
		Labeo boggut; Cirrhina reba	Nadkarni (1962)
Type –VI	The kidneys bilobed and separated for the most part.	*Notopterus; Mastacembelus* (slightly modified)	Kamalaveni (1961) Nadkarni (1962)

Ogawa (1961) classified kidneys of teleosts into five types only. This does not fit in with the types found in fresh water teleosts under study. The occurrence of variation in the typical form of fish kidneys has not been understood properly. Walter and Sayles (1949) have opined that "the typical shape of the kidneys of fishes is modified to conform to the presence of swim-bladder which in fishes is a bed fellow of these organs." Ogawa (1961) has suggested that fusion has a relation to phylogeny and evolution. In the light of the present observations the author is inclined to agree with this view but also believes that ecological aspect in shaping the kidney structure cannot be ignored.

Since kidneys develop as a pair of structures from the nephrotomes, a completely separate pair of kidneys should be therefore considered to be a primitive condition and a complete fusion of kidneys as a derived condition. From this point of view the kidneys of *Botia birdi* are more advanced than those of other species shown in Table 2. Whether the fusion of kidneys is of any advantage to the fish is not clear unless the fusion is a result of economy of space available within the fish body.

Whereas the pronephros of the larval teleost is functional, it becomes non-functional as an excretory organ in the adult. Guital (1901, 1902), however, has found that the pronephros retains its renal function throughout life in a few telcosts e.g. *Fierasfer, Zoarces* and*Lepidogaster*. With these exceptions the mosonephric nature of the adult functional kidneys of teleosts is an established fact. These mesonephric kidneys with the exception of *Amia* and *Lepisosteus* lack nephrostomes. Kerr (1919) asserted that since the posterior definitive kidneys of all gnathostomous fishes is derived from the entire caudal portion of the nephrotomic plate, it should be called as opisthonesphros, representing both the embryonic mesonephros plus the caudal nephrogenic material which in higher vertebrates forms the metanephros (Hickman and Trump 1969).

Summary and Conclusions

The kidneys of six species of fresh water teleosts from Kashmir valley of the Indian union have been studied. Comparison of the morphology of kidney of the fishes under study with other Indian fresh water studied earlier by Kamalaveni (1961) and Nadkarni (1962) has revealed that fresh water fishes of India can be classified into six distinct types which have been named as follows:

1. Schizothoracine type
2. Cyprinine type
3. Botine type
4. Silurine type
5. Carpine type
6. Notopterine type

The study has confirmed the view expressed by Ogawa (1961) that the kidney structure is related with phylogeny and evolution. The present author believes that ecology also plays an important role in shaping the kidney form. Except in *Botia birdi* a clear distinction between the head kidney and the kidney proper has also been found.

References

Audige, J. (1910). Arch. Zool. Exp. et gen., Ser. 5, 4 : 275-624. Contribution a l 'e' lude des reins des poissons t' eleost' ee'n.

Balfour, F.M. and W.K. Parker (1882). Phil. Trans., London, 173: 359. On the structure and development of Lepidosteus.

Bridge, T.W. (1932). "Cambridge Natural History" Macmillan and Co., London.

Goodrich, E.S. (1930). "Studies on the structure and development of vertabrates" Macmillan and Co., London.

Hickman, C.P. and B.F. Trump (1969). In "Fish Physiology" (Hoar and Randall eds.). Vol. I. Chap. 2. 91-297. Acad. Press N.Y. and London.

Hyrtl, C.J. (1855). S.B. Akad. Wiss. Wien., Math. Natur Wiss. Kl., Wien 13d., 12, 5: 179. Ueber den Zussammenhan, der geschlachts und Harnwer, Kzengaberden Ganoiden.

Kamalaveni, S (1961). Record. Ind. Mus. Delhi, 59:83-118. Morphology of the urinogenital system in some Indian teleostean fishes.

Kaul, B.L. (1971). Studies on the functional Anatomy of urinogenital system of some Kashmir fishes. Ph. D. Thesis submitted to Kashmir University.

Kingsley, J.S. (1926). "Comparative Anatomy of Vertebrates" P. Blakiston' and Co., Philadelphia.

Lal, M.B. (1963). Ichthyologica (2) 2 : 45-50. Studies on the Urinogenital organs of the Mahaseer *Tor putitora*.

Lankester, R. (1909). "Treatise of Zoology" Part. IX (Vertebrata Craniata), Adam and Charles Black, London.

Muller, J. (1846). Abh. K. Akad. Wiss. Berlin, 1846. Uber d. Ben U. Grenzen der Ganoiden.

Nadkarni V.B. and M.A. Hooli (1962). Proc. Second. All-Ind. Cong. Zool., Varanasi. Urinogenital system in some teleosts.

Ogawa, M. (1961). Sci. Repts. Tokyo, Yoiker Deigaku, Sect. B. 10 (149) : 61-68. Comparative study on the external shape of the teleostean kidney with relation to phylogeny.

Owen, B.B. (1938). Biol. Bull. Woods Hole, 74: 349-363. A simple teleost kidney in the genus *Cyclothne*.

Parker T.J. and W.A. Haswell (1940). "A Text Book of Zoology" Vol. II, Macmillan and Co. London.

Sedgwick, A. (1905). "A Students Text Book of Zoology". Vol. 2, Swan Sonnaschein, London.

Sharma, S. (1969). The annals of Zoology, v (2): 111-114 Haemopoeisis in the kidney of *Notopterus notopterus*.

Chapter 9

A Comparative Study of the Renal Organs of Freshwater Teleostean Fish Part-II—Histology

*B.L. Kaul**

Abstract

Histological studies of the renal organs of the six fresh water teleosts from Kashmir (India) waters have been made. The fish studied include *Schizothorax esocinus* (Heckel), *Oreinus plagiostomus* (Heckel), *Cyprinus carpio specularis* Linn., *Crossocheilus latius* Heckel, *Botia birdi* Chowdhry and *Glyptothorax kashmirensis* Hora. Although the microscopic structure of the renal organs conforms to the structure of such organs in other teleostean fishes yet there are variations in size and structure related to ecology. Large sized giant glomeruli have been observed in *G. kashmirensis*. It has also been observed that smaller the size of the renal corpuscles larger their number and thus greater the area of excretion. Smaller glomeruli are therefore of greater advantage to the fish.

Introduction

Despite the important role played by the kidneys of fresh water teleosts in excretion and osmoregulation scant attention has been paid to the study of their microscopic structure. The earliest studies are those of Muller (1846), Balfour and Parker (1882). Studies made during the present century include those of Marshall and Smith (1930), Marshall (1934), Pfeiffer (1933), Grafflin (1937), Guyton (1935), Ford (1958), Ogawa (1961, 1962), Kamalaveni (1961), Lal (1963), Hickman (1965), Capreol and Sutherland (1968), Sharma (1969) and Kaul (1971, 1974).

The present work is a detailed histological study of the kidneys, ureters and urinary bladder of six fresh water fishes from Kashmir (India). The species studied include *Schizothorax esocinus* (Heckel) *Oreinus plagiostomus* (Heckel), *Cyprinus carpio specularis* Linn., *Crossocheilus latius diplocheilus* Heckel, *Botia birdi* Chowdhry and *Glyptothorax kashmirensis* Hora.

Materials and Methods

The teleosts studied were collected from the river Jhelum and its tributaries and from the famous Dal and Wular lakes in the Kashmir valley (India). Freshly caught

* Department of Zoology, Govt. College for Women, Parade Ground, Jammu (India) Pin 180001.

fish were pithed and dissected in order to remove their kidneys and associated organs i.e. ureters and urinary bladder. These were fixed in Bouin's fixative and formol-alcohol. Sections were later cut 5-7 microns thick and stained in Mallory's triple stain and haematoxyline-eosin.

Observations

Histologically the kidney of a teleost has two distinct parts, the 'head-kidney' without nephrons and the kidney proper with nephrons. A nephron consists of the renal corpuscles (vascular glomerulus in bowman's capsule), a neck segment, the proximal convoluted tubule, intermediate segment and the distal convoluted tubule. Besides, the kidney proper contains a variable amount of intertubular tissue. The associated renal organs include the ureters and the urinary bladder.

1. Schizothorax esocinus (Heckel)

(i) Head Kidney

It is composed of lymphoid tissue and there are neither any glomeruli nor any renal tubules in head kidney.

(ii) Kidney

The kidney is composed of glomeruli, and uriniferous tubules surrounded by lymphoid tissue. The lymphoid tissue consists of granular leucocytes, lymphocytes and lymphoblasts (Fig. 1). The glomerulus is enclosed in a bowman's capsule which is more or less rounded in shape. The glomerulus is well vascularized and the capillaries are in the form of a branched tree, while the substance of the glomerulus contains a large number of R.B.C's shown as black dots diagramatically (fig. 8b). The number of renal corpuscles is the largest in the middle thick and fused part of the kidneys being more concentrated towards the sides than in the centre. The anterior and the posterior parts of fused kidneys possess fewer glomeruli. Occasionally some glomeruli lie closer forming small groups but with separate vascular poles. The bowman's capsule is a little sac of simple squamous epithelial cells. The squamous epithelium is also reflected on to the glomerulus (Fig. 8b). The bowman's capsule shows an invagination at the vascular pole of the glomerulus (fig. 8b). A few R.B.C.'s were observed in the bowman's capsule.

The glomeruli range in size from 45-70 microns x 35-50 microns, the majority being of the intermediate size. They number 9-10 on an average per field at 15x × 10x in zeiss monocular microscope. In the anterior and posterior parts the average number of glomeruli of 15x × 10x is 2-3 per field.

The uriniferous tubule is differentiated into neck segment, proximal convoluted segment, intermediate segment and distal convoluted segment.

There is a short neck segment. It is made up of cuboidal cells, having more or less spherical, basal nuclei. The lumen is small and the average outer and inner diameter is 32 microns and 11 microns respectively. There is a moderate brush border present.

The proximal convoluted tubule has two distinct parts (Fig. 2). The first part is

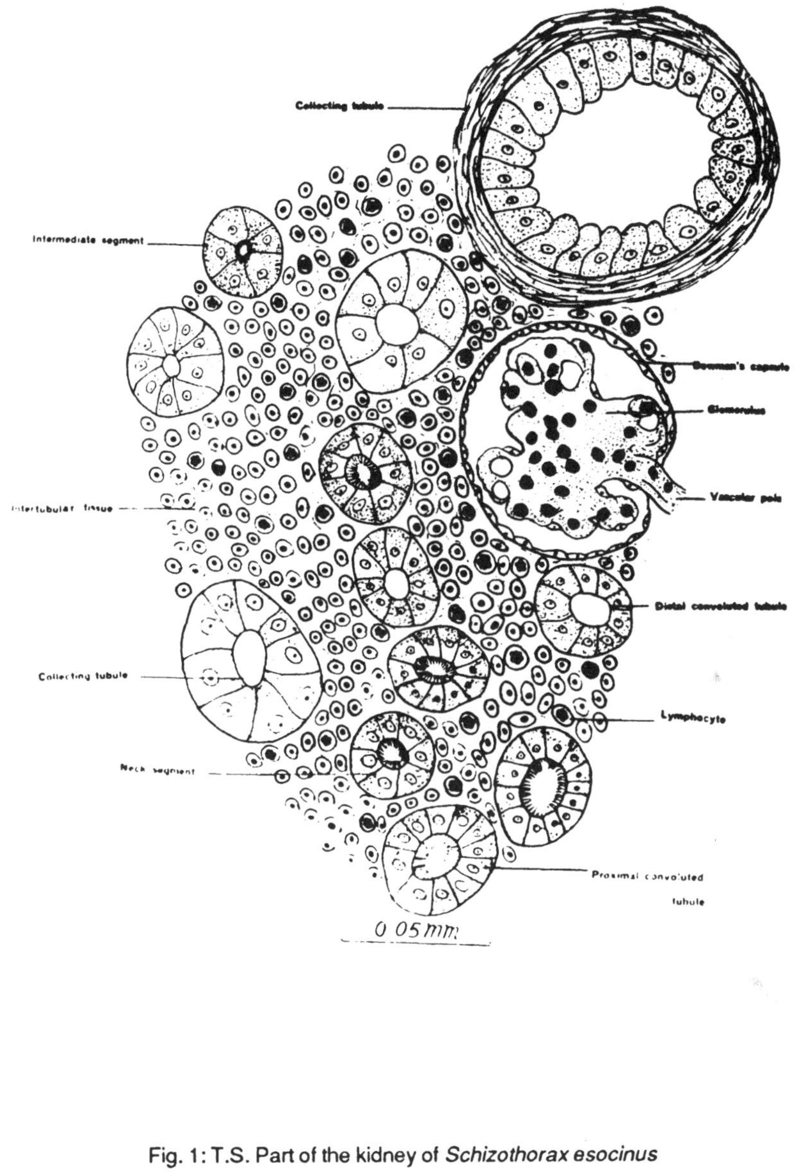

Fig. 1: T.S. Part of the kidney of *Schizothorax esocinus*

circular in outline; it is made up of cells with faint boundaries, granular cytoplasm and oval or oblong nuclei each having a small nucleolus. The external and the internal diameters of this portion on an average are 42 microns and 16 microns respectively. The brush border is about 5 microns in height. The second portion is wider with a narrow lumen. The cells are glandular, have oval nuclei each having a small nucleolus. The external diameter is more than that of the first portion but the internal diameter is lesser. The brush border is lower in height than the first portion being about 3 microns.

The intermediate segment is present between the proximal convoluted segment and distal convoluted segment. It has a narrow lumen with a narrow brush border.

The distal convoluted segment is made of large cells with centrally placed nuclei. They are without a brush border. The average external diameter is 39 microns and the internal diameter is 13 microns.

The initial collecting tubule (Fig. 1) is made up of large cells with large central nuclei. The lumen is wide. The larger collecting tubules are invested with a connective tissue sheath.

The Juxtaglomerular complex was not seen in this fish.

The Corpuscle of Stannius is composed of groups of low columnar cells, covered over by a fibrous capsule.

(iii) Ureter

It has a simple lumen lined by 3-7 cells thick stratified epithelium. There is a deeply staining connective tissue surrounding the stratified epithelium. It contains elastic fibres and blood vessels interspersed in it. In the distal part the ureter is seen to possess a 4-5 cell thick stratified epithelium lining the lumen; while the connective tissue part contains two distinct layers, (a) the connective tissue proper with elastic fibres lying internally and (b) a thick layer of circular muscles lying under the adventitia.

(iv) Urinary Bladder

The wall of the bladder is composed of five layers, which from inside to out are:

(a) Endothelial lining,

(b) Submucosa,

(c) Muscularis of circular and longitudinal muscles,

(d) Tunica adventitia, and

(e) Outer peritoneum.

The endothelium is composed of 6-10 layers of stratified epithelium. The cells have prominent nuclei. Below lies the submucosa with elastic fibres. There is a layer of circular muscles forming a ring and a layer of longitudinal muscle bands interspersed by connective tissue. The adventitia is covered over by peritoneum.

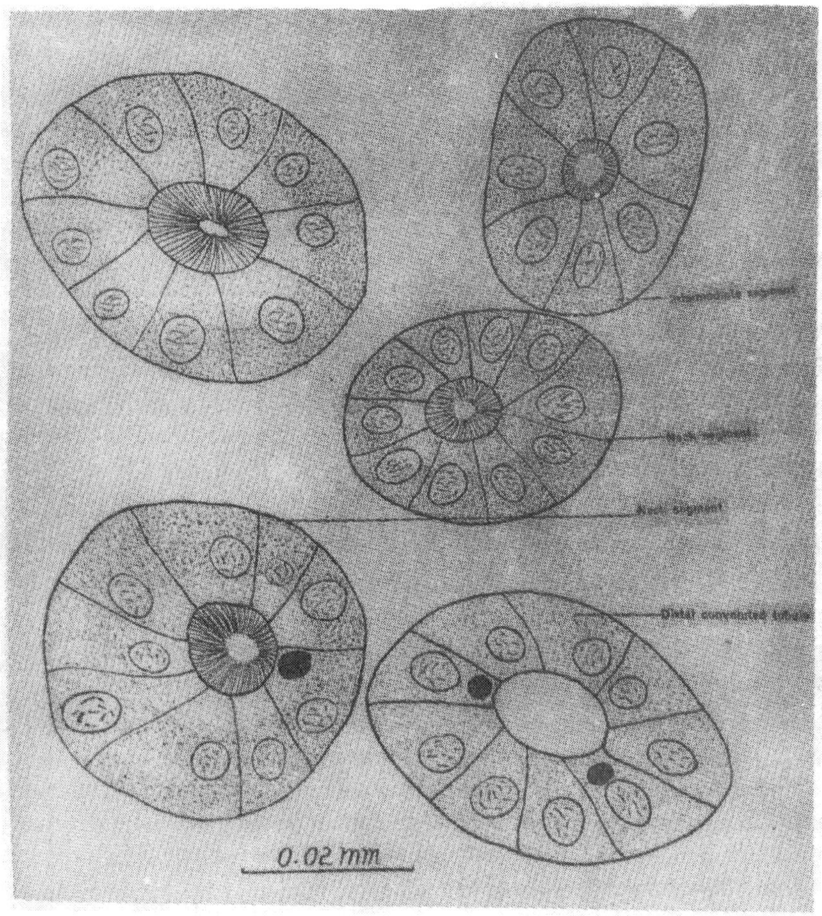

Fig. 2: T.S. of renal tubules of *S. esocinus*

2. Oreinus plagiostomus (Heckel)

(i) Head Kidney

It is composed of lymphoid tissue without any glomeruli or uriniferous tubules.

(ii) Kidney

The kidney is composed of nephrons surrounded by lymphoid tissue (Fig. 3; 9a). The glomerulus is enclosed in a more or less oval bowman's capsule. The bowman's capsule is lined by squamous epithelial cells of which only a few could be seen. The glomerulus also has squamous epithelium of the bowman's capsule reflected on it. The renal corpuscles are densely distributed in the thick fused region of the kidneys. Their number is much less in the anterior and posterior parts. The bowman's capsule is only somewhat turned inwards at the vascular pole. The glomerulus contains a few demarcated lobe-like structures, the R.B.C's lying interspersed in it.

The glomeruli range in size from 40 microns-75 microns × 48-60 microns. The larger number of geomeruli have a size nearer to the lower range. The number of glomeruli per field at 15 x X 10x is 8-9 in the thick middle part, 2 in the anterior part and 3 in the posterior part (Table 1).

The uriniferous tubule presents four clearly differentiated parts, viz, a neck segment, a proximal convoluted segment, an intermediate segment and a distal convoluted part.

There is a short neck segment with a narrow lumen and with a brush border. It has an average external diameter of 36 microns and an internal diameter of 10 microns. The cells are cuboidal with spherical central nuclei.

The proximal convoluted tubule has large and cuboidal cells with abundant cytoplasm. The nuclei are likewise large with prominent nucleoli. The brush border is about 5 microns in height. The external diameter is on an average 40 microns and the internal diameter is about 14 microns (Fig. 3) shows that the proximal convoluted tubule in T.S. as well as in L.S. in which it is apparent that some parts have narrow lumina while others are with wider lumina.

The intermediate segment is made up of large cuboidal cells with spherical central nuclei. The lumen is extremely narrow with a brush border.

The distal convoluted tubule is made up of tall cylindrical cuboidal cells with prominent central nuclei. There is no brush border and lumen is about the same in size as of the proximal convoluted tubule. The average external diameter is 55 microns and the internal diameter about 15 microns.

The initial collecting tubule is wider than the distal convoluted tubule and is made up of large cells. The larger collecting tubules are invested with a layer of fibrous connective tissue (Fig. 3).

The Juxtaglameruler complex could not be traced.

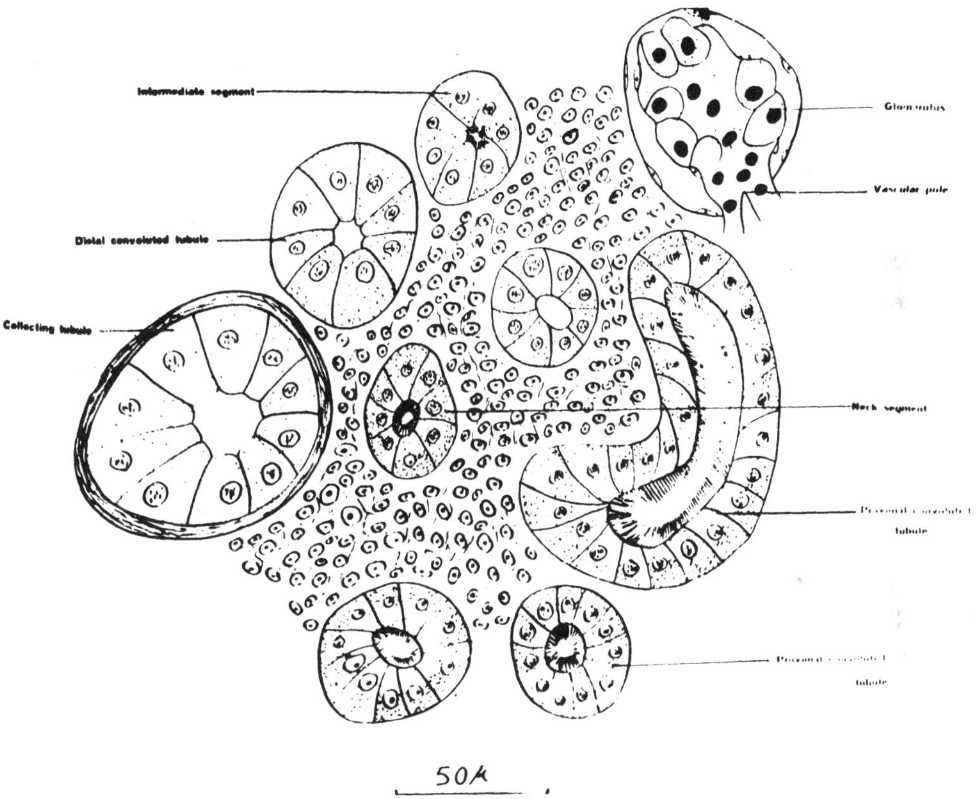

Fig. 3: T.S. Part of the kidney of *Oreinus plagiostomus*

(iii) Ureter

It has a lumen lined by sinuous stratified epithelium 2-5 cells thick. The cells with large nuclei are compactly arranged. The submucosa lying below is a fibrous type of connective tissue. The circular muscles are in the form of a ring. In the lower portion of the ureter nearest to the urinary bladder are present longitudinal muscles outside the circular muscles. There are a few blood vessels present in the connective tissue. The adventitia surrounds the ureter.

(iv) Urinary Bladder

The wall of the urinary bladder is composed of five layers which are from inside to outside as follows:

(a) Endothelial lining,

(b) Submucosa,

(c) Muscularis of circular and longitudinal muscles.

(d) Tunica adventitia, and

(e) Outer peritoneum.

The endothelium is made up of 3-5 or more layers of stratified epithelium. Among the epithelial cells are sometimes present secretory goblet cells (Fig. 12). The stratified epithelial cells have large nuclei and are compactly packed together. Next to the endothelial layer is present submucosa of connective tissue with elastic fibres. This is followed by layers of circular and longitudinal muscles with loose connective tissue interspersed between them. This is present an advertitia covered over by peritoneum.

The urinary bladder of *Oreinus plagiostomus* is often bifurcated internally into two cavities, each showing the same histological elements. A thick common advertitia invests both the cavities (Fig. 12).

In one serial of sections of urinary bladder sections of some unidentified helminth parasite were observed. The endothelial lining of the urinary bladder was found eroded.

3. Cyprinus carpio specularis Linn.

(i) Head Kidney

It does not contain nephrons and is made up only of lymphoid tissue.

(ii) Kidney

It is composed of nephrons surrounded by lymphoid tissue (Fig. 4). The renal corpuscles are rounded in shape. The bowman's capsule enclosing the glomerulus is lined by squamous epithelium reflected on to the glomerulus (Fig. 10b). The thick fused part of the kidney has greater number of renal corpuscles than any other part. The glomeruli range in size from 60-70 microns × 75 microns. The number of glomeruli per field at 15 x × 10x is 5-7 in the thick middle part and 1-3 in the anterior and posterior parts (Table 1). The glomeruli are uniformly distributed in thick middle part and no groups were observed. Each glomerulus has many capillary loops grouped into 6-8 lobes (Fig. 10b).

The uriniferous tubule is composed of a neck segment, proximal convoluted segment, intermediate segment and distal convoluted segment.

Whereas the neck segment, proximal convoluted segment and the intermediate segment possess a brush border the distal convoluted tubule is without it (Fig. 10a).

The short neck segment is made up of cylindrical cells. It is provided with a 3 microns tall brush border. The average external and internal diameters are 50 microns and 14 microns respectively (Fig. 4).

The proximal convoluted tubule has a wide lumen and is made up of cuboidal and cylindrical cells having large centrally placed nuclei. While the brush border in the region is 5 microns tall, the average external and internal diameters are 55 and 15 microns respectively. Besides, the intercellular walls between cells are distinct.

Between the proximal convoluted tubule and the distal convoluted tubule an intermediate segment with well marked brush border is present. It is made up of low columnar epithelium with spherical central nuclei, and has a narrow brush border (Fig. 4). The average external and internal diameters are 40 and 15 microns respectively.

The distal convoluted tubule is made up of columnar epithelial cells with distinct cell membranes. The lumen is large and the average external and internal diameters of distal convoluted tubule are 60 and 20 microns respectively.

Basement membrane surrounds all the types of renal tubules mentioned above. The initial collecting tubules are narrower than the distal collecting tubules. These are made up of distinct simple columnar epithelial cells, with basal nuclei.

Groups of large collecting tubules covered with fibrous connective tissue were observed in the peripheral parts of the kidney (Fig. 10a). These tubules are made up of tall columnar epithelium with basal nuclei and deeply staining nucleoli. The lumina of these tubules are quite large and without brush borders.

Juxtaglomerular complex was not observed.

(iii) Ureter

It is lined with a 2-5 layers thick stratified epithelium. Next to the epithelium lies a submucosa of fibrous connective tissue. The lower part of the ureter has a muscularis is composed of circular and longtiudinal muscles. There is an adventitia which makes outer sheath of the ureter.

(iv) Urinary Bladder

The wall of the urinary bladder has many villi-like epithelial projections making the lumen markedly sinuous. It is composed of five layers which from inside out are:

(a) Endothelial lining,

(b) Submucosa,

(c) Muscularis of circular and longitudinal muscles,

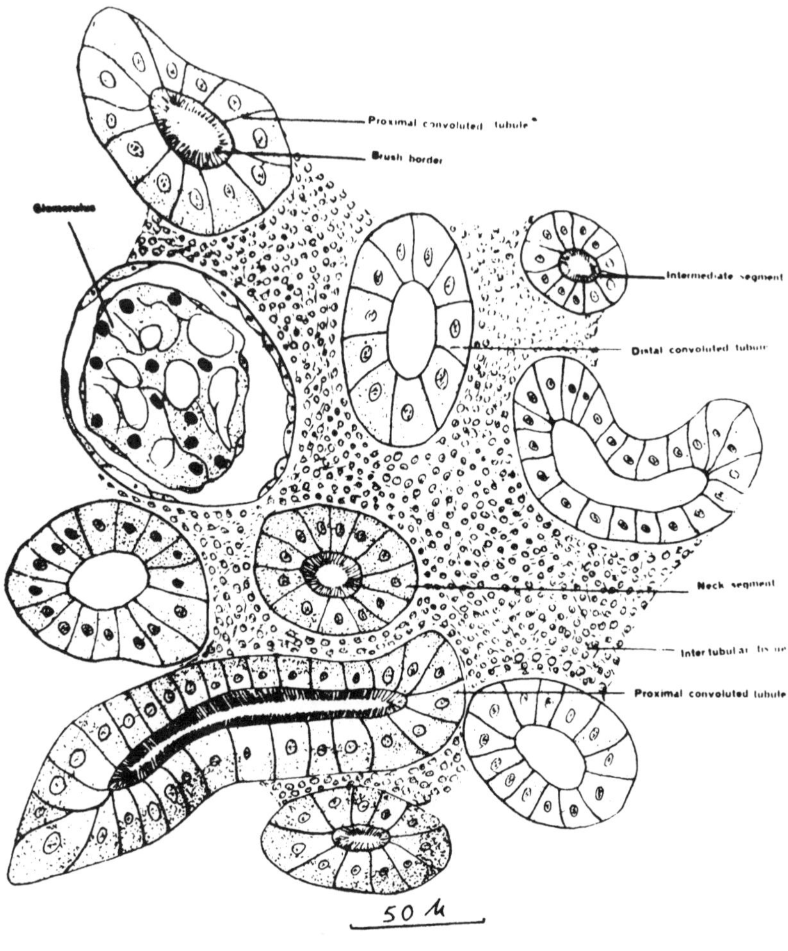

Fig. 4: T.S. Part of the kidney of *Cyprinus carpio*

(d) Tunica adventitia, and

(e) Outer peritoneum.

The endothelium consists of 3-6 or more layers of stratified epithelium. The epithelial cells are small. Capillaries and R.B.C.'s lie scattered among the epithelial cells. Goblet cells in the endothelium were not seen in this species. The submucosa lying below the endothelium is fibrous which stains deeply. There are broad bands of circular muscles interspersed with connective tissue and thin bands of longitudinal muscles. Blood vessels and groups of R.B.C.'s are usually found in the connective tissue between and outside the bands of muscles. The adventitia surrounding the muscularis is often thick and there is a thin peritioneal layer present.

4. Crossocheilus latius diplocheilus (Heckel)

(i) Head Kidney

It is composed of only lymphoid tissue without any nephrons.

(ii) Kidney

The fused kidney consists of nephrons surrounded by lymphoid tissue. The renal corpuscles are more numerous in the thick middle part, being oval in outline. The bowman's capsule is lined by squamous epithelium which is reflected on to the glomerulus. The glomeruli are well formed and each with 3-6 demarcated lobules. A few glomeruli are seen in the anterior and posterior parts of the kidneys.

The glomeruli range in size from 45-85 microns X 35-60 microns in size, majority being 45-40 microns in size. There is little gap between the bowman's capsule and the glomerulus (Fig. 5). The number of glomeruli per field at 15 x × 10x magnification is 3-7 in the thick middle part and 1-2 in the anterior and posterior parts (table 1). The glomeruli tend to get grouped in twos or threes, but have separate vascular poles.

The renal tubule is differentiated into neck segment, proximal convoluted segment, intermediate segment and distal convoluted segment (Fig. 5).

The neck segment is made up of cuboidal epithelial cells with large central nuclei. The average external and internal diameters are 25 and 8 microns respectively. The brush border is 2-5 microns tall.

The proximal convoluted tubule has circular outline with a brush border about 4-5 microns in height. The lumen is wide. The average external and internal diameters are 47 and 16 microns respectively. The cells are cuboidal and cylindrical epithelium with medium size central nuclei.

The intermediate segment is narrower than distal segment. The cells have well marked boundries and there is a low brush border only 3 microns in height. The average external and internal diameters are 28 and 8 microns respectively.

The distal convoluted segment is without a brush border. It is made up of cells with well marked boundries and central nuclei. There is a well marked basement membrane around it. The lumen is narrow and the average external and internal diameters are 50 and 12 microns respectively.

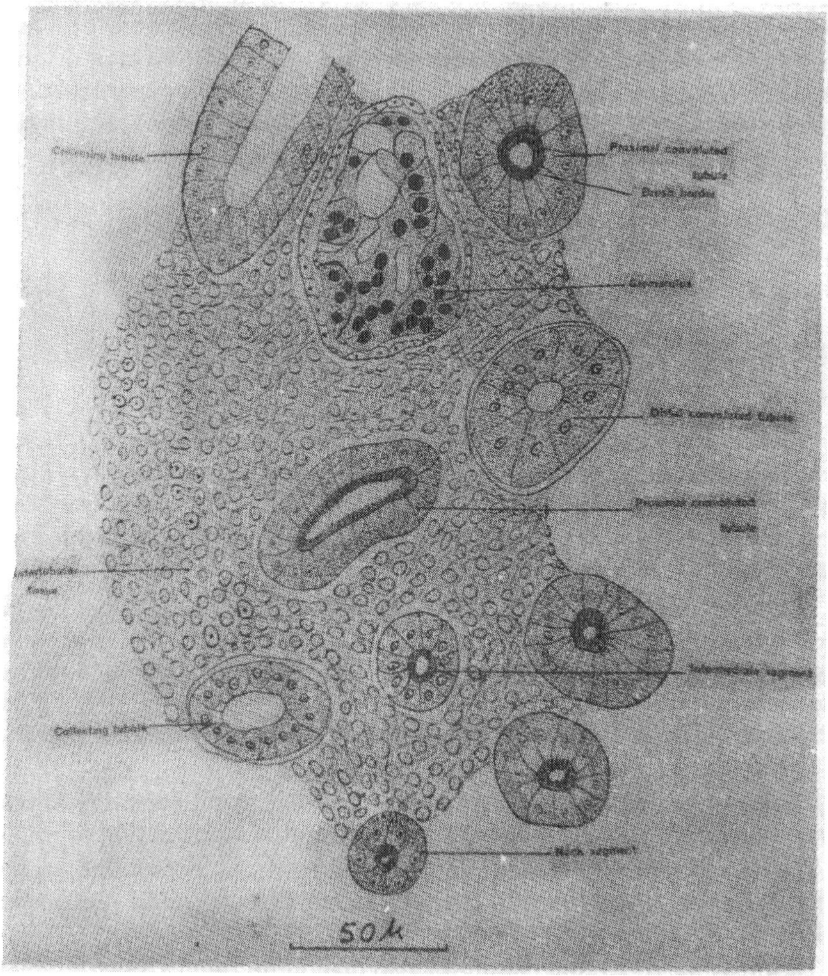

Fig. 5: T.S. Part of the kidney of *Crossocheilus latius*

The initial collecting tubule (Fig. 5) is made up of columnar epithelial cells. The larger collecting tubules are lined by tall columnar epithelium.

No Juxtaglomerular complex was observed.

(iii) Ureter

It has an irregular lumen lined with 2 or 5 layers of stratified epithelium. Next to it is a submucosa which stains lightly. The circular and longitudinal muscles are present. There is a thin layer of adventitia. Goblet cells lie scattered towards the internal part of the endothelium.

(iv) Urinary Bladder

The proximal part of the urinary bladder contains two cavities. But the distal part has a single lumen.

Internally the wall of the bladder has villi-like projections and it consists of the following layers:

(a) Endothelium

(b) Submucosa,

(c) Muscularis of circular and longitudinal muscles,

(d) Adventitia, and

(e) Peritoneum

The endothelium is composed of 3-6 layers of stratified epithelial cells, with deeply staining and prominent nuclei. Numerous secretary vesicles (goblet cells) lie embedded in the epithelium. The secretory vesicles are larger than the epithelial cells. No R.B.C's were observed among the epithelium. The submucosa consists of fibrous connective tissue, with blood capillaries. In some cases secretory cells designated here as mucous cells have also been observed. Both the circular and longitudinal muscles are present in the form of bands interspersed with connective tissue. Thin adventitia and peritoneum mark the outer boundaries. Lymphoid masses and blood vessels were frequently come across in sections.

5. Botia birdi Choudhry

(i) Head Kidney

It is absent.

(ii) Kidney

The glomeruli and renal tubules are well formed in this species. The renal corpuscles are rounded or slightly oval. The bowman's capsule is lined with squamous epithelium, also reflected as a visceral layer on the glomerulus. The urinary pole is distinctly visible (Fig. 11a) and is lined with well marked ciliated epithelial cells. It leads into the neck segment. Though glomeruli are more dense in the thick middle part of kidney they are peripherally disposed. Just a few glomeruli

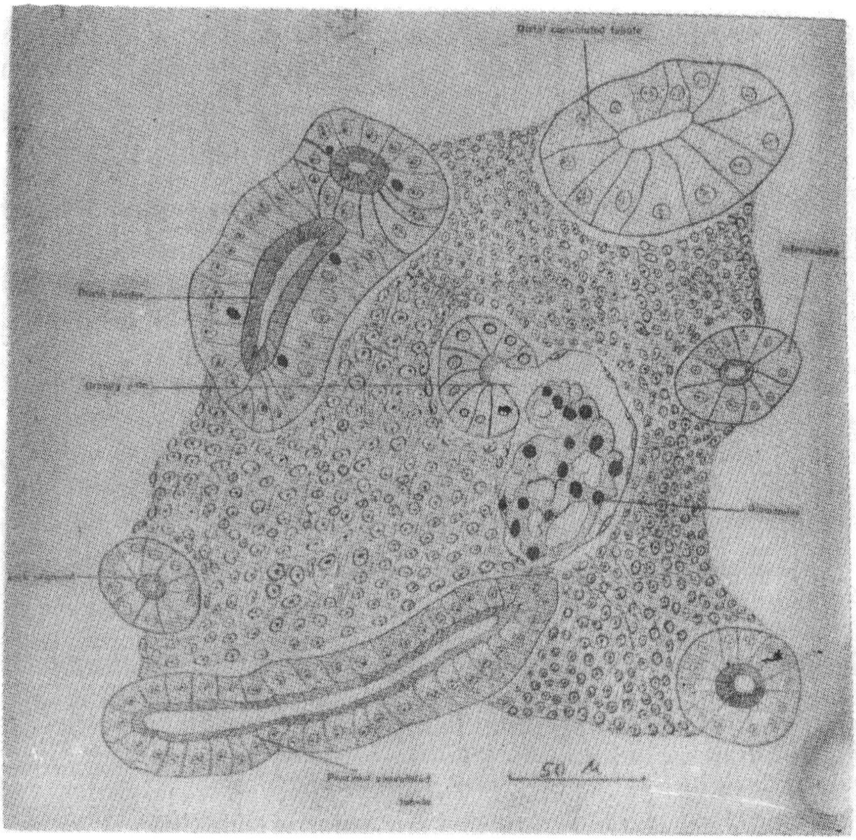

Fig. 6: T.S. Part of the kidney of *Botia birdi*

and renal tubules are present in the other parts of the kidneys. The middle part connecting the two kidneys only, has a few glomeruli.

Glomeruli range in size from 75-80 microns X 35-55 microns. The majority of glomeruli, however measure 65 X 40 microns. The number of glomeruli per field at low power is 4 to 5. A typical glomerulus has a tree-like capillary net-work and is lobulated and interspersed with R.B.C's (Fig. 11a). The anterior part of kidney has no or 1 glomerulus per field. It is mostly composed of lymphoid tissue. The posterior part has 1 or 2 glomeruli per field (Table 1).

The renal tubule is marked into neck segment, proximal convoluted tubule, intermediate segment and distal convoluted segment.

The neck segment is made up of cuboidal cells with large basal nyclei. It measures 30 and 12 microns respectively, in external and internal diameters.

The proximal convoluted segment has a large lumen than the neck segment with a taller brush border. It is lined with cuboidal and cylindrical epithelium having large nuclei and deeply staining nucleoli. It has average external and internal diameters of 45 and 15 microns respectively. The brush border is 5 microns tall. The proximal convoluted tubule is much longer than the distal convoluted tubule (Fig. 6).

The intermediate segment is made up of cuboidal and cylindrical cells with large basal nuclei. The brush border is present and lumen is quite narrow. On an average it measures 35 and 10 microns in external and internal diameters respectively.

The distal convoluted segment is larger. The cells are tall and cylindrical with prominent basal nuclei and distinct boundaries (Fig. 6). It measures about 60 and 12 microns respectively in its external and internal diameters. The lumen is narrow and large collecting tubules are covered with muscular capsule (Fig. 6).

Juxtaglomerular complex was not observed.

(iii) Ureter

It has a simple narrow lumen lined by 2-5 cells thick layer of stratified epithelial cells. Below this layer lies submucosa made up of fibrous connective tissue. The outer adventitia is composed of loose connective tissue and has a few blood vessels in it.

(iv) Urinary Bladder

The urinary bladder has a wide lumen lined by stratified epithelium thrown into small projections. The wall is composed of five layers, which are from inside to outside as follows :

(a) Endothelium,

(b) Submucosa,

(c) Muscularis of circular and longitudinal muscles,

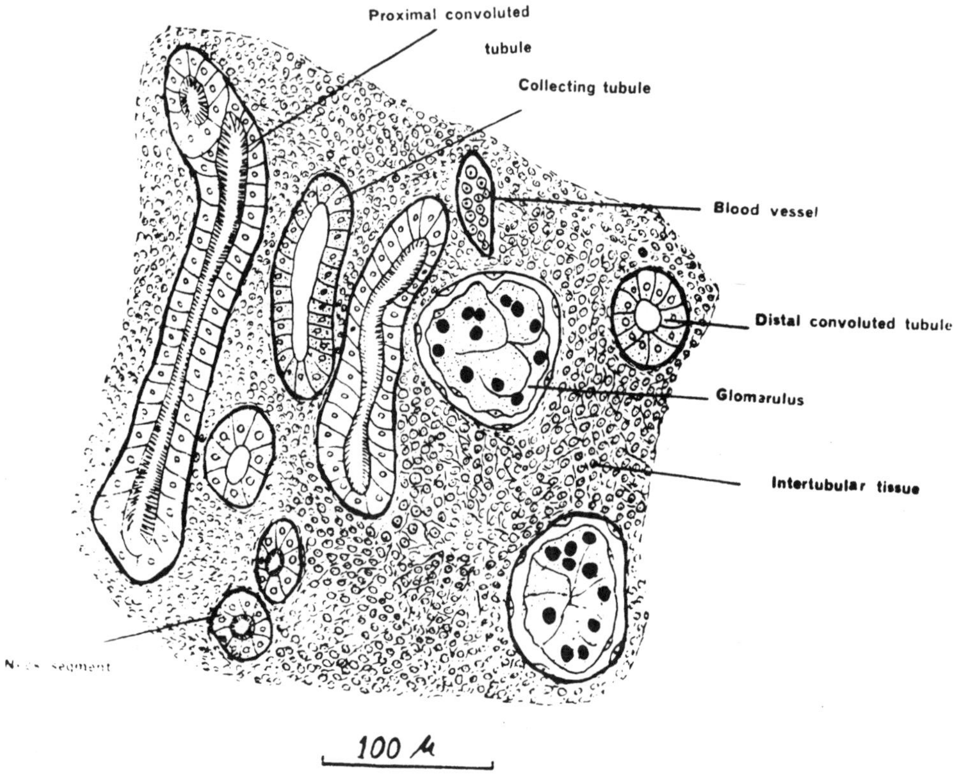

Fig. 7: T.S. Part of the kidney of *Glyptothorax kashmirensis*

(d) Advertitia, and

(e) Peritoneum

The endothelium consists of 2-5 cells thick stratified epithelium. There is an underlying fibrous connective tissue submucosa in which muscle fibres are also present. The muscles are of two types, the inner circular muscles and outer longitudinal muscles, interspersed by connective tissue. The outer adventitia is made up of loose connective tissue. The outer most layer is a thin peritoneum

6. Glyptothorax kashmirensis Hora

(i) Head Kidney

It consists of a mass of lymphoid tissue without nephrons.

(ii) Kidney

The kidneys are composed of nephrons and lymphoid tissue. The glomeruli are rounded, oval and even almond shaped. The bowman's capsule is lined with squamous epithelium which is also reflected on the glomerulus. The glomeruli are lobulated and situated more towards the margins of the kidney. The thick wide anterior part possesses more glomeruli than any part of the kidney.

The glomeruli range in size from 50-100 microns X 50-92 microns. Some glomeruli are much larger and are termed here as "giant glomeruli" (Fig. 11b). They are regularly found in all the sections and range in size from 140-155 microns X 80 microns. Giant glomeruli may occur in groups of two or more at a place. The number of both types of glomeruli taken together at 15 x X 10x is 4-5 per field (Table 1). In the posterior part the number is 1 or 2.

The renal tubule is differentiated into neck segment, the proximal convoluted segment and distal convoluted segment.

The neck segment has columnar epithelial cells with large nuclei. The average external and internal diameters are 27 and 6 microns respectively. The brush border is 2.5 microns high. The cilia of the neck segment extend into the urinary pole of the glomerulus (Fig. 7).

The proximal tubule consists of cuboidal cells with indistinct boundaries but with large nuclei and deeply staining nucleoli. The cytoplasm is granular and the part of the cell bordering on the lumen has a rugged appearance. The brush border is 3.5 microns in height. The average external and internal diameters are 47 and 20 microns respectively.

The distal convoluted tubule is smaller than the proximal convoluted tubule. It has a narrow lumen and is without a brush border. The cells are cuboidal and with clear boundaries. The nuclei are heavily chromatic and the average external and internal diameters are 45 and 15 microns respectively.

The intermediate segment is not present in this species.

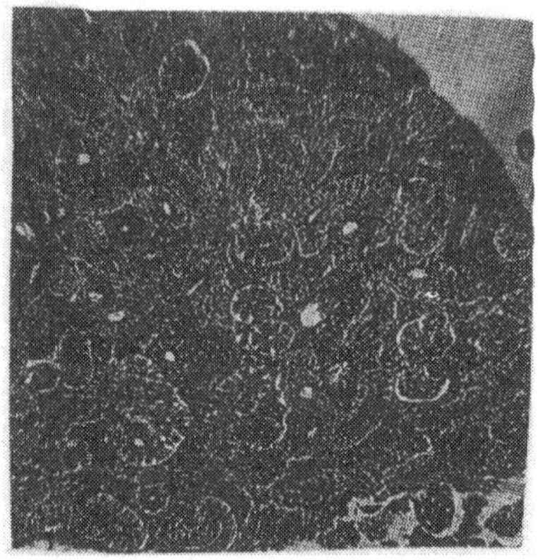

a

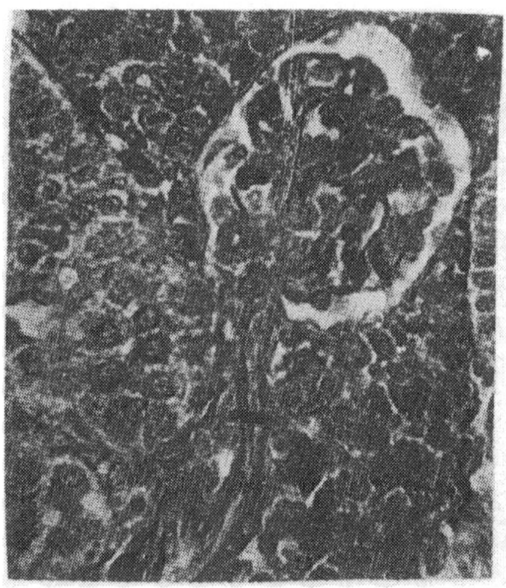

b

Fig. 8: (a) Photomicrograph of T.S. of the kidney of *S. esocinus*
(b) Photomicrograph of glomerulus of *S. esocinus*

The initial collecting tubules are wide and are made up of simple columnar epithelial cells, having a clear cytoplasm.

(iii) Ureter

The ureter has stratified epithelial cells lining a simple lumen. They are surrounded by circular and longitudinal muscles. There is a thick adventitia and a thin layer of peritoneum.

(iv) Urinary bladder

The lumen of the urinary bladder is lined by stratified epithelium which is not thrown into villi. The wall is composed of:-

(a) Endothelium,

(b) Submucosa,

(c) Muscularis of circular and longitudinal muscles,

(d) Adventitia and

(e) Peritoneum.

There are 2-5 layers of stratified epithelium. The epithelial cells are small and compactly arranged with quite prominent nuclei. Goblet cells are absent. The submucosa consists of fibrous connective tissue having many blood capillaries. The bands of circular muscles are not continuous being interrupted by intervening connective tissue. The longitudinal muscles form a continuous row. The adventitia is generally a thick layer covered over by a layer of peritoneum.

R.B.C's turned orange by the Mallory's triple stain were observed in the lumen and also in the tissue of the urinary bladder.

Discussion

A typical teleostean kidney tubule consists of:

(1) Bowman's capsule and glomerulus

(2) a neck segment typically ciliated;

(3) a proximal convoluted tubule (with brush border) having two cytologically distinct regions;

(4) a ciliated intermediate segment and

(5) a distal convoluted tubule without a brush border.

The author's findings regarding the histology of teleostean kidney agree in most points with the observations of Marshall and Smith (1930), Marshall (1934), Edwards (1928, 1929 and 1935), Grafflin (1937), Guyton (1935), Nash (1931), Ford (1958), Moore (1933) and Ogawa (1962).

Except in *Glyptothorax kashmirensis* in which the intermediate segment could not be differentiated, the structure of the kidney tubule in all other species studied by the author conforms to the typical teleostean type described above. According to Hickman and Trump (1969) intermediate segment has variously been reported as being present or absent. In their opinion it can best be regarded as a specialization of the second part of the proximal convoluted tubule, which specialization is absent in *Glyptothorax.*

Ogawa (1962) observed that ciliated epithelium of the neck segment may sometimes continue into the inner surface of the urinary pole of the glomerular capsule for some distance. This was also observed by the author in case of *Glyptothorax kashmirensis* and *Botia birdi*. The neck segment varies in length in fresh water teleosts. According to Hickman and Trump (1969) the neck segment is very short in some fresh water teleosts while it is quite long in case of some. The neck segments as observed in *Glyptothorax* and *Botia* are long, and short in all the other fishes studied. The distal convoluted tubule has been identified by the present author in all the fresh water teleosts studied. Contrary to the observations of Ogawa (1962) who could not recognise it in three fresh water fishes viz. *Onchorhynchus masou, Salvelinus pluvius,* and *Plecoglossus altivelis.*

The proximal convoluted tubule was invariably found by the present author in all the species studied to be composed of large cuboidal cells. The brush border however varies in height. In *Oreinus plagiostomus* and *Schizothorax esocinus* it is about 5 microns in height. In *Cyprinus carpio specularis* the brush border is 5-6 microns tall. In *Botia birdi,* it is about 8 microns tall, while in *Crossocheilus latius* it is 3 microns tall. It is only 2.5 microns in case of *Glyptothorax kashmirensis*. The universal occurrence of brush border, large basal nuclei and abundant mitochondria along the basal membrane in the teleostean proximal convoluted tubule suggests that as in other vertebrates, it functions in active solute reabsorption and secretion. This view is held by many physiologists including Hickman (1965) and Hickman and Trump (1969).

As in studies in other fishes by different authors the distal convoluted tubule lacks a brush border in all the fishes studied by the present author.

The precise function performed by the distal convoluted tubule is not understood in absence of definitive investigations on the subject. In amphibians micropuncture studies have revealed that the distal convoluted tubule is the site of urine dilution in these animals. Since fresh water fishes are well known for copious amounts of urine secreted by their kidneys, it may be assumed that the distal convoluted tubule is the site of urine dilution in freshwater teleosts as well.

The glomeruli vary in shape, size and number in various species. The present author observed that the glomeruli are more concentrated in the thick fused part whereas their number is much less else-where. Ogawa (1962) states that nephrons tend to be concentrated in the posterior part of the kidneys. He further states that fishes which are less active possess smaller number of glomeruli compared to the active and fast swimming ones which possess more numerous and better developed glomeruli. This would, however, hold true only for freshwater fishes. In primitive fishes the glomeruli are large but not so numerous.

A summary of the average size and number of glomeruli (per field in Zeiss microscope) in the species studied by the present author is given in Table. 1.

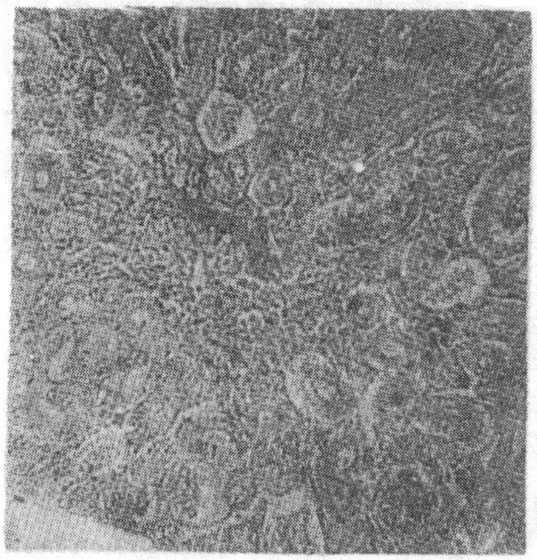

a

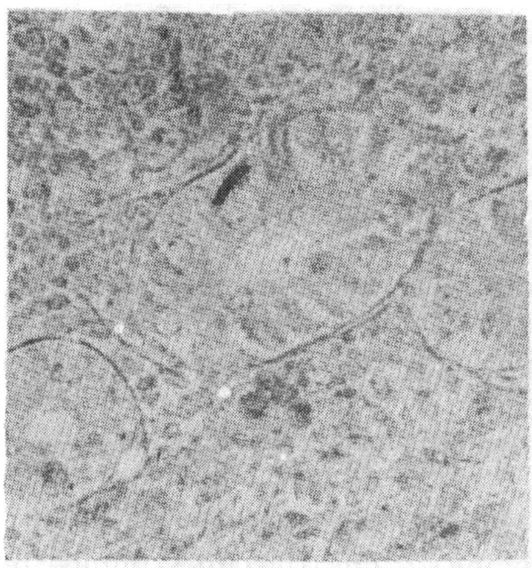

b

Fig. 9 (a) Photomicrograph of T.S. of the kidney of *O. plagiostomus*
(b) Photomicrograph of renal tubules of *O. plagiostomus*

Table 1 : Comparative study of size and number of glomeruli of some Kashmir fishes

Species	*Average size*	*Average number per field at 15 x × 10x in the mesial part*	*Average number per field at 15x × 10x in the anterior and posterior part*
Schizothorax esocinus	40 x 60 microns	9-10	2-3
Oreinus plagiostomus	57.5 x 67.5 microns	8-9	3
Cyprinus carpio	65 x 75 microns	5-7	1-3
Botia birdi	77.5 x 45 microns	4–5	1–2
Crossocheilus latius	65 x 47.5 microns	3-7	1-2
Glyptothorax kashmirensis	(i) 75 x 66 microns (ii) 147.5 x 80 microns	4-5	1-2

The renal corpuscles (glomerulus and bowman's capsule) of *Schizothorax esocinus* and *Cyprinus carpio* are more or less round in shape. In *Oreinus plagiostomus* and *Crossocheilus latius* they are oval in shape. In *Botia birdi* they are round or oval in shape while in *Glyptothorax kashmirensis* they are round, oval or even almond shaped.

From Table 1 it is obvious that the smallest renal corpuscles are found in *Schizothorax esocinus* (and incidentally the number is the largest) and the largest in *Glyptothorax kashmirensis.* Smaller the size of the glomeruli the larger their number and thus the greater the area of excretion. Smaller glomeruli would thus be more advantageous to the fish. The number and size of renal corpuscles is directly related to the salt content of the environment and therefore to the volume of urine excreted.

Schizothorax esocinus and *Orienus plagiostomus* inhabit clean fresh waters of river Jhelum and some fast running streams. The largest number of glomeruli present in their kidneys is a clear indication of presence of small amounts of salts present in the water they inhabit. In streams and rivers the salts are washed away due to the running state ofwater. The concentration of the body fiuids of the fish being more than the concentration of the surrounding water, the latter flows into the body of the fish. In order to get rid of the large quantities of water the number of glomeruli is large in the fish living in fresh water with less salt concentration.

Cyprinus carpio, Botia birdi and *Crossocheilus latius* are essentially inhabitants of lakes in Kashmir. The concentration of salts in lakes is much more than in the streams and rivers. The problem of water entering into the body of the fish is much less in lakes. This explains the presence of smaller number of glomeruli in these three fishes.

Glyptothorax kashmirensis is a small torrential stream fish. The concentration of

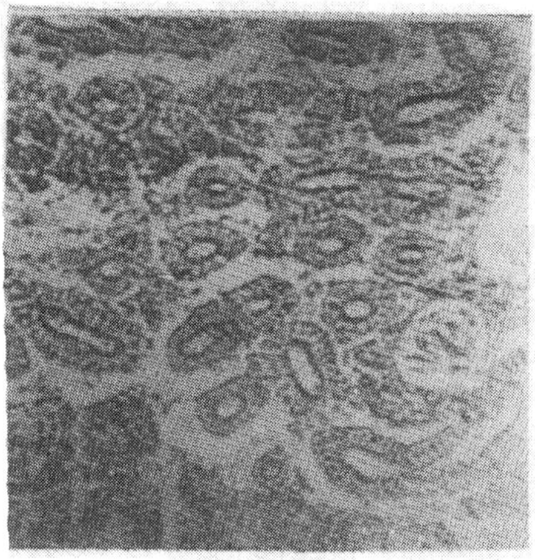

a

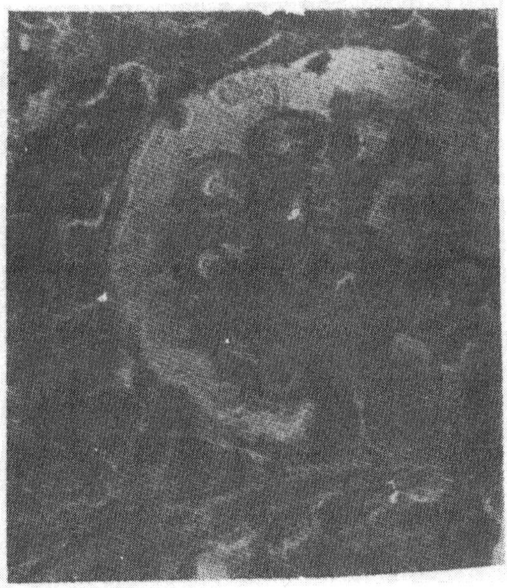

b

Fig. 10: (a) Photomicrograph of T.S. of the kidney of *C. carpio* (Highly magnified)
(b) Photomicrograph of glomerulus of *C. carpio*

salts in torrential streams in the least because of the very fast running state of waters in them and therefore the inflow of water through the body of this little fish is highest amongst all. *Glyptothorax* has solved the problem of getting rid of water not by increasing the number of renal corpuscles and renal tubules (because of the limited space available within the kidneys) but by increasing the size and efficiency of renal corpuscles.

By increasing the number of renal corpuscles, the number of renal tubules would be automatically increased. It would therefore result in an increase in the reabsorption of large amounts of filtrate and that was not required. The fish therefore chose to increase the size of the renal corpuscles, and also provided its neck segment with enormous number of cilia projecting into the urinary pole of the glomerulus for rapid filtration. The occurrence of "giant glomeruli" in *Glyptothorax*, in addition to the large glomeruli cannot otherwise be explained. The presence of some almond shaped "giant glomeruli" is interesting. Large well vascularized glomeruli are also found according to Hickman and Trump (1969) in the kidney of yellow bull head, *Ictalurus natalis.*

It is also a well established fact that the kidneys of fresh water bony fishes have many glomeruli compared to the marine bony fishes with glomerular kidneys. The glomeruli of the fresh water teleosts are larger than those of the marine teleosts.

Romer and Grove (1935) postulated their theory of fresh water origin of vertebrates based on the palaeontological and anatomical evidence. The kidney structure of fishes favours osmotic regulation in fresh water medium and this fact supports the theory of fresh water origin of vertebrates. Both the fossil record and kidney structure and function of sharks and marine teleosts suggest that their ancestors lived in inland waters. Sharks and marine teleosts invaded salt water independently and evolved different methods of combating the dangers of high salt concentration.

Marshall and Smith (1930) studied the problem of osmoregulation in fishes and suggested for the first time that the glomerular kidney was an evolutionary adaptation to fresh water. The glomerulus, it is believed, develops in response to the necessity for filtering off large amounts of fresh water as in the fresh water fishes. In marine teleosts a large number of glomeruli would be rather harmful. So in order to conserve water the number of glomeruli in them is small. In some marine teleosts glomeruli have disappeared altogether. Such a-glomerular kidneys are found in Lophobranch (Hout 1987; Verne 1921-22; Edwards 1928) and in goose-fish, *Lophius piscatorius* (Marshall and Grafflin 1928; Edwards 1928) and in toad-fish, *Opsanus tau* (Marshall 1929). It was found by Grafflin (1931) and Defries (1932) that the aglomerular kidneys have only proximal convoluted tubule with brush border, obviously to reabsorb salts in order to keep the body fluids at a higher concentration. It is an interesting fact that Grafflin (1933) observed the presence of glomeruli in *Myoxocephalus octodecimspinosus* but found them absent in a closely related species *M. scorpius.* Grafflin's view that in *M. scorpius* the glomeruli were lost with the growing age of the fish has not been substantiated by the observations of Forster (1953) who concluded that the aglomerular condition is randomly distributed within the species of the fish.

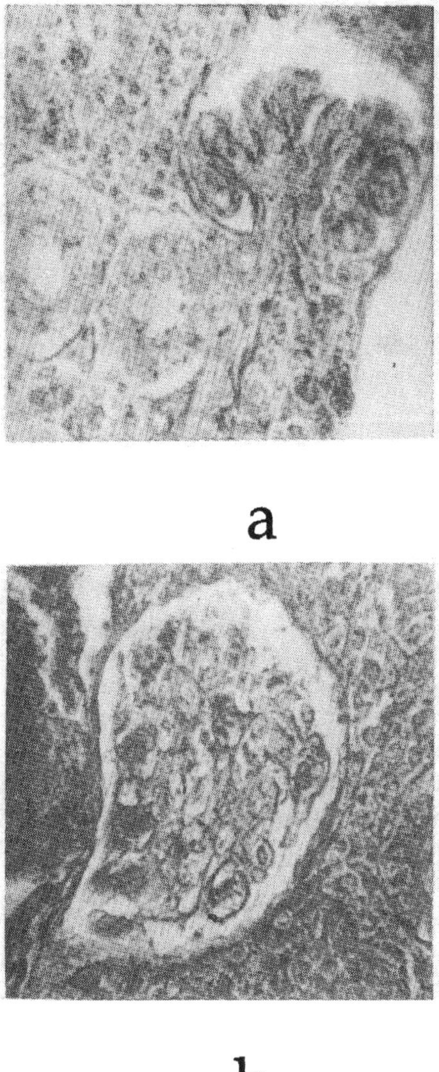

Fig. 11: (a) Photomicrograph of glomerulus of *B. birdi*
(b) Photomicrograph of giant glomerulus of *G. kashmirensis*

With the apparent exception of the aglomerular fresh water pipe fish(*Microphis boaja*) reported by Grafflin .(1937) fresh water teleost kidneys are universally glomerular, although the kidneys of embryonic and perhaps juvenile forms also are sometimes aglomerular (Audige 1910). In *Microphis boaja* the nephron appears to consist of only two regions, one which appears to represent the second proximal segment in fresh water teleosts, and a collecting duct system (Hickman and Trump 1969). None of the species studied by the present author are aglomerular. The aglomerular nature of *Microphis boaja* is difficult to explain, except on the basis of its recent incurssion into fresh water, as all its relations are in the sea. A large part of nitrogenous wastes is removed in fresh water teleosts by the gills as ammonia and urea, and the gills may be termed as accessory excretory organs.

Nash (1931) is of the opinion that total filteration by the kidneys depends on three factors:-

(1) the number of glomeruli,

(2) the size of glomeruli,

(3) the degree of branching of capillaries within the glomerular tuft.

The present author is also of the opinion that the number, size and degree of branching capillaries determine the amount of urine formation.

Capillary branchings were observed in *Schizothorax esocinus*. The large glomeruli of *Glyptothorax kashmirensis* with elaborate branchings tend to be oval or even almond shaped. The author is of the opinion that the degree of branchings of capillaries within the glomerulus determines the shape of the glomerulus. The rounded glomeruli are less efficient (having lesser capillary loops) compared to oval glomeruli (which can accommodate a larger number of capillary loops). Nash (1931) further states, "in a comparison of different species it seemed to be true that larger the glomerulus, the more numerous were the branchings of its capillaries", being maximum in giant glomeruli.

The glomeruli contain numerous capillary loops frequently grouped into two or more (and sometimes many lobules) (Hickman 1965). During the present studies the author could distinguish 6-8 lobules in the glomerulus of *Cyprinus carpio specularis*. Lobulated nature of the capillary loops in the glomorulus was also observed in *Glyptothorax kashmirensis*. Whether lobulated nature of the glomerular capillaries is of any advantage to the animal is not clear.

Moore (1933) observed double and polyglomerular arrangements arising from singular vascular poles in certain freshwater species. While no such arrangements were observed by the author during the present studies, groups of two or three glomeruli with separate vascular poles were observed in serial sections in the kidneys of *Schizothorax* and *Crossocheilus*.

The peripheral disposition of the glomeruli of *Botia birdi* resembles the similar condition observed by Ogawa (1962) in *Plecoglossus altivelis*.

The intermediate segment between the proximal convoluted tubule and distal convoluted tubule has been found in all the fishes studied by the author except in *Glyptothorax kashmirensis*. Edwards (1933) and Marshall (1934) have described the

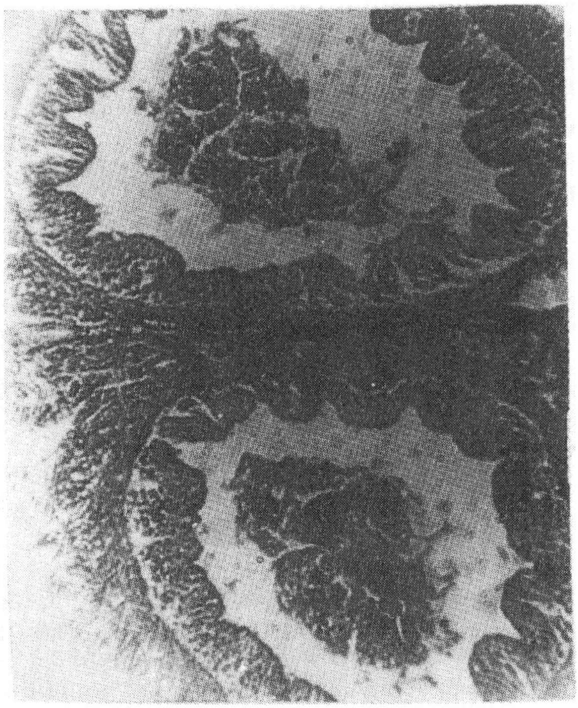

Fig. 12: Photomicrograph of T.S. Urinary bladder of *O. plagiostomus*

occurrence of a long ciliated intermediate segment in the nephron of the Gold fish *(Carassius auratus)*, *Danio malabaricus* and in *Amiurus nebulosus*. They have suggested that the intermediate segment of fishes, undergoes a change to the intermediate segment of amphibians and raptiles, and still further develops to the thin segment of Henle's loop in mammals, which is formed by the distal convoluted tubule. The observations of the present author regarding the r/R ratio (r stands for lumen of the second portion of proximal segment and R stands for its diameter) do not agree with the results of Ogawa (1962) except in case of *Cyprinus carpio specularis*, in which case the ratio comes to 0.29. According to Ogawa the r/R is 0.29 or less in typical fresh water fish, being 0.22-0.27 in cyprinid fishes, 0.32 or more in land locked or sea originated fishes and 0.24-0.33 in marine fishes. Ogawa further states that the ratio is 0.35 or more in sluggish fishes. If this rule is applied to Kashmir fishes, then all of them shall have to be categorised as sluggish fishes although that is not the case.

Juxtaglomerular cells were described by Sutherland (1966) and Capreol and Sutherland (1968) in some fishes. These are granular epitheloid cells having a sporadic occurrence in some species around afferent arterioles (Capreol and Sutherland 1968) and in others extend some distance from the glomerulus (Lagios 1968). Hormone renin is believed to be secreted by the juxtaglomerular cells, in teleost fishes. The demonstration of juxtaglomerular granules according to Sutherland and Hartroft (1968) requires special staining techniques. The present author was not able to identify these granules using Mallory's triple stain.

The histologic study of the ureters and urinary bladder has revealed a simple and similar structure in all the six species of teleosts studied. The same histologic elements namely the stratified epithalium, lamina propria and muscles are found in both the structures. The only apparent difference lies in the increase of smooth circular muscles of the urinary bladder resulting in a thicker wall. The structural similarity of the ureters and the urinary bladder is, therefore, a sufficient proof in favour of the view that the latter is a dilation of the former.

Summary and Conclusions

Histological studies of the renal organs of six freshwater teleosts from Kashmir (India) have revealed that their structure, with minor variations, conforms to the structure of such similar organs in other freshwater teleostean fish. Large sized 'giant glomeruli', however, have been observed in *Glyptothorax kashmirensis*. It has also come to light that smaller size of the renal corpuscles larger their number and thus greater the area of excretion. Smaller glomeruli are thus more advantageous to the fish.

References

Audige, J (1910). Arch. Zool. Exp. et Gen., Ser, 5, 4 : 275-624, Contribution a l'e' lude des reins des poissons te' leoste'e'n.

Balfour, F.M. (1881). "Comparative Embryology" Vol-2-London.

Capreol, S.V., and L.E. Sutherland (1968). Canadian Journal of Zoology. 46 : 249-255. Comparative Morph. of Juxtaglomerular cells in fish.

Defries, A. (1932) Anat. Rec, 54: 185-195.

Edwards, J.G. (1928) Am. J. Anat. 42: 75-108. Studies on aglomerular and glomerular kidneys. I. Anatomical.

—(1929). Anat. Record 44 : 15-28. Studies on aglomerular and glomerular kidneys II. Cytological.

Ford, P. (1958). Can. J. Zool., 36: 45-47. Studies on the development of the kidney of the pacific pink salmon (Oncorhynchus gorbuscha) II. Variation in glomerular count.

Forster, R.P. (1953). J. Cell Comp. Physiol., 42-487-510.

Grafflin, A.L. (1937a). Anat. Record 68-287-304. The structure of nephron in fishes.

Guyton, J.S. (1935). Anat. Record. 63: 212-226. Structure of the nephron of the South American Lung fish *Lapidosiren paradoxa*.

Hickman, C.P. (1965). Trans. Roy. Soc. Canada. Vol. III Series IV: 213-236. Studies on renal function in fresh water teleost fish.

Hickman, C.P. and B.F. Trump (1969). In "Fish Physiology" (Hoar and Randall eds). Vol. I. Chap. 2. 91-227. Acad. Press N.Y. and London.

Hout, E. (1897) C.R. Acad. Sci, 124-1462. Sur les Capsules surranales, les reins. le tissue lymphoide des poissons lophobranchs.

Kamalaveni, S. (1961 a). Record. Ind. Mus. Delhi, 59 : 83-118 Morphology of the urinogenital system in some Indian Teleostean fishes.

Kaul, B.L. (1971). Ph. D. Thesis Univ. of Kashmir. Studies on the Functional Anatomy of urinogenital system of some Kashmir fishes.

Kaul, B.L. (1974). Curr. Sci. 43(8): 256. A note on the presence of 'Giant Glomeruli' in a sisorid cat fish *Glyptothorax kashmirensis*.

Lagois, M.C. (1968). Gen. Comp. Endocrinol. 11: 248-250. Granular epithaloid cell involution in renal arteries of a euryhaline fish, *Cymatogester* adapted to hypotonic salinities.

Lal, M.B. (1963). Ichthyologica (2) 1-2: 45-50. Studies on the urinogenital organs of the Mahaseer *Tor putitora*.

Marshall E.K. Jr (1930) Am. J. Physiol., 94: 1-10. A comparison of the function of the glomerular and aglomerular kidney.

Marshall, E.K. Jr. and H.W. Smith (1930) Biol. Bull., 59: 135. The glomerular development of the vertebrate kidney in relation to habitat.

Marshall, E.K. Jr (1934). Physiol, Revs. 14 : 133-159. The comparative physiology of the kidney in relation to theories of renal secretion.

Moore, R.A. (1933). Contrib. Franz Theodore Stone Lab., Ohio St. Univ., 5 : 1-34. Morphology of the kidney of Ohio fishes.

Nash, J. (1931) Am-J-Anat., 47: 425-445. The number and size of glomeruli in the kidneys of fishes with observations on the morphology of the renal tubules of fishes.

Ogawa M. (1962). Sci. Repts. Siatama Univ. Ser. B. 4 (2) : 107-129. Comparative study of the internal structure of the teleostean kidney.

Pfeiffer, C.A. (1933). J. Morph. 54: 459-472. The anatomy and Blood supply of the urinogenital system of *Lapidosteus platystomus*.

Sharma, S. (1969). The annals of Zoology, V (2), 111-114. Haemopoeisis in the kidney of *Notopterus notopterus*. Can. J. Zool. 46: 257-263.

Sutherland, L.E. and Hartroft, P.M. (1968). Comparative Morphology of Juxtaglomerular Cells. II. The presence of Juxtaglomerular cells in embryos.

Verne, J. (1921-22). Arch. de Anat. microscopique, 18 : 357-407.

Chapter 10

Benthic-Macroinvertebrates as Indicators of Aquatic Environment

*Usha Moza**

Abstract

The need for biological approach in monitoring environment quality has became an accepted fact. The benthic macro-invertebrate community has been used frequently to assess the aquatic environment. A wide variety of procedures have been developed to evaluate the benthic data, so that this parameter too can be expressed in simple quantifiable terms and understood by layman. Various concepts like saprobian system, indicator organisms, indicator communities have been thoroughly developed with regard to riverine system of developed countries. But the problem regarding biological parameters and their indices is that these can not be universalized like chemical parameters, so each ecologically different region has to develop its own biotic approach to problem. In India where river ecology has changed a lot due to accelerated pace of development with an ever increasing population demand for the necessaties, the concept of biological monitoring is gaining popularity. Various agencies have been framed to evaluate the rivers with context to pollution both organic & inorganic. Accordingly most of the Indian rivers have been evaluated chemically, but so far biological evaluation has not been well documented. The basis for biological approach is to have list of organisms according to "Saprobian system". An attempt is herewith made to enlist the indicator organisms in relation to Indian rivers.

Introduction

The Biotope of any aquatic ecosystem is indicative of its trophic status. The total biotope is subject to change both in biomass and quality due to anthropogenic activities and the changes it reflects are the changes in aquatic environment of the given system at the given time. In view of this fact, the biological means form an important component of the environment monitoring surveillance. It is not feasible economically or otherwise to monitor all the biotic groups simultaneously and so far it has not been possible to identify the most useful component of aquatic ecosystem for intense monitoring, but literature survey indicates that use of benthic macroinvertebrates to be more prevalent for assessing aquatic environment quality especially in rivers. (Fig. 1).

Macro-invertebrates are preferred over other biotic indices particularly in flowing waters because of the following reasons.

1. Cosmopolitant distribution.

* Central Inland Capture Fisheries Research Institute, Karnal (Haryana).

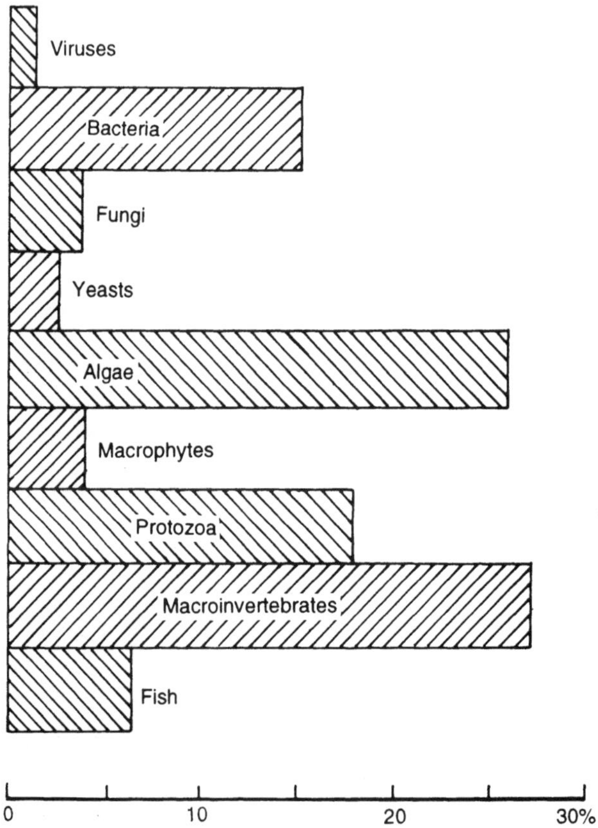

Fig. 1: Percentage distribution of taxa recommended for use as indicators based on a literature survey

2. Easy sampling due to suitable size and no elaborate equipment required for sampling.
3. No preservation problems.
4. Availability of identification key for most groups.
5. A High hysteresis value—because of
   (i) Sedantary habitates
   (ii) Occupation of similar niche
   (iii) Low mobility
   (iv) Long life cycle comparatively. All these factors do not enable them to escape deleterious substances that enter the environment, thereby being more prone to integrated pollutional effects both localized and for longer periods than *planktons* and *fish*, which are drifting and mobile organisms.
6. Exhibit heterogeneity-different phyla exhibit wide range of responses to environment changes especially organic.
7. Community as a whole is likely to be sensitive indicator.
8. Ready accumulation of all types of pollutants both water soluble a fat soluble like organochlorine, pesticides etc. which settle at bottom with organic debris.
9. Application of numerical methods especially species diversity index valid to large extent.

Benthic Sampling Methods

Sampling benthic macroinvertebrates although easy is not universal, depends on the current, velocity/flow, depth and substratum of aquatic ecosystem. Various methods used are as follows.

(i) **Kick sampling** (Hynes 1960): Used in fast flowing streams with large stony substratum. A given area for a given time is disturbed by vigorous movement of feet and the fauna gets collected in the net placed at the bottom.

(ii) **Surber sampling** (Welch 1948): Used in shallow water with gravel substratum. The apparatus is combined net and sampling quadrat. The area within quadrat sampler is disturbed by hand and organisms get collected in net.

(iii) **Shovel sampling** : Using in low land river or water bodies having shallow depth and muddy substratum. These are essentially nets mounted on robust frame with a strong-cutting edge at front.

(iv) **Dredge sampling** : This type of sampling is feasible in low land rivers/ lakes/reservoirs having considerable depth with soft bottom. Sampling apparatus in Fig. 2.

Sampling Frequency : Investigations on the number of samples of macroinvertebrate necessary to provide reliable estimation although well documented by

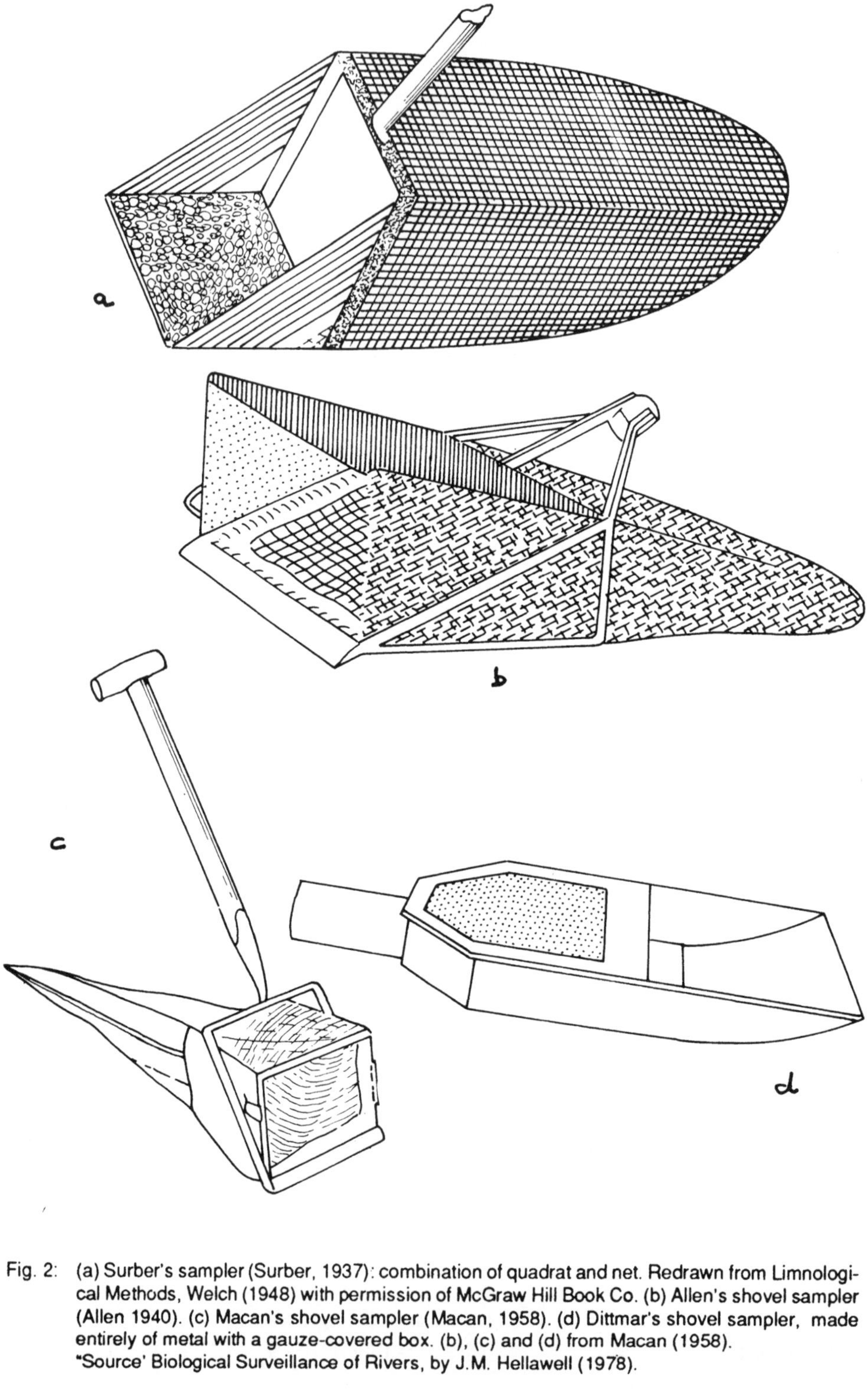

Fig. 2: (a) Surber's sampler (Surber, 1937): combination of quadrat and net. Redrawn from Limnological Methods, Welch (1948) with permission of McGraw Hill Book Co. (b) Allen's shovel sampler (Allen 1940). (c) Macan's shovel sampler (Macan, 1958). (d) Dittmar's shovel sampler, made entirely of metal with a gauze-covered box. (b), (c) and (d) from Macan (1958).
'Source' Biological Surveillance of Rivers, by J.M. Hellawell (1978).

Hellawell (1978) depend more on the fauna characteristics, seasonal factors and life history stages. It is necessary to conduct pilot survey to obtain estimates of number of sample, required for a given degree of precision with confidence.

Analysis System

The first attempt at utilizing the benthos to provide information on water quality was the classic work of Kolkwitz and Marrson (1909) and their "saprobian system". They developed list of organism associated with various zones of pollution. These zones ranged from oligosaprobic having clean water. α & β mesosaprobic to polysaprobic having low dissolved oxygen, high BOD and decomposable organic matter (Fig. 3). Saprobian system zones were considered as centres of optimum growth and development of organisms. System was based on effects of organic matter decomposition in smooth flowing streams and focussed on tolerance of organisms to low dissolved oxygen. The system introduced the concept of indicator organisms.

Oligo-saprobic	Poly-saprobic	α Meso-saprobic	β Meso-saprobic	Oligoo – saprobic

Kolkwitz and Marsson, 1908, 1909

Indicator organism : The organisms which has narrow range of tolerance and remain restricted in a specialized or restricted environment conditions.

Depending on the species response to organic enrichment, 3 types of species were distinguished (Hawks 1962).

(i) **Saprophobic species :** The non-tolerant species, mostly present in clean waters which eventually get eliminated in succession due to organic enrichment, *Hydrophilia* sp:, *Limnophilid* sp;, (Caddis worms) *Ephemerella* sp:, *Ecdyonurus* sp: (may fly larva), *Nemoura* (stone fly nymph), *Gammarus pulex* (fresh water shrimp), *Pisidium* sp: (Bivalve), *Chaetogaster* (Oligochaeta).

(ii) **Saproxenic species:** These have wide range of tolerance to organic enrichment and are present mostly in mesosaprobic zone sometimes in clean waters also. *Asellus aquaticus;* (shrimp) *Helobdella* sp; *Eropdella* sp; (leech) *Baetis rhodani*, *Lymnaea pereger*, *Sphaerium* sp.

(iii) **Saprophilic species :** Present in organically rich waters, but as degree of enrichment increases these also get eliminated. *Chironomous* sp., *Tubifex tubifex*, *Eristalis tenax*, *Physa acuta*. Hynes (1960) showed that entire community is indicator of any ecological change. Thus the concept of *Indicator communities* was developed. Stone flies, May flies, Caddis worms, and some shrimps were indicator communities of clean waters and Oligochaetae, Pulmonata snails and chironomid larvae were shown to be indicators of high turbidity, suspended solids & organic enrichment. Comparison of species between oligosaprobic and polysaprobic zone showed that species composition differ both qualitatively and quantitatively between the two. Oligosaprobic zones possess many species, represented by few individual

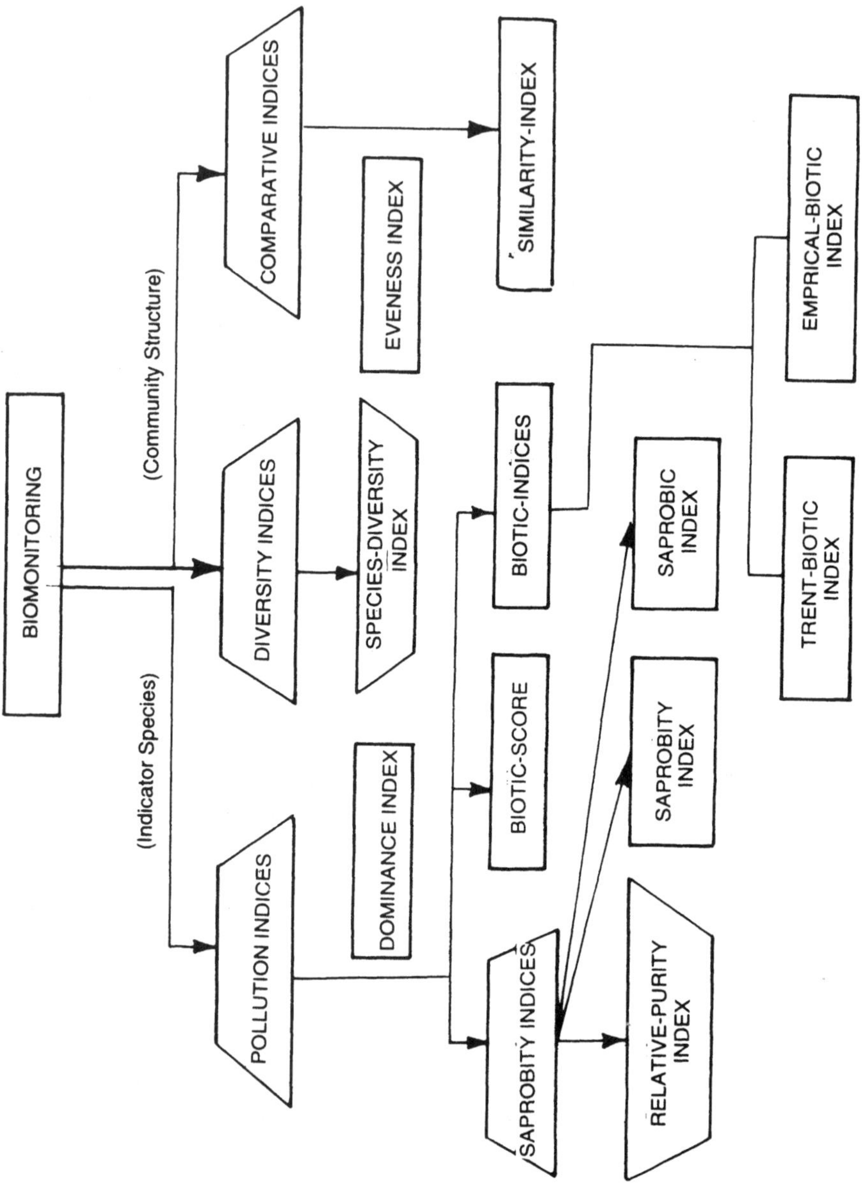
BIOMONITORING
(Community Structure)
(Indicator Species)
COMPARATIVE INDICES
DIVERSITY INDICES
POLLUTION INDICES
SIMILARITY-INDEX
EVENESS INDEX
SPECIES-DIVERSITY INDEX
DOMINANCE INDEX
BIOTIC-INDICES
BIOTIC-SCORE
SAPROBITY INDICES
EMPRICAL-BIOTIC INDEX
TRENT-BIOTIC INDEX
SAPROBIC INDEX
SAPROBITY INDEX
RELATIVE-PURITY INDEX

Fig. 3

in given area-compared to less species but more individual in polluted waters, thus the concept of *community diversity* were developed. Quantification of community structure resulted in *diversity* and *similarity indices.*

Analytical Methods

Numerous indices have been developed to interpret the basic data, but so far no unitary method could be developed because of different response of indicator value to different type of perturbations. (Table 1).

Table 1: Biocoenotic Responses of Indicator Value

A.	The appearance of disappearance of individual species from a community, i.e. a change in species list.
B.	A reduction in the numbers of species of taxa present in the community i.e., a reduction in the diversity.
C.	A change in the population of individual species within the community usually : i.e. change in species diversity.
D.	A change in proportional species composition of the community. i.e. change in similarity

Basically there are 3 approaches—(a) Pollution indices (b) Diversity indices (c) Comparative indices. Former is based on the concept of *Indicator species* and the latter two on *community structure,* its diversity and stability. The 3 approaches between themselves comprise all forms of analysis used in evaluation of Benthic Biota (Fig. 4).

Pollution Indices: All pollution indices are derived on the observation that there is progressive loss of clean water with increasing pollution load-usually organic and the difference between the species occurring above and below the point of discharge of pollutant gives indication of the system. In this type of indices following points were taken into account. 1. Broad tolerance of large invertebrates (Type A of Table 1) 2. Overall diversity of fauna (Type B), but there is no provision on relative abundance (Type C & D of Table 1). Various saprobic indexes have been developed to measure this indices based on three common factors (Table 2). (a) Numerical value is given to each indicator species based on its position in scale of general response to given environment condition (Table 2.1) (b) Value is weighed by the indicator value of taxon (Table 2.2) (c) Weighting is given according to abundance of taxon (Table 2.3).

Diversity indices : Diversity indices are based on the concept that community structure undergo change due to change in environment, thus the diversity of species in a community has relation to stability of the ecosystem and to establish underlying relationship, many diversity indices have emerged (Table 3). Diversity indices is used to assess the intensity of change due to ecological stress. Stress can

Table 2 : Commonly used Pollution Indices & Methods of Analysis (After Hellawell 1986)

	Indices			*Rating*	*Requirement*
1.	Relative purity index (Knoop 1954)	$I = \frac{\sum (o+b)}{\sum (o+b+a+p)}$	o = oligosaprobic b = beta-mesosaprobic a = alpha-mesosaprobic p = polysaprobic	0-7 point scale	Requires relative abundance information
2.	Saprobity index (Pantle and Buck 1955)	$I = \frac{\sum sh}{\sum h}$	s = degree of saprobity h = abundance (Rare = 1, frequent = 3, abundant = 5)	0-4 oligo = 1 poly = 4	- do -
3.	Saprobic index (Zelinka & Marvan 1961)	$I = \frac{\sum ahg}{\sum hg}$	a = saprobic valency h = abundance g = indicator value	0-5 0 = clean water 5 = high 1 = Rare	- do -
4.	Trent Biotic Index (Woodiwiss, 1964) (Refer Table 2.1)			0-10 Grade assigned by reference to presence of key texa	Adequate species list required
5.	Empirical Biotic index (Chutter 1972) (Refer Table 2.2)	$I = \frac{a \times b}{h}$	a = No. of individuals b = quality value h = total no. of samples	0-10 Polluted water = 10	Both qualitative and quantitative list of species.
6.	Biotic Score (Chandler 1970) (Refer Table 2.3)			0 -100	- do -

***Table 2.1 : Trent Biotic Index**

Key indicator Groups	*Diversity of Fauna*	*Total number of groups (see Part 2) present*					
		0-1	*2-5*	*6-10*	*11-15*	*16+*	
Column No : 1	*2*	*3*	*4*	*5*	*6*	*7*	*Line No:*
				Biotic Index			
Plecoptera nymphs present	More than one species	-	VII	VIII	IX	X	1
	One species only	-	VI	VII	VII	IX	2
Ephemeroptera nymphs present	More than one species	-	VI	VII	VIII	IX	3
	One species only	-	V	VI	VII	VIII	4
Trichoptera larvae present	More than one species+	-	V	VI	VII	VIII	5
	One species only+	IV	IV	V	VI	VII	6
Gammarus present	All above species absent	III	IV	V	VI	VII	7
Asellus present	All above species absent	II	III	IV	V	VI	8
Tubificid worms and/or Red Chironomid larvae present	All above species absent	I	II	III	IV	-	9
All above types absent	Some organisms such as *Eristalis tenax* not requiring dissolved oxygen may be present	0	I	II	-	-	10

* *Source:* "Biological Surveillance of Rivers" by J. M. Hellawell, 1978.

* Baetis rhodani excluded

\+ Baetis rhodani (Ephem.) is counted in this section for the purpose of classification.

Part 2 Groups

The term 'Group' here denotes the limit of identification which can be reached without resorting to lengthy techniques. Groups are as follows:

1. Each species of Platyhelminthes (flatworms)
2. Annelida (worms) excluding *Nais*
3. *Nais* (worms)
4. Each species of Hirudinea (leeches)
5. Each species of Mollusca (snails)
6. Each species of Crustacea (log-louse, shrimps)
7. Each species of Plecoptera (stone-fly)
8. Each genus of Ephemeroptera (may fly) excluding *Baetis rhodani*
9. *Baetis rhodani* (may fly)
10. Each family of Trichoptera (caddis-fly)
11. Each species of Neuroptera Larvae (alder-fly)
12. Family Chironomidae (midge larvae) except *Chironomus thummi (= riparious)*
13. *Chironomus thummi* (blood worms)
14. Family Simulidae (black-fly larvae)
15. Each species of other fly larvae
16. Each species of Coleoptera (beetles and beetle larvae)
17. Each species of Hydracarina (water mites)

Table 2.2 : Empirical Biotic Index (Quality Value) of Benthic Genera (Chutter 1972)

Taxon	*Quality Value*	*Taxon*	*Quality Value*
Hydra spp.	6	*Amphipsyche scottae*	3
Planarians	3	*Cheumatopsyche thomasseti*	2
Rhabdocoelida	10	*Macronema capense*	3
Tardigrada	7	*Hydroptila* spp.	2
Nematoda (except Mermithidae)	7	*Orthotrichia* spp.	2
Hirudinea	7	*Stenelmis* spp.	3
Choroterpes spp	1	*Psychoda* spp.	10
Caenidae	1	Culicidae	10
Micronecta spp	4	*Chironomus* spp.	10
Baeites harrisoni (only)	6	*Eristalis* spp.	10
B. harrisoni with 1 (other species)	4	Ancylidae	4
B. harrisoni (2 other species)	2		
B. harrisoni (3 other species)	0		

Quality value for other texa is 0, except Cladocera and Copepoda.

Empirical Biotic Index

Species	*a*	*b*	*h*
Tubifex tubifex	2	8	16
Dendrocoelum lacteum	5	3	15
Asellus aquaticus	10	0	0
Gammarus pulex	30	0	0
Baetis rhodani	20	6	120
Erpobdella octoculata	2	7	14
Limnaea stagnalis	10	0	0

h = a = 79 ab = 165

$$\text{Index} = \frac{ab}{h} = \frac{165}{79} = 2.09$$

Interpretation of the Index-Values

Value	*Interpretation*
0-2	Clean, unpolluted waters.
2-4	Waters slightly enriched owing to naturally-occurring organic matter or high quality effluents containing a little organic matter. Chemically similar to 0-2.
4-7	Enriched waters, the higher the Biotic Index value the greater the enrichment. Obvious increase in BOD and nitrogenous compounds in the water and rather wide diurnal fluctuations in dissolved oxygen are to be expected.
7-10	Polluted waters with great increases in the concentrations of substances associated with organic pollution.

Table 2.3: Biotic Score (Chandler 1970) (After Hellawell, 1978)

Species	*No.*	*Abundance*	*Score*
Baetis rhodani	20	Common	48
Gammarus pulex	30	Common	40
Dendrocoelum lacteum	5	Few	33
Limnaea stagnalis	10	Few	28
Asellus aquaticus	10	Few	22
Eropdella octoculata	2	Present	24
Tubifex tubifex	2	Present	22
		Biotic Score	217

Note : The individual score of each species is determined in relation to abundance in the sample. Sum of individual scores gives the Biotic score.

be of any type, therefore it obviates inherent weekness of most pollution indices. The indices are based on 3 concepts of Biococnotic response, are more objective, as these are derived mathematically from quantitative data. Wilhm (1970) attempted to relate species diversity index (Shanon-Weaver 1964) to water quality and hypothesized that value greater than 3.0 indicate clean condition, 1.0 indicate severe pollution, value in between indicate, moderate pollution.

Comparative Indices : Comparative indices are based on the assessment of two or more community structures from polluted and non-polluted stretch of a water body, mostly used in riverine surveillance based on concept D of Bio-coenotic response. Several indices are available, but better comparisons are made with comparative indices in which the relative or absolute abundance of species are utilized-Table 4).

Table 3

		Commonly Used	*Diversity Indices*
Species richness index (Margalef 1951)	(d)	$d = S\text{-}1/\log 2^N$	Ratio between total no. of organism in community: No. of sp: present in ecosystem
Concentration of dominance (Simpson 1949)	(c)	$c = (n_1 / N)^2$	Measure of sp: distribution in a community
Diversity Index (Shannon & Weaver 1964)	(H)	$H = \sum_{r=1}^{s} P_r \log_2 P_r$	
Evenness Index (Pielou 1969)	(J)	$J = \frac{H}{\log_2 s}$	
	s =	No. of species	Ratio between sp. diversity
	N =	No. of organisms	to sp. packing
	P_r =	Proposition of individuals in r^{th} species	

Table 4

Similarity Indices	
Coefficient of similarity (Jaccard, 1912)	$\frac{J}{a+b-j}$
Quotient of similarity (Sorenson 1948)	2j/a+b
	j = No. of sp. common to both ecosystem a = No. of sp. in ecosystem a b = No. of sp. in ecosystem b

Performance of Indices : Many systems and several indices have been proposed for the analysis of benthic data. But it is unlikely that any single index can satisfy all requirements. Hawks (1962) found that by using relative frequency of benthic organisms and their classification in the saprobian system of Kolkwitz and Marsson (1908, 1909), a useful pollutional grading could be derived. Attempts have also been made to compare several of the pollution indices to calibrate them against conventional measure of pollution such as BOD (Table 5). The Biotic indices no doubt give direct information of water quality, the limitation lies in the fact that index value/ quality factor, score system for indicator organisms can not be same at national or international level, because of the geographical distribution and range. Species on the threshold of geographical range are likely to respond more sharply to environmental stress than when same species is present within the range. Quality factor assigned is mainly towards organic waste and should summarise tolerance for wide variety of pollutants. As such development of regional biotic index value may be more useful.

Compared to pollution indices: Diversity indices are more objective as it does not make use of autecological information (information regarding response of the individual taxa) and is unaffected by sample size Edwards *et. al.* (1975). Diversity index especially that of Shanon andWeaver (1964) has been observed to be most sensitive of systems for assessing environment quality of river through benthic community structure (Hawks 1962, Hellawell 1986). The main draw back of the method is that although high diversity indicates good quality ecosystem, low diversity does not necessarily indicate poor quality, because low diversity may be caused by other stresses also. Many torrential head streams with a high quality water have a low diversity because of severe physical conditions (Archiabald 1972). Since biotic index and diversity index have different Biocoenotic responses, comparison of both should prove a comprehensive evaluation of the aquatic environment.

Conclusion

Biological methods of assessing aquatic ecosystem started since early part of this century through the works of Kolkwitz & Marrson (1908, 1909) in Europe and Richardson (1928) in America. Towards the end of fifties, biological surveillance of

Table 5 : Comparison between BOD-Characteristics and Various Grading of Pollution. After Marstrand 1973.

mg/1 BOD	*General Characteristics*	*Desig-nation*	*Class : based on Carpenter (1928)*	*Trent (Woodiwiss, 1964)*	*Lothians (Graham, 1965)*	*Biotic score (Chandler, 1970)*
2	Salmonids Plecopterans Ephemeropterans Trichopterans Amphipods Coarse fish.	Very Clean	A	9-10	1	900+
2-3	All above groups represented Coarse fish.	Clean	AB	7-10	1	500-900
2-3	Above groups restricted except Amphipods Trichopterans and Baetids Few fish.	Clean	B	6-8	1-2	300-500
3-5	Trichopterans restricted; Baetids rare; *Asellus* dominant Molluscs and leeches present	Fairly clean	C	5-6	2-4	110-400
5-10	As above, but no Baetids. leeches present	Doubtful	C	3-5	2-4	45-300
5-10	No fish, other-wise as above No fish.	Doubtful	CD	2-4	4-5	15-80
10+	Oligochaeta and red chironomids only	Bad	D	1-3	4-5	9-20
10+	None of above; only air breathing *Eristalis tenax*	Bad	D	0-1	6	0-10

Source : "Biological Surveillance of Rivers" by J.M. Hellawell 1978.

rivers became established as an integral part of monitoring water quality. Few authorities like Bristol Avon River Board (Bielby 1960); Trent River Board (Woodwiss 1964); Luthians Purification Board (Graham 1963); Carolina division of environmental management (Lenat *et. al* 1977), South African rivers (Chutter 1972), Canadian rivers (Clark 1977), developed biological methods for classifying the

Table 6 : Pollution Indicator Macro-Invertebrates (Indian Rivers)

Genera	*Water-Quality*	*Class*	*Tolerance*
Heptagenia sp; *Epeorus* sp; (May-fly nymphs) *Hydropsyche* sp; *Rhyacophila* sp; (Caddis fly) nymphs) *Psephenus* sp; (Beetle larvae), *Perla* sp; (Stone-fly nymphs). *Anax* sp; (Dragon fly nymph). *(Eropdella* sp; (Leech)	pH = 8.3 to 9.0 D.O. = 7.8 to 9.8 BOD = Nil	Clean water Xenosaprobic	Do not tolerate low dissolved oxygen and high temperature
Trichoptera, Plecoptera larvae, *Ephemeroptera* nymph. *Nais* sp., *Chaetogaster* sp. (Oligo) *Lamelliden marginalis, Melania striatella tuberculata, Lymnaea* sp., *Pila globosa* sp. (Mollusca) *Gammarus pulex* (Amphipod)	pH = 6.4 to 10 D.O. = 7.2 to 9.2 BOD = 0 to 3	Clean water Oligosaprobic	Taxa are pollution sensitive. Do not tolerate organic load.
Earthworm, *Aelosoma* sp: (Oligochaeta) *Brachycerea* sp. (diptera). *M. striatella, Valvata sp., Gyraulus.* sp., *V. bengalensis* (Mollusca. *Baetis* sp., *Rhantaticus* sp., *Berosus* sp., *Macrogomphus* nymphs	pH = 7.9 to 8.1 D.O. = 7.0 to 8.4 BOD = 3-8	Slighted Polluted β mesosaprobic	Organisms tolerate moderate amount of pollution
Dixa sp., *Tipula* sp., *Chironomus* sp. (Diptera) *Nais* sp., *Aelosoma* sp. (Oligochaeta) *Rhantaticus* sp., *Berosus* sp., (Beetle larvae) *Physa* sp. *Amnicola* sp. *V. bengalensis* (Mollusca)	pH = 7.0 to 8.0 D.O. = 6.9 to 7.2 BOD = 5 to 10	Moderately Polluted α mesosaprobic	Organisms tolerate fair amount of organic pollution.
Tubifex sp., *Limnodrilus hoffmeisteri, Branchiura* sp. (Oligochaeta) *Chironomus tendipiformis, Eristalis tenax* (Diptera)	pH = 7.0 D.O. = nil - 4.8 BOD 10	Heavily Polluted Polysaprobic	Organisms tolerate anaerobic conditions
Limnodrilus hoffmeisteri. Branchiura sp. (Oligochaeta)	pH = 6.0 D.O. = Nil to traces Electric conduc-tivity } 2250-3000	Polluted & anaerobic Isosaprobic (Waste disposable sites)	Macro-invertebrates absent fauna present Occasionally

Values expressed D. O. = ppm
BOD = mg/1
EC = micro-mhos/cm at 25 C

pollutional condition of rivers. These biological surveillance led to development of various indices-especially regarding benthic composition (Woodwiss 1964), Grahm 1965), Chandler 1970, Chutter 1972). These indices has been assessed from time to time and found valid regarding evaluation of Environment quality (Clarke 1977, Leppakoski 1977), Wilhm & Dorris 1968, Philips & Segar (1986).

In India the environmental monitoring of aquatic ecosystem at national level is a developing concept and many agencies are involved like, Central Water Pollution Control Board, New Delhi; National Environmental Engineering Research Institute, Nagpur; Indian Standard Institute, New Delhi. Recently Inland Capture Fisheries Research Institute, Barrackpore with a different objective has taken the biological surveillance of main rivers and its tributaries. Ganga Action Plan has come into existance since 1983 to work in a comprehensive manner on the assessment of trophic status and pollutional level of river Ganga, but so far no biotic index or quality score is developed for indicator organisms present in Indian rivers. Some of the rivers like Ganga, Yamuna, Gomiti have been classified on the water quality system (Singh & Misra 1987), and few benthic indicators identified within these rivers systems (Jhingran 1989). An attempt is herewith made to compare various chemical parameters, pollutional grading to the benthic fauna present in different rivers of India (Table 6), from the works of Ahmad & Singh (1989), Jhingran (1974), Jhingran (1991), Krishnamoorthi & Sarkar (1979) Singh (1988), Moza (1992) Avon 1989), to enlist the indicator organisms according to saprobian system, so as to frame the first step in the use of benthic macro-invertebrates in environment pollution studies with contest to Indian Riverine system.

References

Ahmad, S.H.; and Singh, A.K. 1989. *J. Environ. Biol.* 10(3) 283-91.

Anon 1989. Brief Report on G.A.P. Schemes vis-a-vis water quality of River Ganga 1983-88. Bkp. Publication.

Archibald, R.E.M. 1972. *Water Research* 6 : 1229-1238.

Bielby, G.H. 1960. J. *Inst. Sew. Purif.* 3 298-301.

Chandler, J.R. 1970. *Water Pollution Control* (4) 415-422.

Chutter, F.M. 1972. *Water Research* 6 : 19-30.

Clarke, R. Mev. 1986. In "Biological indicators of freshwater pollution and Environmental Monitoring" pp 133-141.

Graham, T.R. 1965. Annual Report Lothian River Purification Board.

Hawks, H.A. 1962. "River pollution II causes & effects" Published by Butterworth, London.

Hellawell, J.M. 1978. "Biological Surveillance of Rivers". Publication Water Research Centre, London.

Hellawell, J.M. 1986. "Biological indicators of freshwater pollution and Environmental Monitoring" Published by Elsevier Applied Science, London.

Hynes, H.B.N. 1964. *Chem. Ind.* 435-436.

Jaccard, P. 1912. *New. Phytol.* 11: 37-50.

Jhingran, V.G. 1974. "Fish and Fisheries of India" Pub. Hindustan Publication Corporation.

Jhingran, A.G. 1989a. A Synopsis CICFRI Barrackpore 1-26.

—1991. National Conference on Aquatic Sciences in India. Jawahar Lal Nehru University, New Delhi.

Knoop, H. 1961. *Gewasserkudle. Mitt.* 5 66-73.

Krishnamoorthi, K.P. and Sarkar, R. 1979 *Proc. Symp. Environ. Biol.* 133-135.

Kolkwitz, R. and Marsson M. 1908. *Ber. dt. bot. Gen.* 26, 505-519.

—1909. *Int. Rev. Hydrobiol.* 2, 126-152.

Lenat, D.R. ; Smock, L.A. Penrose, D.L. 1977. "Biological Monitoring for Environmental Effects" pp. 97-109.

Leppakoski, E. 1977. "River Pollution II causes and effects, pp. 125-132.

Margalef, R. 1951. *P. Inst. Biol. Apl.* 9: 5-27.

Moza, U. 1992. Course Material for FAO Trainees CICFRI Publication 1-6.

Pantle, R.; Buck, H. 1955. *Gas-4, Wasserfach.* 96. 604.

Philips, J.H.D.; Segar, A.D. 1986. Marine Pollution Bulletin 17 No. 1. 10-17.

Pielou, E.L. 1960. *J. Theor. Biol.* 13 : 131-144.

Richardson, R.E. 1928. *Bull. Ilinois Nat* Hist *Surv.* 17, 385-475.

Shannon, C.E. and Weaver, W. 1964. "The Mathematical theory of communication". Univ. Illinois Press.

Simpson, E.H. 1949. *Nature* 16 : 427-449.

Singh, S. & Misra, G.N. 1987. "Water Pollution" (Conservation & Management) Pub. by: Gonodaya Prakashan, Nainital.

Singh, S.N. 1988. *J. Inland Fish. Soc.* India 20(2) 30-34.

Sorenson, T. 1948. *K. dansk. Vidensk. Selsk.* N.S. 5 : 1-34.

Welch, P.S. 1948. "Limnological methods" McGraw Hill Book Company, London.

Wilhm, J.L. 1970. *Technical Report* No. 7. Institute Water Research, Michigan. State Univ.

Woodiwiss, F.S. 1964. "Trent River Board Classification" Chemy. Ind. 443-447.

Zelinka, M. and Marvon, P. 1966. *Verh. Int. Verein-heor. angew. Limnol* 16, 817-22.

Chapter 11

Effect of Mechanical Stress on Early Embryonic Stages of *Tor tor*

*Kuldeep K. Sharma**

Abstract

The importance of impact of mechanical stress in relation of fishery biology lies in determining the effect of flood conditions on developing eggs. This type of stress may limit fish production. During present studies on eggs of *Tor tor* (Mahaseer), hatching success under stress is observed to be 88% against 94% (in non-stress conditions). The hatching interval is compressed to 80 hrs. in stress conditions against 100 hrs. in non-stress conditions. Mechanical stress has been observed to produce 15% developmental abnormalities in stress conditions (directly bubbled and aerated set 'A') against only 7.5% abnormality in indirectly bubbled aerated set 'B'.

Introduction

Mechanical stress can become an important environmental influence on early life history stages of fish. It can occur in nature because of disturbances due to habitat deterioration, or in managament practices which may involve transportation of seed for transplantation thus vitally affecting the year class strength of fishes. Such type of impact of embryonic stages of freshwater fishes have not been estimated in the past except that of Pommeranz (1974), who have projected this impact on early stages of Place. The importance of this study in relation to fishery biology lies in determining the effect of flooded conditions on developing eggs.

The present studies depict experimental results to determine the effects of mechanical stress created by constant bubbling through water containing eggs and larvae of Mahaseer *(Tor tor)* from Jammu.

Materials and Methods

Fertilized eggs of *Tor tor* were collected from River Tawi with the help of haul nets and sieves and immediately taken to the laboratory in plastic containers. Equal number of eggs were kept in two sets of experimental containers filled with tap water as described by Pommerang (1974). Set 'A' of the experimental containers having eggs which were directly exposed to constant bubbling (of an aerator to create

* Department of Biosciences, University of Jammu, Jammu - 180004.

mechanical stress). Set 'B' container acted as control involving indirect bubbling (to avoid mechanical stress) caused by aerator opening in a gauge core in the cylinder separating the aggs placed in outer jacket of the water column. This arrangement eliminated the direct mechanical stress while maintaining all physico-chemical parameters at the same level as in 'Set A'. The experiment was conducted up to the time of complete hatching and then the larvae were reared to study the subsequent survival in both the groups. The observations were recorded after every 24 hr duration.

Table 1 : Showing the Effect of Mechanical Stress (like Constant Bubbling and Aeration) on the Eggs of *P. Conchonius*

	Mean hatching duration (hrs.)	*Mean hatching success (%)*	*Mean egg mortality (%)*	*Mean alnormal hatch (%)*
Direct bubbling	80	88	12	15
and aeration	(60-85)	(75-90)	(10-15)	(10-20)
Indirect bubbling	100	94	6	7.5
and aeration	(80-100)	(90-96)	(4-10)	(5-8)

Ranges in parentheses.

Results and Discussion

Perusal of Table 1 reveals that 88% hatching success against constant mechanical stress like bubbling and aeration in *Tor tor* indicates a higher resistance of its eggs to such a situation, when compared to the eggs of *Xenentodon cancila* and *Aspidoparia morar* (a sympatric species) exposed to similar conditions where only 58.4% and 74% hatching recorded respectively.

Pommeranz (1974) also indicated similar results in developed eggs of plaice and highlighted that in case of eggs developing under direct mechanical stress register a low survival as compared to conditions of non-direct stress.

It is thus indicated from present findings (when compared to previous ones on *X. cancila*, Malhotra *et. al.*, 1981) that a better fortification in the developing eggs of *Tor tor* may be due to the presence of protective egg envelope. Such an association of survival in relation to protective envelop of developing eggs has earlier been discussed by Blaxter (1969) who has been of the view that in Plaice the higher susceptibility of larvae to mechanical stress (as compared to their eggs) may be due to effective sheltering by the egg shell, while the perivitalline space may act as a shock absorber. This differential survival among eggs in these Jammu fishes may be due to better egg protection on *Tor tor* and *A. morar* as compared to *X. cancila* (Malhotra *et. al.* 1981), which is evidenced by thick egg shell and proportionately wider perivitaline space in *Tor tor* and *A. morar* 0.6mm (0.4 - 0.8) and 0.5mm (0.3-0.7mm) respectively in comparison to that of *X. cancila* 0.06 (0.04-0.09mm).

Zotin (1958) on the other hand, attributed shell strength to activation by calcium contents of water, which during the course of present investigations were found to be higher in the breeding habitats of *Tor tor* (40.5 ppm) as compared to 20.05 ppm in *X. concila* and 37.07 ppm in *A. morar.*

Table 1 reveals that mechanical stress also hastened the hatching process in the developing eggs of *Tor tor* by affecting the pace of embryonic development. 100 per cent hatching success in *Tor tor* has been achieved after 80 hrs. in the directly aerated 'Set A' while in non-directly aerated 'Set B' 100% hatching occurred after 100 hr. From these observations it is thus concluded that mechanical stress advances the hatching time but hatching success is retarded. In *Tor tor* the mechanical stress has also been observed to induce deformities in newly hatched larvae particularly in directly aerated and bubbled set 'A', where 15% larvae are recorded as abnormal suffering from deformed yolk-sac; scoliosis; lordosis and kyphosis. In indirectly aerated and bubbled set only 7.5% of the hatchings are observed with similar afflications. All these observations indicate that mechanical stress has adverse effect on the hatching success in *Tor tor,* wherein such effects are also manifested in the forms of abnormal hatching. Matlak (1970) reported the misshapen heads in carp fry are attributable to mechanical disturbance to which developing eggs are exposed. Ulanowicz (1974) also reported that much of the damage induced by exposure to mechanical stress in developing stages is not immediate but manifests itself later in form of structural deformities in developing juvenlies. The present findings in *Tor tor* are in agreement to the views of Ulanowicz (*op. cit.*).

Generally *Tor tor* breeds in River Tawi and Jhajjar nallaha. As per observations this fish breeds in the water column only when these are flooded. The water bodies are gravelly packed with sandy deposition and silt laiden water. The eggs were found floating and were also washed ashore when water turbulances were caused by flooding conditions. These stray turbulants caused by stray current always effect the early development of *Tor tor.* It is quite clear from the above findings that these mechanical disturbances can cause appreciable mortality in the early developing stages thereby strongly affecting the year class strength of *Tor tor.*

It is thus concluded from the present findings that :

1. Mechanical stress causes mortality during early embryonic development.
2. It advances the hatching duration but hatching success is retarded.
3. Frequency of deformities increases in the hatchings as a result of this stress.

References

Blaxter, J.H.S. 1969. Development : Eggs and larvae In : Fish Physiology, Vol. III. W.S. Hoar and D.J. Randall (eds.) Acedemic Press, New York and London, 178-252.

Malhotra, Y.R. Kuldeep Sharma and Chhatterpati Sharma, 1981. Effect of Mechanical stress and varying temperature on embryonic stages os *Xenentodon cancila* and *Aspidoparia morar* from freshwaters of Jammu. Proc. IV. All India Symposium of Environmental Biologists, 95-102.

Maltak, O. 1970. Vorlanfinger Bericht uber Missbildung des Kopfes baim karp fenstnch, Acta Hydrobiol. 12 (4) : 391-398.

Pommeranz, T. 1974. Resistance of Plaice eggs to mechanical stress and light. In: "The Early Life History of Fishes, pp. 765. Springer-Verlag Heidelberg, Berlin, N.Y. ed. by J.H.S. Blaxter.

Ulanowicz, R.E. 1974. The mechanical affects of water flow of Fish eggs and larvae in fisheries and Energy production a symposium, ed. by S.B. Saile, Published by Lexington–Books, Lexington. M.A. U.S.A.

Zotin, A.J. 1953. The mechanism of hardening of the salmonid egg membrane after fertilization or spontaneous activation. J. Embryo. Exp. Morph., *6*: 546-568.

Key words: Mechnical stress; Sympatric, Fortication, Hatching success, Hatching interval, Deformities.

Chapter 12

Icthyofauna of the Sector of Kaveri River in Head Region

*M.N. Madhyastha and S. Murugan**

Each nation has different geographical regions and differing agro-climatic zones. India being a tropical country has diverse ecological systems of which we are proud of Himalayan and Western Ghat ecosystem and which are primarily mountain ranges of great significance. The natural condition and topographic variations brings in series of changes of which precipitation rate of rain is one resulting in origin of streams and rivers. The advantage of snow clad mountains of Himalayan range, which supply a constant flow of waters in the stream is absent in the rivers of Western Ghat regions. The monsoon source is the source of supply for the rivers and streams originating along Western Ghat range.

There are a few major rivers originating along Western Ghat range flowing eastwards whereas only a few smaller rivers flow westward of the hill range. These rivers provide a series of micro habitats which support abode for a variety of fishes and other organisms. The lotic system with differential physico–chemical characters turns into lentic system in some of the cases and then provides different ecological niches.

With the advent of technological development every ecosystem is changing fast and there are immense influences on the flora and fauna of the ecosystem. The changes brought in may be : (i) food related, (ii) environmental related which in itself may be natural or man-made, (iii) technology related and (iv) co-inhabitant related.

The last reason mentioned may be due to the impact of food fish and ornamental fish introduced from an alien country (competition for food and space, transmission or introduction of new diseases).

Even the indigenous species are under severe ecological crisis from modification of river system, lakes, ponds and other habitations due to hydropower, flood control, river basin development and such other environmental aberrations. Development and mangement plans should protect the indigenous species and biodiversity.

* Department of Biosciences, Mangalore University, Mangalagangotri-574 199. Karnataka State, India.

Table 1 : Ichthyofauna

A total of 30 species of finfishes were recorded from river Kaveri and its tributaries. The recorded sp. of finfishes are listed below.

	Name of the Species	*Kaveri River*	*Harangi Reser.*	*Hemavathy River*	*Res.*	*Yagachi River*
1.	*Notopterus notopterus*	X	√	X	√	√
2.	*Danio aequipinnatus*	√	X	X	X	X
3.	*Rasbora daniconius*	√	X	X	X	X
4.	*R. caverii*	X	X	X	X	√
5.	*Barilius gatensis*	√	X	X	X	√
6.	*Cyprinus carpio carpio*	X	X	√	√	X
7.	*Cirrhinus mrigala*	X	√	X	√	X
8.	*Puntius conchonius*	√	√	√	√	√
9.	*P. sarana sarana*	X	√	X	X	X
10.	*P. arulius arulius*	X	√	X	X	X
11.	*Labeo rohita*	X	X	X	√	X
12.	*L. calbasu*	X	√	X	√	X
13.	*L. fimbriatus*	X	√	X	√	X
14.	*Tor mussullah*	X	√	X	X	X
15.	*Catla catla*	X	√	X	√	√
16.	*Lepidocephalichthys*	√	X	X	X	X
17.	*Mystus armatus*	X	X	√	√	X
18.	*M. punctatus*	X	X	X	√	X
19.	*M. cavasius*	X	√	X	√	X
20.	*Callichrous bimaculatus*	X	√	X	√	X
21.	*Wallago attu*	X	√	X	√	X
22.	*Channa marulius*	X	X	X	√	X
23.	*C. puntatus*	X	X	√	√	√
24.	*C. orientalis*	X	X	X	X	√
25.	*Chanda nama*	X	X	X	√	X
26.	*Glassogobius giuris giuris*	√	√	√	√	X
27.	*G. biocellatus*	X	X	√	√	X
28.	*Discognathus lamta*	√	X	X	X	X
29.	*Mastacembelus armatus*	X	X	X	√	X
30.	*Poecilia reticulata*	√	X	√	X	√

River Kaveri is one of the east flowing rivers originating at Western Ghats. It has many tributaries of which Harangi and Hemavathy apart from the main Kaveri has been explored to some extent. Like all the rivers in the country river Kaveri too have been subjected to the onslaught of rapid industrialisation and urbanisation. The entire Kaveri riverine system together with its delta and mangrove regions is subjected to intense use and anthropogenic influences.

30 species of fin fishes were found from the river Kaveri and its tributaries in the head region. The details of the species composition and its special variation has been

depicted in Table 1. *Puntius conchorius* was observed in abundance at all the riverine stretches. The presence of *Tor khudree* only at Harangi reservoir revealed that the environmental conditions favoured its enhanced growth. A noteworthy feature noticed is the presence of a high population density of *Poecilia reticulata.* Though the water quality of this stretch was poor, this species appeared in both the years of observation. The fish disappeared after April only to reappear during pre-monsoon season.

Table 2 : Systematic List of Ichthyofauna of the Hemavathy Reservoir

*Scientific Name**	*Local Name*	*Occurrence*	*Seasonal Abundance*
Family : Notopteridae			
Notopterus notopterus	Chappalimeenu Chamari; Pappasi Olagu thatte	C	Monsoon
Family : Cyprinidae			
Cyprinus carpio communis	Paremeenu	C	Premonsoon & monsoon
Puntius conchorius	Koracha meenu Gende meenu; Gidpakke machhalu	A	Monsoon
Labeo rohita	Rohu meenu	C	Premonsoon & Monsoon
Labeo fimbriatus	Kemmeenu; Gandu meenu	C	Monsoon
Cirrhinus mrigala	Mrigal meenu	C	Premonsoon & Monsoon
Catla catla	Dodda gende; Catla meenu	C	Premonsoon & Monsoon
Family : Bagridae			
Mystus armatus	Girlu	S	Monsoon
Mystus cavasius	Girlu ; Naaru girlu Meese girlu, Nayi girlu	S	Monsoon
Mystus punctatus	Haddu meenu Kettalu	S	Monsoon
Family : Siluridae			
Ompok bimaculatus	Gooddalumeenu ; Bale meenu	S	Premonsoon & Monsoon
Wallago attu	Kembarimeenu; Godle Bale meenu	S	Premonsoon & Monsoon
Family : Channidae			
Channa marulius	Hoovina meenu	C	Premonsoon & Monsoon
Channa punctatus	Bill Korava	C	Premonsoon & Monsoon

(Contd.)

Scientific Name	*Local Name*	*Occurrence*	*Seasonal Abundance*
Channa gachua	Hola Korava Kallu Korava	C	Monsoon and postmonsoon
Family : Chandidae *Chanda nama*	Sipari meenu Gajina meenu	R	Monsoon
Family : Gobidae *Glossogobius giuris giuris*	Nattikannu korava Bangisidda meenu	R	Premonsoon
Glossogobius biocellatus	Nattikannu	R	Premonsoon
Family : Mastacembelidae *Mastacembelus armatus* armatus	Havu batti Haavumeenu	S	Monsoon and Postmonsoon

A = Abundant C = Common R = Rare S = Seasonal

Hemavathy reservoir was taken for indepth seasonal study. The systematic list of Ichthyofauna is given in Table 2.

Riverine ecosystem is a highly perturbed environment. Riverine fishes, early life stages, fish spawning, hatching and nursery habitats are of signal importance. Further the extensive physical differences such as back waters, static shallow, swift and sluggish, lentic situations, situation of main channel also determine the flow, velocity, depth etc. Patterns of larval distribution, nest building; habitat selection by larvae and juveniles presumably benefits the population by increasing food availability and survival chance and deceasing predation.

Sedimentation primarily impacts lithophils, phytophils, phytolithophils and nest building. There is a positive correlation between species richness of juvenile fishes and both per cent silt in the substrate and slow current indicating the value of backwater habitats to early life stages.

Protecting the biota of the rivers has become difficult task due to development activity undertaken by different sectors. Habitat animal compatibility is lost in many cases resulting in pushing some of the species into endangered list or threatened list. Environmental stress has increased to a great extent and animals have failed to adapt to different changed ecological niches. Further concentrations of chemical substances in the ambient environment has lead to the phenomenon of bio-accumulation and biological magnification. Many a time short-term external exposure of a substance has resulted in long-term internal exposure if the toxicant remains in the body.

Thus there is an urgent need to protect the threatened or endangered fragile habitat and attempts should be made to protect each endangered habitat, thus protecting the species inhabiting there.

Chapter 13

Biology of Indian Belone, *Xenentodon Cancila* (Hamilton)-A Freshwater Fish from Jammu Waters of J & K State. I - Periodization in Life History of Xenentodon Cancila (Hamilton).

*Subash Chander Gupta & Kuldeep K. Sharma**

Abstract

In the present communication, the periodization of *Xenentodon cancila* (an inhabitant of slow running streams of Jammu), raised in the laboratory, from the eggs collected from nature, is described & discussed. Transparent fertilized eggs measure about 2.01-2.55 mm in diameter, found attached to macrovegetation with adhesive fibres, collected during March-April. Characteristic morphological and diagnostic features at different stages of development based on melanophore occurrence & other morphic events in periodization viz: cleavage stages and post-cleavage stages, fry stages, post-fry and fingerling stages, have been elucidated and established.

Introduction

Identification of early developmental stages, assigning them to their species, determining their ages and morphic attainments at different developmental stages (periodization) is a taxing problem in the present state of knowledge particularly on the biology of Indian freshwater fishes. Simultaneously, this important aspect is of immense practical application in the developing fields of fishery biology, particularly in this country. The present existing informations on the biology of Indian fishes is in no way comparable to the extensive knowledge accumulated as a result of years of well planned and systematically conducted research on the importance of fishes of the western countries. The little we know of a few fishes of this subcontinent is mostly from isolated observations. Thus a great deal of further work is required before any practical use could be made of the knowledge.

Although considerable investigations have been conducted on fishes during recent years but no reliable information exists regarding the identification of early life history stages of Indian freshwater fishes. Until the recent works of Malhotra *et al* (1980), & Gupta (1982), no information was available on this aspect of fishery biology from the region. Though *Xenentodon cancila* is an important food fish, yet it is unfortunate that this fish lacked the attention of fish biologists or larval fish

* Department of Biosciences, University of Jammu, Jammu-180001, J & K.

specialists towards such studies. Thus need was felt to investigate the life history of *X. cancila* so as to enable the larval fish ecologists or any other fish biologists to identify the fish larvae which form an important and interesting branch of fishery biology.

Though a few workers have generated informations on the eggs and larval forms of Indian freshwater fishes, but one of these workers have tried to project a single character or a group of characters that could be useful in larval identification of these species.

Xenentodon cancila is a very peculiar fish, as far as its life history and early development compared to other fish species is concerned, because in *X. cancila*, most of life history stages (inlcuding pigmentations) are passed within the egg and the larvae is generally born as an adult, with well developed mouth and jaws, otherwise formed later in the larval stages of other fish species. Thus with a view to elucidate important morpho-characters which help in identification of early development stages in the life history of a freshwater belone, *Xenentodon cancila,* the work on its life history has been investigated in the laboratory. *Xenentodon cancila* inhabits small lotic stream of Jammu (and occasionally available in River Tawi also). It exhibits peculiar breeding habits as it breeds among equatic vegetation from March to April every year and is categorized as phytophillous, characteristically different from other fishes (viz : Cyprinids) in having a toothed beak.

Materials and Methods

Eggs of *Xenentodon cancila* were collected during March-April from the side pool of a stream, a tributary of Ranbir canal at Gho-Manasan. The sidepool is large, deep, with muddy bottom, highly vegetative with so many plants and has its communication with the mainstream; provides an ideal place for spawning of *X. cancila*, as the eggs and adults are found in the sidepool. The eggs were found attached (in bunches or isolated) to submerged parts of plants with the help of microfibres. The vegetation was taken out, examined for eggs, eggs one by one were detached from the plants with the help of foreceps, then transferred to a container/containing water & brought to laboratory; a few however were fixed in formaldehyde/70% alcohol in the field itself for studies.

In the laboratory, transparent fertilized eggs were separated from translucent unfertilized ones. The former were then maintained in different hatching troughs & subsequent developmental studies made at the room temperature. Most of the collected eggs were at varying stages of development and thus the time of their spawning and initiation of fertilization in the eggs could not be ascertained The developing eggs were isolated according to their developmental stages, recording the intervening period between the successive stages until they hatched. The larvae were then transferred to fish-hatching incubator (a modified version of B.O.D. specially fabricated for the studies, with water circulation arrangements alongwith cooling & heating mechanisms) maintained at 22°C-24°C, and were examined every 24-hours for any morphic attainments. The periodization (time duration mentioned against various developmental stages) has been determined post-hatching. As the larvae exhaust, they were provided with different foods for exogenous feeding in the laboratory during their post larval stages. The larvae were fed twice a day (i.e. once in the early morning before sun rise and next in the late evening after sun set) with artificial food and natural planktonic food.

Observations

I. Embryonic Development

From within the eggs collected, the following cleave stages of development have been recognized.

(a) Cleavage Stages

(i) Fertilized Egg. (Fig. 1)

Fertilized eggs are transparent, spherical, adhesive, highly yolked, with very hard shells, measure about 2.01-2.55 mm in diameter, and have an inner vitellina membrane (covering the yolk) and an outer zona-radiata or perivitelline membrane. Separation of vitelline membrane from the yolk facilitates the zygote to rotate within the egg shell. Initially, the yolk is creamish-white in colour which changes into golden-yellow, as the develepment proceeds. The presence of micro-fibres on the egg membrane for attachment to vegetation is a unique feature, observed in this freshwater fish.

(ii) Blastodisc formation (Fig. 2)

This is an early stage in development in which cleavage has not yet commenced but cytoplasm has segregated to one pole of yolk. The blastodisc is golden yellow whereas rest of yolk is light-yellow & occupies about half of the space of egg.

(iii) Morula Stages or early Gastrula Stage (Fig. 3)

Blastodermal cells small, numerous, tightly packed, consisting of about half of the yolk mass. The germinal disc has spread over the yolk.

(iv) Gastrulation (Fig. 4)

Blastodermal cells invade about 1/2 of yolk, a small portion of which is uncovered, forming a germ-ring at one pole of egg.

(b) Post-cleavage stages

The eggs kept for further developmental stages in the troughs in the laboratory exhibit the following developmental stages :

(v) Initiation of Embryonic ridge (Fig. 5)

(After 8 hours of Blastodisc formation).

Gastrulation complete, yolk completely invaded by blastodermal cells, thus blastopore closed. Initiation of embryo is in the form of a longitudinal ridge, with a slightly bulging anterior and narrow-posterior end.

(vi) Differentiation of cephalic and caudal ends; optic-lobes;

4-Myotomes and kupffer's vesicle. (Fig. 6) (after 20-hours of blastodisc formation).

Embryo is well defined with head and tail ends, elongate; broad anteriorly, narrow posteriorly. Optic lobes indicated on head; 4-myotomes appear in the lower half of body, and a kupffer's vesicle appears in the caudal portion. Perivitelline space has become wide at this stage; yolk occupies about 2/3 of egg proper. The embryo is creamish-white, but yolk is yellowish in colour.

(vii) 8-12 Myotomc stage (Fig. 7). (after 20 hours of blastodisc formation).

Embryo elongates further, occupies about 1/3 of circumference of yolk, yolk-mass has enlarged and surrounds the kupffer's vesicle in the caudal region. Myotome number increased to 8-12; perivitelline space further widens.

(viii) 14-Myotome stage (Fig. 8). (After 20-hours of blastodisc formation).

There is not much change except for the addition of 2 more myotomes over the previous stage. In the caudal region, kupffer's vesicle disappears but yolk-mass has become more broader than in the previous stage.

(ix) Enlargement of Yolk-mass in the caudal portion (Fig. 9)

(after 24-hours of blastodisc formation).

All the characters are similar to the previous stage except for that the yolk-mass in the caudal region has become broader & massive. Perivitelline space wider than the previous stage. The enlargement of yolk-mass (surrounding the kupffer's vesicle, when present) in the caudal region appears to have been associated with the formation of yolk-sac.

(x) Yolk (Vitelline vessels) circulation stage (Fig. 10) (after 60 hours of blastodisc formation).

Larva much elongated, head broad and blunt. Eyes massive, brain rudiments appear. Further enlargement of yolk mass now takes the form of a yolk-sac; yolk (vitelline vessels) circulates in the form of a network of vessels on the entire yolk-sac, start from the region below pectoral fin on both the sides of body in different directions, runs continuously with the same speed, unites with the vessels coming from the anus and then enters the mouth.

Body well pigmented with stellate melanophores, a few melanophores appears on head and eyes. Dorsal melanophores are arranged in rows on either side of mid-dorsal line extending upto posterior 1/3 of body. Numerous stellate melanophores scattered all round on the yolk-sac.

(xi) Tail-free stage (Fig. 11) (after 96-hours of blastodisc formation)

Embryo 'C' shaped, broad anteriorly, tapering posteriorly. Head adherent to yolk-sac, tail now free from yolk-sac and heterocercal. Snout obtuse and eyes lateral on head.

Dorsal melanophores arranged in the form of 3 bands almost in the middle of body; scattered laterally, almost extending upto caudal end. Yolk-sac bears numerous melanophores all round, golden yellow in colour but embryo itself is yellowish.

(xii) Appearance of Mouth, pectoral fin and larval fin fold (Fig.12)

(after 144 hours of blastodisc formation).

Embryo curled over the yolk-sac; mouth cleft indicated, pectoral fins appear on the sides of body just posterior to head. Larval finfold appears both dorsally and ventrally in the region which is free from yolk-sac, circular around caudal end. Eyes massive with lens, bulging out on head. Orbital head free from yolk-sac. Body thick and opaque.

Melanophores mostly dotted. Head melanophores restricted to the mid brain region as a band. Dorsal melanophores numerous dots, arranged in 9 definite bands, start from a little posterior of pectoral fin, extending laterally upto posterior 1/3 of caudal end. Melanophores on the ventral side of body adjoining yolk-sac are arranged in slaunting bars (demarcating the position of myotomes), extending over the yolk-sac. Fine black pigments scattered in groups mostly on the ventral side of yolk-sac. Eyes pigmented with brown melanophores.

(xiii) Jaw-formation stage (Fig. 13). (after 168 hours of blastodisc formation).

Jaws formed, larva long, head more broad and massive Yolk-vessels abundant.

Body entirely covered with fine stellate melanophores. Body pigments are mostly of three types i.e. yellow, orange and black. Yellow and black pigments are scattered along the body from snout to caudal end; Yellow pigments small and dotted, mostly occupy lateral sides of body, less numerous as compared to black pigments; black pigments large and fine stellate, occupy mostly the dossal surface of body; a few yellow pigments also scattered on head. Orange pigments are small & dotted, scattered on the egg giving orange tinge; a few black pigments also scattered on egg. Eyes bear black pigments.

(xiv) Differentiation of Dorsal and Anal fins (Fig. 14) (after 180-hours of blastodisc formation).

Dorsal and anal fins indicated opposite each other in the posterior 1/3 of body. Head & caudal ends blunt. Pectoral fin elongates further and shows occasional movements.

Body melanophores increase in number, a few scattered melanophores added almost in the middle of body on the dorsal surface. Melanophores on yolk-sac decrease in number.

(xv) Larval Movement Stage (Fig. 15) (After 216 hours of blastodisc formation).

Larva now 'U' shaped, shows occasional wriggling movements accompanied by continuous movements of pectoral fin. Indistinct rays appear in caudal fin.

Head melanophores lobate after the shape of underlying brain lobes, and increase numerically. A few melanophores noticed on snout. Dorsal melanophores numerous on either side of mid-dorsal line. Ventro-visceral melanophores dotted or stellate, increase numerically & have started scattering laterally upwards.

(xvi) Appearance of fin-rays (Fig. 16) (after 288 hours of blastodisc formation)

Larva coiled on itself around the entire yolk-sac, tail has extended over the posterior end of head. Lower jaw much longer than the upper and blunt. Opercular cleft indicated. Pre-orbital head flattened (depressed). Countable rays appear in median fins, 6-8 rays in dorsal, anal and caudal fin each. Body broad, thick and opaque; wriggling movements of body with continuous beating of pectoral fins have become frequent. Myotomes 58, i.e. 32 pre-anals and 26 post-anals; pre-anals prominent and distinct than post-anals.

Dotted melanophores appear for the first time on the tip of lower jaw; and a few stellate melanophores on snout. Dorsal lobation of head melanophores (large & stellate) prominent; large & stellate in the anterior & posterior 1/3 of body; small dotted in the middle of body. Lateral-line well pigmented. Ventro-visceral melanophores numerous, mostly stellate, extending upto lateral-line. Dorsal melanophores alongwith lateral-line and ventro-viscerals merge at caudal peduncle. Dorsal, anal and caudal rays pigmented at their bases also. Pigmentation dense at the point of junction of yolk-sac and the body; fine stellate melanophores in the posterior 1/3 of body occupy entire lateral & dorsal sides; demarcating few myotomes in the region.

(xvii) Differentiation of heart and pre-anal ventral fin-fold (Fig. 17) (after 336 hours of blastodisc formation)

Embryo ring-like, coiling being in one plane. Mouth cleft reaches upto anterior edge of eye orbit. Ventrally, bones indicated on throat. A tubular heart formed just posterior to head between yolk-sac & embryo. Auditory-vesicle and anus indicated. Jail has extended beyond the posterior end of head upto the level of half of yolk-sac. Pre-anal fin fold prominent. Yolk-sac much reduced. Eyes massive, occupying 1/2 the height of head. Dorsal and anal fins now have 12-14 rays each; and caudal with 12 branched rays.

Melanophores numerous small, dotted on the pre-orbital head (densely pigmented), on the tip of upper and lower jaws (densely present) extending upto anterior tip of eye orbit; a few small, dots ventrally on throat, and between posterior end of yolk-sac and anus on the mid-ventral line, and on yolk-sac. Internal melanophores on head noticeable. Dorsal melanophores stellate, arranged in 2-3 rows on either side of mid-dorsal line, extending upto tip of caudal peduncle and diffused in the posterior 1/3 of body; and scattered downwards in the posterior half of body (posterior to yolk-sac), reaching upto lateral-line. Ventro-visceral melanophores numerous, arranged on either side of mid-ventral line. Few myotomes are also marked by melanophores on their surfaces in the posterior half of body.

The embryo is now at the verge of hatching. Wriggling and jerking movements of embryo become very much rapid and foceful, accompanied by continuous movements of pectoral fins, as the time of hatching nears. Then ultimately movements of embryo burst the egg membrane thus affecting the hatching of larva.

After hatching, the hatchling come to surface after bursting the egg, swims there for sometime with the flickering of tail; then go to the bottom, lie there in the normal posture (ventral side touching bottom & dorsal side facing upwards). The tail continuously flickers while the larva rests at the bottom. Mostly these larvae remain on the surface in a slaunting position (the mouth on the surface and tail towards bottom).

II. Larval Development

Morphometric data and the distribution of melanophores during larval stages are depicted in Tables 1 and 2.

(a) Fry Stages

(i) Larva just after hatching (Fig. 18)

Body of larva long, elongated, broad anteriorly tapering posteriorly; thick and opaque. Yolk-sac oblong and transparent. Mouth well formed, blunt with thick jaws, lower jaw longer than upper; gap of mouth reaches below anterior extremity of eye. Eyes massive, dorsal, occupying 1/2 the height of head which is flexed downwards. Pre-orbital head flattened; auditory-vesicles massive; jaw and opercular bones formed; opercular cleft distinct. Pre-anal length of body almost twice post-anal length, later equals the yolk-sac length. Pre-anal ventral fin-fold broad, extending upto posterior 1/3 of yolk-sac. Dorsal and anal fins with 16 rays each, continuous with round caudal fin and originate opposite each other just posterior to anus. Uncountable rays appear in broad and circular pectoral fin. The path of yolk-circulation is same but these vessels now don't enter the mouth (head) as the mouth is free from yolk-sac but they remain circulating around the yolk-sac & then enter the body near heart region. Heart seen beating continuously in living larvae.

Body highly pigmented with dotted or stellate melanophores. Melanophores on jaws extend upto maxilla, even little beyond the posterior border of eye. Lower jaw pigmented on upper and ventral sides with numerous pigments, a few pigments noticed on the inner side also. Post-anal part of body highly pigmented with small, stellate melanophores. Lateral-line melanophores extend from the level of posterior 1/3 of yolk-sac, deep seated; rest of body pigmentation is superficial. Melanophores below lateral-line are somewhat larger than those above it, arranged on the surface of myotomes starting from posterior 1/3 of yolk-sac upto caudal end; but in the post-anal region, melanophores on the upper half also mark the surface of myotomes. Fine black melanophores also scattered on the area adjoining the yolk-sac and body. Yolk-sac pigmented only on the lateral sides, marking the path of yolk-circulation; ventrally yolk-sac unpigmented but a streak marks its distinction from the neck. Body melanophores brown, yolk-sac bears black melanophores.

(ii) 24-hours post hatching Stage (Fig. 19)

4-6 rays appear in pectoral fin. Myotomes now 60, i.e. 36 pre-anal, 24 post-anals. (2 myotomes appear to have been added to pre-anal part, thus anus shifted posteriorly by four myotomes). Swim-bladder rudiments appear. Head and body longer; snout less obtuse; pre-orbital head less flattened. Heart beats about 220 times per minute.

Melanophores sufficiently increase in number, pattern remain the same. Lateral-line melanophores increase numerically. Melanophores lining the lower surface of myotomes are faint and diffused, very small sized in the caudal region. Pectoral fin pigmented at the base. Melanophores at the junction of yolk-sac. and body increase numerically and extend over the yolk-sac. A group of internal melanophores noticable at the posterior border of eye, continuous in a streak, upto the base of pectoral fin. Heart demarcated by pigmented streak, Ventro-visceral melanophores dense.

(iii) 48-hours post hatching Stage (Fig. 20)

Dorsal finfold completely absorbed, pre-anal ventral finfold still persists but reduced in breadth; Pectoral fin now with 6-8 rays, caudal with 14 rays. Head straight, trunk well demarcated by a depression. Nasal sacs appear. Larvae show reduction in body length. Heart beats at 235 times/minute.

Head and dorsal melanophores merge. Pre-orbital head densely pigmented with melanophores arranged in a semi-circle around anterior half of nasal sacs; Jaw bones (upper & lower) densely pigmented. 3-4 distinct stellate melanophores grouped together on the operculum just near the base of pectoral fin. Melanophores increase numerically on operculum & ventrally on lower jaw.

(iv) 72-hours post hatching Stage (Fig. 21)

Head and body long, elongate, narrower than previous Stage. Yolk-sac much reduced. Pectoral fin now with 8 rays, dorsal and anal rays branched. Teeth appear, 4-5 on either side on upper jaw & 3-4 on either side on lower jaw. Ventrally branchiostegal rays appear. Oesophagus appears as tubular structure, continuous with alimentary canal upto anus.

Head well pigmented. Pectoral rays pigmented. Dorsal melanophores extended downwards, join lateral-line in the posterior half, very less in the anterior half. Melanophores decrease on yolk-sac. Interal pigmentation visible on the surface of swim-bladder.

(v) 96-hours post hatching Stage (Fig. 22)

4 pairs of gills indicated. Opercular cleft distinct & open. Upper jaw distinctly smaller than the lower and obtuse. Teeth in a row of 6-7 on either side of both jaws. Larvae show slight reduction in body length.

Gill cleft well pigmented, jaws densely pigmented on their inner side. Melanophores increase numerically on operculum & ventrally on heart but decrease on general body; small, dotted and less numerous on snout.

(vi) 216-hours post hatching Stage (Fig. 23)

Beak formed on head. Teeth increase in number, 14 in a row on either side in both the jaws (smaller in outer row). Pre-orbital head half the length of head; eyes massive, dorsal, bulging out on head, occupying about 3/4th of height of head. Gape of mouth narrow. Auditory-vesicle quite massive. Heart clearly distinct.

Snout (beak) densely pigmented with stellate melanophores; dotted in the inter-orbital space. Melanophores fine, stellate increase in number on head but decrease in number on heart, yolk-sac & rest of body. Ventrally, melanophores on jaw extend over pre-orbital head. Internal melanophores visible on head and auditory vesicle. Melanophores appear on nasal sacs and nasal bones.

B. Post Fry Stages

(vii) 264–hours post hatching Stage (Fig. 24)

Yolk–sac completely absorbed. Swim bladder well developed, present between vertebral column and alimentary canal; alimentary canal distinct opening at anus. There are 16 teeth on either side in both the jaws. Dorsal and anal fins with 17 branched rays; caudal fin complete with 15 rays (adult form).

There is not much change in pigmentation except for that internal melanophores on operculum and gill arches visible. Dorsal melanophores increase in number; those present laterally in the anterior half (below lateral-line) above the surface of alimentary canal and swim bladder are less numerous. Pre-anal ventro-visceral melanophores loosely arranged, post-anal ventro-viscerals dense.

(viii) 384– hours post hatching Stage (Fig. 25)

Ventral fin indicated close to anal fin almost in the posterior 1/3 of body, excluding caudal fin. Dorsal and anal fins now separated from caudal fin. Teeth 24 on either side in both the jaws. Pectoral fin now with 8–10 rays.

Melanophores fine, stellate, larger than previous stage. Lateral line melanophores distinct, deep seated in the posterior half of body. Aorta–ventralis demarcated by pigmented streak, a few scattered on the surface of heart; a few black medio-ventral melanophores appear. Internal pigmentation visible on swim-bladder and alimentary canal, and also on operculum starting from posterior extremity of eye upto pectoral fin base.

(ix) 456–hours post hatching stage (Fig. 26)

Body long and narrow; head elongates further. Eyes some what longer than round. Teeth 26 on either side in both the jaws. Pre-anal ventral fin fold still persists.

A general increase in overall pigmentation. Eyes black with white lens. Melanophores stellate on the body but small, dotted in the caudal region. Pattern of pigmentation remains unchanged.

(x) 504–hours post hatching Stages (Fig. 27)

Pectoral fin complete with 11 branched rays (adult form). Brain lobes clearly distinct. Swim bladder long, present as transparent chamber between vertebral column & alimentary canal.

Melanophores slightly decrease in number above the surface of swim-bladder & alimentary canal. Medioventral melanophores increase numerically. Pectoral and caudal rays well pigmented.

(xi) 624–hours post hatching Stage (Fig. 28)

One spine added to dorsal and anal fins each (1/17); ventral fin with 3–4 rays; pre-anal ventral finfold absorbed. Snout length more than twice and head length twice the previous stage. Teeth 46–48 on either side in both the jaws and hooked inwards. Operculum, throat, head & jaw bones prominent. Swim bladder elongate, alimentary canal long. Pre-anal length twice the post-anal length which is almost equals the orbital length (from tip of snout upto posterior extremity of eye). Myotomes 'Z' shaped, the number remains unchanged.

Body highly pigmented with small dotted or stellate melanophores. Dorsal labation of head melanophores distinct. Eyes pigmented on the anterior, dorsal & posterior borders. Melanophores increase on operculum, swim-bladder & above the surface of alimentary canal, and ventrally on aorta-ventralis & surface of heart, but medio-ventral melanophores decrease in number. A group of melanophores near the base of pectoral fin merge with head melanophores. Gill filaments well pigmented. Lateral-line highly pigmented with numerous small, dotted melanophores from the posterior end of operculum upto caudal peduncle.

C. Fingerling Stage

(xii) 720-hours post hatching stage (Fig. 29)

All the full complements of an adult fish are formed here, viz: ventral fin complete with 6 rays, dorsal fin complete with 2/15, and anal fin complete with 2/16. A longitudinal groove along the upper surface of head & a longitudinal groove along the dorsal surface of body on either side of mid-dorsal line along the entire length of body. Teeth 66-72 on either side in both the jaws i.e. a row of large, sharp, widely separated ones in both jaws (with an external row of numerous fine ones), lower jaw longer than upper, maxilla reaches beneath the first third of eye orbit. Dorsal fin commences opposite the anal, rather more than twice as far from the anterior extremity of orbit as it is from the posterior extremity of tail; tail has become homocercal, caudal furcate. Nasal sacs somewhat triangular. Body entirely covered with fine black dots or brown stellate melanophores, melanophores rather dense on dorsal surface. Head Melanophores fine, large as compared to rest of body and extend down on the operculum. Snout and jaws densely pigmented on the dorsal and ventral surfaces, and also on the sides. Nasal sacs surrounded by stellate melanophores; arc of fine stellate melanophores surrounding the eye on its anterior, dorsal & posterior edges. Operculum well pigmented. Pre-opercular and opercular bones bear a patch of melanophores each. Lateral-line melanophores and most of black dotted melanophores are deep seated while stellate melanophores are superficial. All the fins except for ventral fin are highly pigmented. Ventrally, a few stellate melanophores scattered on throat; heart melanophores indistinct. Melanophores on swim-bladder, gills and above the surface of alimentary canal indistinct; medio-ventral melanophores still present.

Discussion

In *Xenentodon* cancila, pigmentation is not a distinctive feature in life history, because most of the pigmentation is passed within the egg (in the embryonic stages) which otherwise appears in larval stages of other fish species like *P. conchonius, A. morar, T. tor & B. bendelisis* (Malhotra *et. al.,* 1980 & Gupta, 1982), and the larva is generally born as an adult. The pigmentation may be used as an instrument of aging of *X. cancila,* but other pigmentations which may be used in the identification of larvae of *X. cancila* during different development stages include the following.

Though the melanophores on heart appear in the larvae of *X. cancila* at 24-hours post hatching stage, they make their appearance at 48-hours stage in *T. tor* 72-hours stage in *P. conchonius* & 120 hours stage in *B. bendelisis* (Gupta, 1982), and in *A. morar* at their 72-hours post hatching stage (Malhorta, *et. al.* 1980). Since a semicircle of melanophores around nasal sacs appear in the larvae of *X. cancila* at 48-hours stage, they appear at 456-hours, 720-hours and 744-hours stages in *P. conchonius, B. bendelisis* and *T. tor* respectively (op. cit.) at their 648-hours, 1008-hours and 1104 - hours post hatching stages respectively in which they ultimately form a complete arc around nasal sacs but in *X. cancila* of present investigations, there is no complete arc of melanophores formed at any stage during development. A distinct group of few melanophores appears at 48-hours stage in *X. cancila* on the operculum but in *P. conchonius* (op. cit.) they appear as scattered pigments on the operculum at 312-hours post hatching stage arranging themselves in a distinct row at 768 hours stage. In *T. tor* (op. cit.), they appear at 120-hours stage, a few of them are distinct & are arranged in the form of an arc on the operculum at 288-hours post hatching stage. Their simple occurrence is reported in *B. bendelisis* and *A. morar* (op. cit.) at their 192-hours & 376-hours stages respectively. While *X. cancila* is characterized by the appearance of swim-bladder melanophores at its 72-hours post-hatching stage; *P. conchonuis, A. morar* & *T. tor* are characterized by its appearance at their 48-hours stage, but at 120-hours stage in case of *B. bendelisis.* The appearance of melanophores on gills is achieved at 96-hours post-hatching stage in case of *X. cancila* but this appearance is achieved at 120-hours stage in *T. tor* and at 216-hours & 304-hours stages in *B. bendelisis* and *A. morar* respectively, not reported in *P. conchonius.* Circumorbital completion of melanophores is yet another character which may help in larval aging of *X. cancila* at their 624-720 hours post hatching stage; otherwise in *B. bendelisis, T. tor* & *P. conchonius,* it gets completed at 720-hours stage; 840-hours & 1200-hours stages respectively. The merger of dorsal and head melanophores is noticed at 48-hours stage in *X. cancila* but in *T. tor, B. bendelisis* & *P. conchonius,* it is noticed at 96-hours, 168-hours and 576-hours post hatching stage respectively but not noticed in *A. morar* at any stage during development.

Gupta (1982) while discussing various morphological events during development of 4 cyprinid fishes viz; *P. conchonius, B. bendelisis, T. tor* and *A. morar*, highlighted that a very common feature i.e. larval finfold is quite prominent and is in varying degree of differentiation but in the present case, the larval forms of belone, *X. cancila* are devoid of such type of finfold (gets absorbed during their embryonic stages), thus clearly differentiating them from the larvae of cyprinid fishes. The appearance of such an important character like finfold in early larval forms of fish has also been reported by Ganapati & Raju (1960), Kulkarni (1971), Paramaswaran *et al.* (1969, 1971), Selvaraj *et. al.* (1972), Chakrabarty & Murty (1972) and Desai (1972, 1973).

Present observations reveal that the separation of head and yolk sac is developed simultaneously with the appearance of mouth, pectoral fin and larval finfold in *X. cancila* in their embryonic stages, whereas similar observations have also been made by Parameswaran *et al* (1971) but in the 2 days old larvae of *Nandus nandus* where the above stated event in development do not occur simultaneously. Paradoxically, however Gupta (1982) stated that in cyprinids, the head initially is adherent to yolk-sac and only separates from it a later by gaining prominence. Initiation & formation of mouth, its completion and development of protrusible mouth (in the form of well marked jaws) are distinct morphological events that occur in the development of oral intake apparatus. It is, therefore, seen differently timed in different species and particularly depending upon the capability of larva to feed exogenously. In fishes like *P. conchonius* & *B. bendelisis*, the mouth formation is initiated in 24-hours post hatching stage whereas it is initiated at 48 hours post hatching stage in *T. tor* & *A. morar*. The functionality of mouth also varied to 48-hours stage in *P. conchonius*, 72-hours stage in *A. morar* & *T. tor*, while this important function of mouth do not occur until at 144-hours post hatching stage (op. cit.) in *B. bendelisis*. In *X. cancila*, the mouth cleft and jaws are indicated in the embryonic stages, the teeth on jaws appear at 72-hours post hatching & mouth gets prolonged into a beak—a characteristic feature of adult at 216-hours post hatching stage and this feature becomes very distinctive for the identification of *X-cancila* larvae. Parameswaran *et. al.* (1971), Salvary *et. al.* (1972), Banerji (1974) and Thakur *et al* (1974) also reported the initiation of mouth and its formation in their fish larvae during different development stages.

The first trace of alimentary canal in *X. cancila* has been observed in 72-hours post hatching stage when oesophagus appears as a tubular structure continuous with alimentary canal, opening at anus, but the coiling has not been observed at any stage during development, which may confirm their carnivorous habit as indicated by toothed beak of mouth. The appearance of a long tubular intestine has also been reported by David (1961) in one day old pre-larave of *Botia dayi*. The appearance of alimentary canal as a straight tube has also been reported by Ganepati & Raju (1960) in the newly hatched larvae of Muraenid eel (*Muraena* sp.); Desai & Karamchandani (1967) in 5.6 mm long larvae of *B. bendelisis*; Parameswaran *et. al.* (1969, 1971) in 60-hours old larvae of *Oxygaster bacaila*, 2-days old larvae of *Ompok pabda* and 4 days old larvae of *Nandus nadus:* Selvaraj *et. al.* (1972) in 24-hours old larvae of *Labeo*

Table I : Morphometric data (in mm) by age of *Xenentodon cancila* larvea

Morphic characters	*Larva just after hatching*	*Developmental stages (age) (Hours post-hatching)* 24	*48*	*72*	*96*	*216*	*264*	*386*	*466*	*504*	*624*	*720*
Size interval	8.0	10.5	9.6-10.2	10.4-10.8	9.9	11.5-13.4	13.2	13.8	14.1	14.5	21.9	34.1
Length												
Total	8.0	10.5	9.6	10.6	9.96	12.4	13.2	13.2	14.1	14.5	21.9	34.1
Standard	8.3	9.8	8.9	9.7	9.0	11.5	12.1	12.1	13.1	13.4	20.4	31.6
Pre-anal	5.7	3.0	6.3	6.8	6.3	6.6	9.1	9.8	9.9	10.1	15.9	20.2
Post-anal	3.1	3.4	3.4	3.7	3.6	3.5	4.1	3.9	4.2	4.4	6.0	13.9
Snout	0.4	0.4	0.5	0.7	0.7	1.6	2.2	2.1	2.1	2.4	5.3	13.7
Head	1.8	1.9	1.2	2.3	2.3	3.5	3.7	4.1	4.4	4.4	8.3	21.4
Head depth	1.8	1.8	1.8	0.9	1.2	9.9	1.1	1.1	1.1	1.8	1.7	3.4
Eye diameter	0.6	0.6	0.6	0.6	0.7	0.7	0.7	6.7	9.9	0.8	1.2	2.4
Greatest body depth	1.5	1.5	1.5	1.2	1.4	0.8	0.9	6.9	8.8	8.9	1.4	3.4
Yolk-sac length	3.2	5.0	5.0	5.3	4.4	2.1	-	-	-	-	-	-
Yolk-sac depth	1.0	0.9	1.8	0.7	8.7	0.2	-	-	-	-	-	-
Pre-dorsal length	4.8	6.0	6.4	6.8	6.7	0.6	9.6	9.9	10.0	10.1	16.1	23.2
Anal fin to caudal peduncle	-	-	-	-	-	-	-	6.3	8.8	0.9	1.4	2.1
Eye to dorsal surface of head	0.1	0.1	0.1	0.1	0.1	0.00	0.07	0.1	0.1	0.1	0.1	0.1
Caudal peduncle depth	0.3	0.3	0.3	0.4	8.4	9.4	0.4	0.4	0.5	0.5	0.7	1.6
Swim- bladder length	-	-	0.9	1.2	1.5	-	1.7	1.7	1.9	2.2	3.8	-
Swim-bladder depth	-	-	0.2	0.3	9.3	-	0.3	5.4	8.3	0.3	0.4	-

Table 2 : Number and Position of Melanophores During Different Developmental Stages of *Xenentodon cancila larvae*

Developmental				*Dorsal line region*			*Fin*				
stages (Age)	*Swim bladder*	*Yolk-sac*	*Head region*	*Pre-dorsal*	*Fin-fold*	*Caudal peduncle*	*Dorsal*	*Pectoral*	*Ventral*	*Anal*	*Caudal*
1	*2*	*3*	*4*	*5*	*6*	*7*	*8*	*9*	*10*	*11*	*12*
Embryo (hrs. after blastodisc formation)											
60	-	Many	Few	Melanophores in 2-3 rows on either side of mid–dorsal line			-	-	-	-	-
96	-	Numerous	Few	Highly pigmented with 3 vertical bands in the middle of body			-	-	-	-	-
144	-	*	Many	Highly pigmented with 9 vertical bands; numerous scattered in the caudal region			-	-	-	-	-
168	-	*	Many	*	*	*	-	-	-	-	-
180	-	Less numerous	Many	*	*	*	-	-	-	-	
216	-	Numerous	Numerous	4-5 rows on eithei side of mid-dorsal line			-	-	-	-	-
288	-	Numerous	Numerous	4-5 rows on either side of mid-dorsal line			Pigmented	-	-	Pigmented	Pigmented
336	-	Numerous	*	2-3 rows on either side of mid-dorsal line			Pigmented	Pigmented	-	Pigmented	*
Larval hrs.(Post-	-										
hatching) Larvae just after hatching	-	Numerous	*	*	*	*	Many	Many	-	Many	*
24	-	Numerous	*	*	*	*	Many	Many	-	Many	*
48	-	Numerous	*	*	*	*	Many	Many	-	Many	*
72	Few	Numerous	*	*	*	*	Many	Many	-	Many	*
96	Few	Numerous	*	*	*	*	Many	Many	-	Many	*
216	Few	Numerous	*	*	*	*	Many	Many	-	Many	*
264	Few	-	*	*	*	*	Many	Many	-	Many	*
384	Few	-	*	*	*	*	Many	Many	-	Many	*
456	Few	-	*	*	*	*	Many	Many	-	Many	*
504	Many	-	*	*	*	*	Many	Many	-	Many	*
624	Many	-	*	*	*	*	Many	Many	-	Many	*
720	Indistinct		-	*	*	*	Many	Many	-	Many	*

* = Highly pigmented.

Contd...

Table 2...... Contd.

Developmental stages (Age)	Lateral-line	Ventro-visceral line: Pre-anal	Ventro-visceral line: Post-anal	Caudal peduncle	Throat region	Heart region	Gills	Medio-Ventral-line
	13	14	15	16	17	18	19	20
Embryo (hrs after blastodisc formation)								
60	-	-	-	-	-	-	-	-
96	-	-	-	-	-	-	-	-
144	-	*	*	Many	-	-	-	-
168	-	*	*	Many	-	-	-	-
180	-	*	*	Many	-	-	-	-
216	-	*	*	Many	-	-	-	-
288	Pigmented	*	*	Numerous	-	-	-	-
336	44-45	*	-	Numerous	Many	-	-	-
Larvae (hrs. post-hatching)								
Larvae just after hatching	Many	*	*	Numerous	Many	-	-	-
24	Many	*	*	Numerous	Numerous	Heart demarcated by pigmented streak	-	-
48	Many	*	*	*	Numerous	Heart demarcated by pigmented streak	-	-
72	Many	*	*	*	*	4-6 melanophores on either side of aorta ventralis; 6 on the surface of heart	-	-
96	Many	*	*	*	*	4-6 melanophores on either side of aorta ventralis; 14-16 on the surface of heart	Few	-
216	Many	*	*	*	*	Melanophores indistinct in aorta ventralis; 6-8 on the surface of heart	Few	-
264	Many	*	*	*	*	Pigmented streak lining the heart	Many	-
386	Many	*	*	*	*	Pigmented streak on the aorta ventralis; 12 scattered on the surface of heart	Many	Few
466	Many	*	*	*	*	Heart lined by pigmented streak	Many	Few
504	Many	*	*	*	*	Heart lined by pigmented streak	Numerous	Few
624	Many	*	*	*	*	12 on either side of aorta ventralis; 16 on the surface of heart	Numerous	Few
720	Many	*	*	*	*	Indistinct	Indistinct	Few

* = Highly pigmented.

boggut: Desai (1973) in 10.70 mm long larvae of *Tor tor*: Thakur *et. al.* (1974) in 6-8 hours old larvae of *Heteropneustes fossilis:* and Benerji (1974) in 15-hours old larvae of *Channa punctatus.*

The gills appear for the first time in a distinct branchial chamber at different times in the life history of fishes. They appear at 96-hours post hatching stage in *X. cancila* but their appearance in *P. conchonufis, T. tor, B. bendelists & A. morar* is differently timed. This time related appearance of gills in fish larvae has also been reported by Kulkarni (1971) in 3-days old larvae of *Tor Khudree,* and Chakrabarty & Murty (1972) in 48-hours old larvae of *Cirrhinus mrigala* and 24-hours old larvae of *Labeo rohita.* The duration of yolk-sac period depends upon the quantity of yolk present in the larvae and is related to the exogenous mode of feeding; and after the complete absorption of yolk-sac when the larvae acquire the post-larval characters. In *Xenentodon cancila,* the yolk-sac gets absorbed at 264-hours stage; although the mouth is well formed at birth (unlike the above mentioned cyprinids), yet it continues to derive its energy from stored food in yolk-sac. Such a yolk-sac absorption has been reported by Salvaraj *et. al.* (1972) in *L. boggut* at 42-hours post hatching stage; Chakrabarty and Murty (1972) in 96-hours old larvae of *C. mrigala, C. catla & L. rohita;* Parameswaran *et. al.* (1971) in 2-days old, 3-days old & 4-days old larvae of *O. pabda, M. tengara & N. nandus* respectively but David (1961) reported its complete absorption in 2-days old larvae of *B. dayi*. Desai (1972 & 1973) reported the yolk sac absorption in larvae of *T. putitora* & *T. tor* longer than 11.30 mm and 12.16 mm stages respectively. On the contrary, Ganapati & Raju (1960) reported the yolk-sac absorption in 4-days old embryo (within the egg) of Muraemid eel, *Muraena* sp.

Summary and Conclusions

In the present communication, an attempt has been made to investigate the life history of *Xenentodon cancila,* a freshwater belone. The periodization in the life history stages during development has been made on the basis of occurance of melanophores and other morphic events (characters) during different developmental stages. Since *X. cancila,* is a unique fish (unlike cyplinids), as most of its life history stages are passed within the egg and the larva is generally born as an adult, with well developed mouth and jaws and other important pigmentations. It becomes then evident that appearance of pigments do not follow any set pattern in different fish species, but may follow a patter of appearance within the same species, and can be of some value only in the aging of larvae. Thus it is evident from the present studies that the various characters occurring at various intervals in fishes varies differently and is independant phenomenon.

Acknowledgements

The authors are thankful to Prof. Y.R. Malhotra for critically evaluating the manuscript. They are also grateful to the CSIR and UGC, New Delhi for financial assistances during the course of studies, and to the University of Jammu, Jammu for providing the necessary facilities to carry on these investigations.

Legend to Figures

Plate-I

Figs. 1-17 : Fertilized egg, cleavage and post-cleavage stages of *Xenentodon Cancila.*

1. Fetilized egg
2. Blastodisc formation
3. Morula or ealy gastrula
4. Gastrulation
5. Initiation of embryonic ridge
6. Differentiation of Cephalic and caudal ends; Optic lobes, 4 myotomes and Kupffer's vesicle.
7. 8-12 myotomes
8. 14 myotomes. Note the disappears of Kupffer's vesicle and the enlargement of yolk-mass in the caudal region.
9. Enlargements of yolk-mass in the caudal portion. Note the further enlargement of yolk-mass in the caudal region.
10. Yolk (vitelline vessels) circulation. Note the circulation of yolk-vessels in the form of a net-work of vessels all around the yolk-sac; formation of eyes; appearance of brain rudiments; and head, dorsal and yolk-sac melanophores.
11. Tail-free. Note the formation of bands of dorsal (dorso lateral) melanophores.
12. Appearance of mouth, pectoral fin & larval finfold. Note the arrangement of head and dorsal (dorso-lateral) melanophores in the form of bands.
13. Jaw-formation
14. Differentiation of dorsal and anal fins. Note the differentiation of dorsal and anal fins, and the movements of pectoral fins.
15. Larval movements.
16. Appearance of fin-rays. Note the appearance of rays in dorsal, anal and caudal fins; and dense pigmentation at the junction of yolk-sac and the body.
17. Differentiation of heart and pre-anal ventral finfold. Note the ring-like coiling of the embryo.

(Scale = 1.0 mm).

(ATF-Attaching fibres; BL-Blastodisc; EM-Embryo; KV-Kupffer's vesicle; MT-Myotomes; NC-Notochord; OL-Optic lobes; OV-Optic vesicle; PVM-Perivitelline membrane; VM-vitelline membrane; Y-Yolk; AF-Anal fin; AR-Rays in anal fin; BR-Brain rudiments; CF-Caudal fin; CR-caudal rays; DR-Dorsal fin; DM-Dorsal Melanophores; DR-Rays in dorsal fin; E-Eyes; FF-Finfold; HM-Head melanophores; Mouth; PF-Pectoral fin; VV-Vitelline; YS-Yolk sac; YSM-Melanophores on the yolk sac.)

PLATE - I

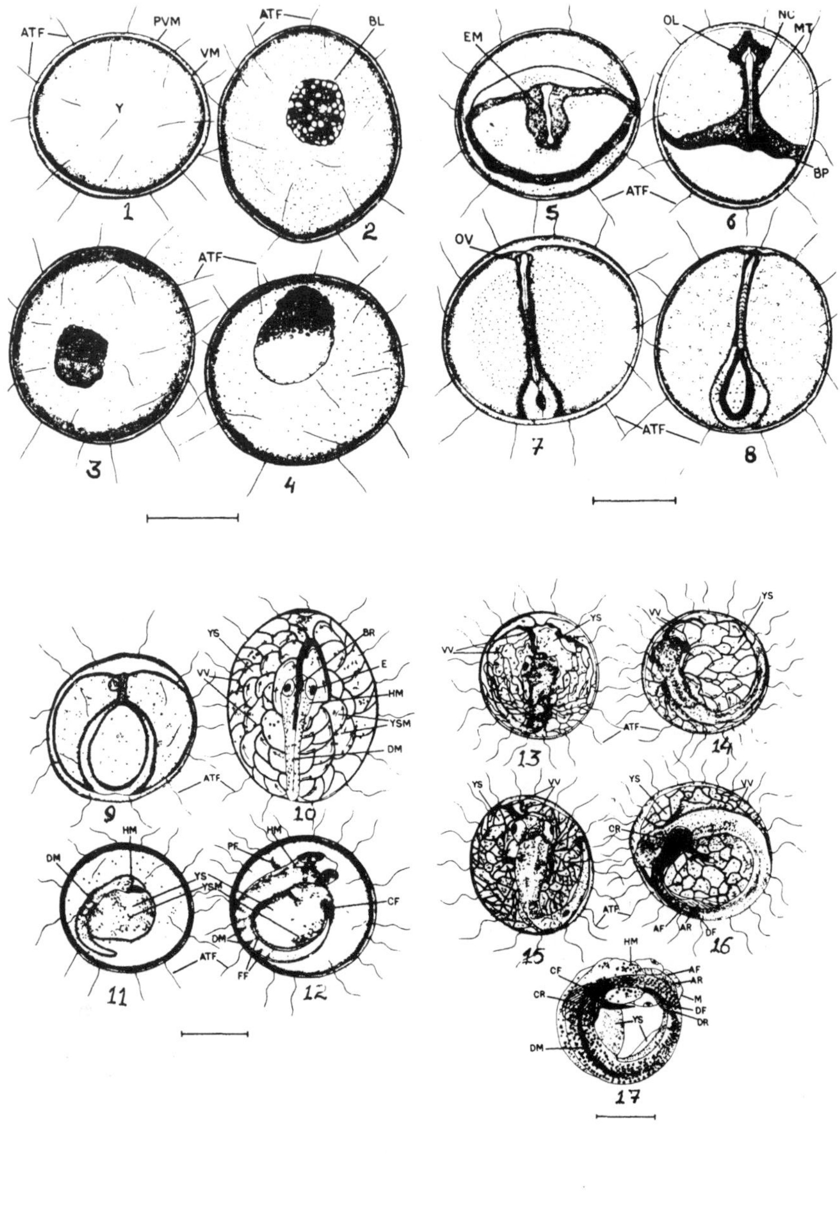
ATF
PVM
VM
Y
BL
EM
OL
NC
MT
BP
OV
YS
BR
E
HM
YSM
DM
VV
PF
CF
FF
CR
AF
AR
DF
M
DR
1
2
3
4
5
6
7
8
9
10
11
12
13
14
15
16
17

Legend to Figures

Plate-II

Figs. 18-20 : Fry stages of *Xenentodon Cancila.*

18. Larva just after hatching.

18A. Lateral view. Note the heterocarcy of tail, differentiation of vent, distinction of lateral-line and Ventro-visceral melanophores and the arrangement of melanophores above the surface of myotomes.

18B. Dorsal view

18C. Ventral view

19. 24-hours post hatching

19A. Lateral view. Note the appearance of pectoral rays and Surin bladder rudiments.

19B. Dorsal view

19C. Ventral view

20. 48-hours post hatching

20A. Lateral view. Note the appearance of nasal sacs, absorption of dorsal finfold and demarcation of trunk by a depression.

20B. Dorsal view

20C. Ventral view

(Scale = 1.5 mm).

(AF-Anal fin; AR-Anal rays; CF-Caudal fin; CR-Caudal rays; DF-Dorsal fin; DM-dorsal melanophores; DR-Dorsal rays; E-Eyes; FF-Finfold; HM-Head melanophores; HM′-Heart melanophores; LLM-Lateral-line melanophores; M-Mouth; MTM-Melanophores on myotomes; PF-Pectoral fin; PR-Pectoral rays; SB-Swim-Bladder; V-Vent; VVM-Ventro-visceral melanophores; YS-Yolk-sac; YSM-Yolk ac Melanophores; NS-Nasal sacs).

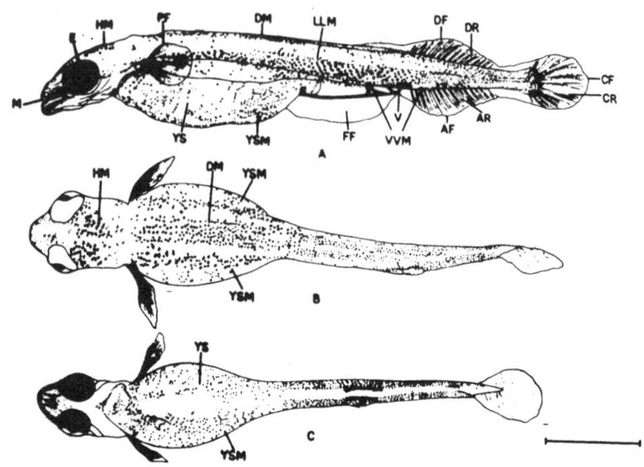
E
HM
PF
DM
LLM
DF
DR
CF
CR
M
YS
YSM
FF
V
VVM
AF
AR
A
HM
DM
YSM
YSM
B
YS
C
YSM

18

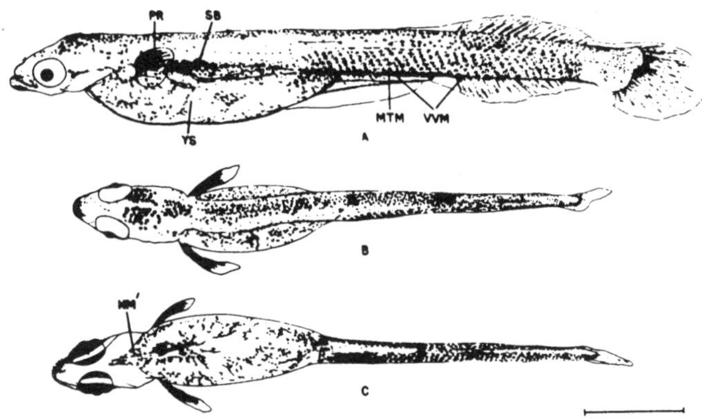
PR
SB
MTM
VVM
YS
A
B
HM'
C

19

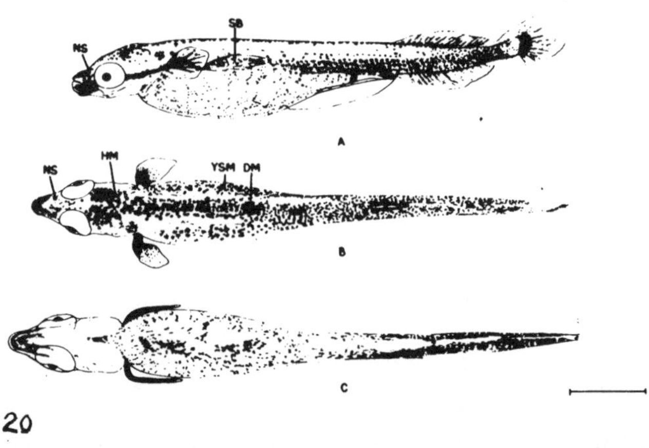
SB
NS
A
HM
NS
YSM
DM
B
C

20

Legend to Figures

Plate-III

Figs. 21-23 : Fry stages of *Xenentodon Cancila.*

21. 72-hours post hatching

21A. Lateral view. Note the appearance of teeth, a tubular Oesophagus continuous with the alimentary canal upto anus, and melanophores on the swim-bladder.

21B. Dorsal view

21C. Ventral view

22. 96-hours post hatching

22A. Lateral view

22B. Dorsal view

22C. Ventral view. Note the appearance of gills and melanophores on them. (Scale = 1.5 mm)

23. 216-hours post hatching

23A. Lateral view. Note the formation of beak.

23B. Dorsal view

23C. Ventral view. Note the distinction of heart. (Scale = 2.5 mm).

(AC-Alimentary canal; BK-Beak; G-Gills; H-Heart; HM - Head melanophores; HM′-heart melanophores; Oes-Oesophagus; SB - Bladder; SBM-Swim-bladder melanophores; TH-Teeth; YS- Yolk sac; YSM-Yolk-sac melanophores).

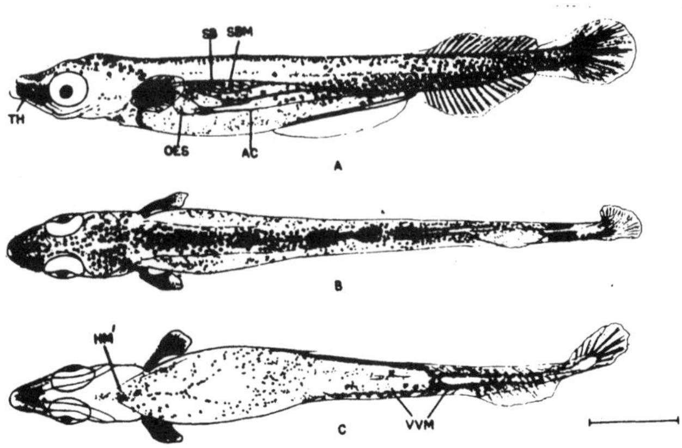

21

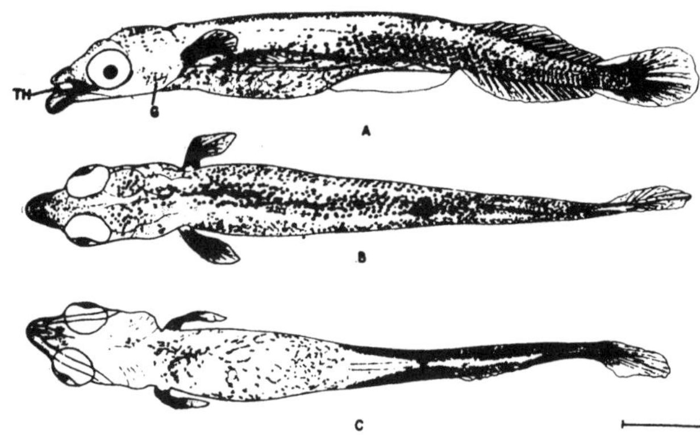

22

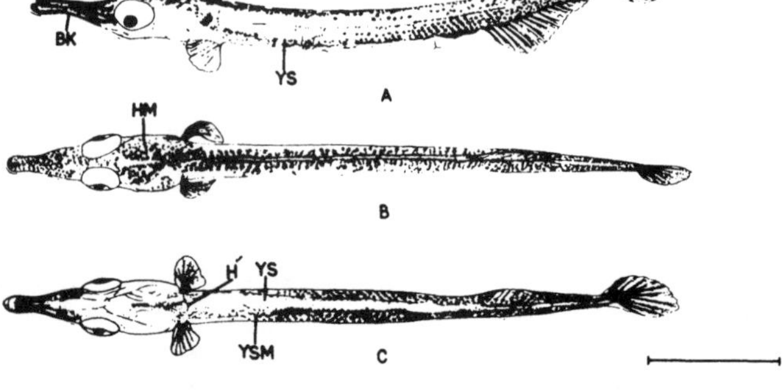

23

Legend to Figures

Plate-IV

Figs. 24-26 : Post-fry stages of *Xenentodon Cancila.*

24. 264-hours post hatching

24A. Lateral view. Note the absorption of Yolk-sac.

24B. Dorsal view

24C. Ventral view

25. 384-hours post hatching

25A. Lateral view. Note the appearance of ventral fin and dense pigmentation of snout.

25B. Dorsal view

25C. Ventral view. Note the appearance of medio-ventral melanophores.

26. 456-hours post hatching

26A. Lateral view

26B. Dorsal view

26C. Ventral view. Note the pigmentation of heart and aorta ventralis

(Scale = 2.5 mm).

(AC-Alimentary canal; AV-Aorta ventralis; G-Gills; HM - Head melanophores; HM′-heart melanophores; MVM-Medio-ventral melanophores; SB-Swim-bladder; VF-Ventral fin).

PLATE - IV

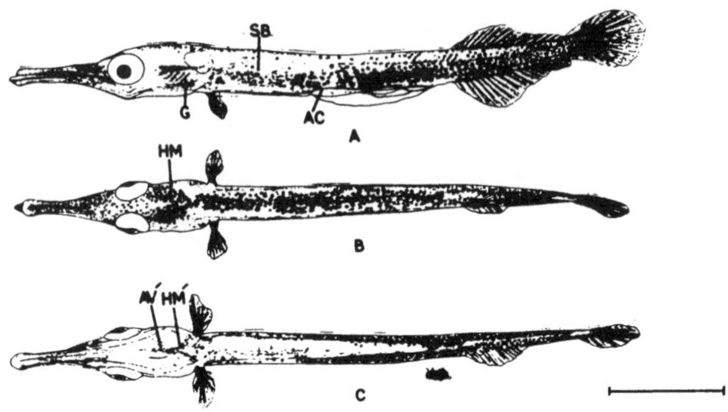

24

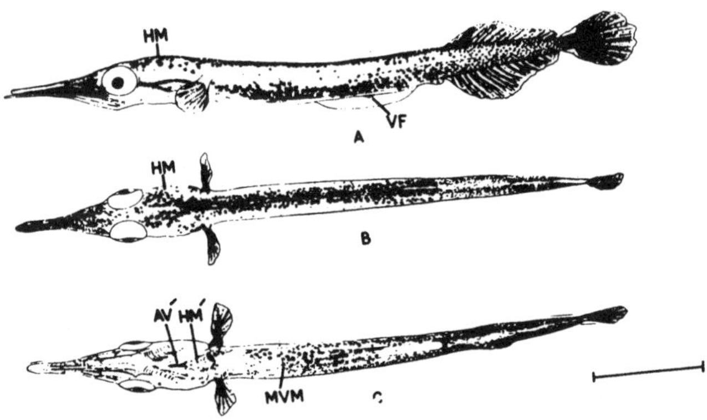

25

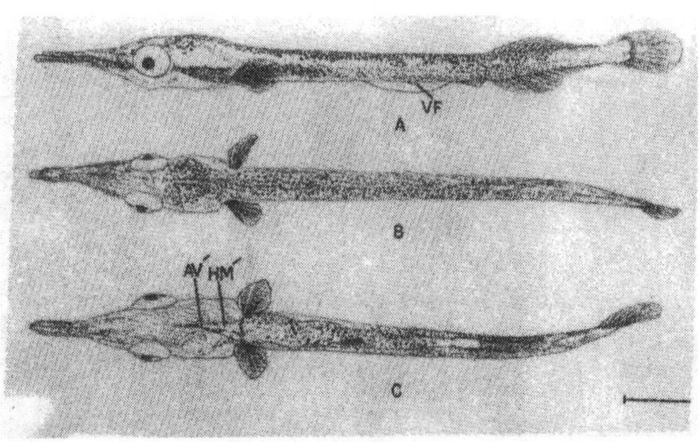

26

Legend to Figures

Plate-V

Figs. 27-29 : Post-fry and fingerling stages of *Xenentodon Cancila.*

27. 504-hours post hatching

27A. Lateral view

27B. Dorsal view

27C. Ventral view (Scale = 2.5 mm)

28. 624-hours post hatching

28A. Lateral view. Note the appearance of a spine in the dorsal and anal fin each; rays in the ventral fin; and absorption of ventral finfold.

28B. Dorsal view

28C. Ventral view (Scale = 4.00 mm)

29. 720-hours post hatching (Fingerling stage)

29A. Lateral view

29B. Dorsal view

29C. Ventral view (Scale = 4.00 mm)

(HM'—Heart melanophores; MVM-medio-ventral melanophores; NS-Nasal sacs; SB-Swim bladder; SBM-Swim bladder melanophores; AS-Anal spine; DS-Dorsal spine; VF-Ventral fin; VR-rays in the ventral fin).

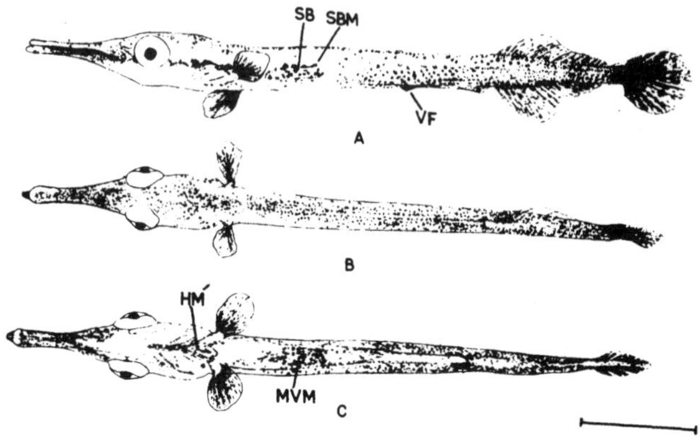

27

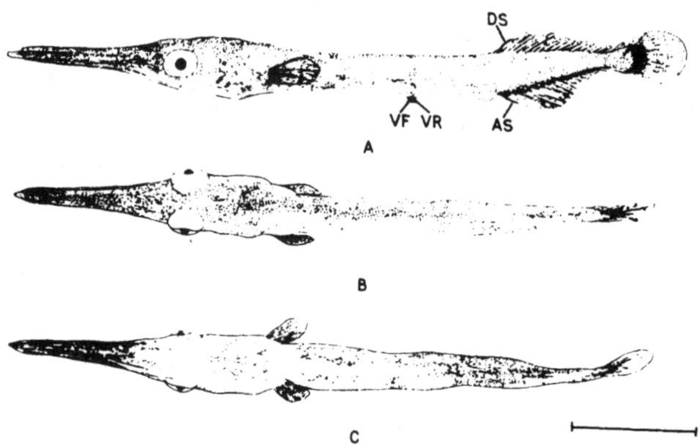

28

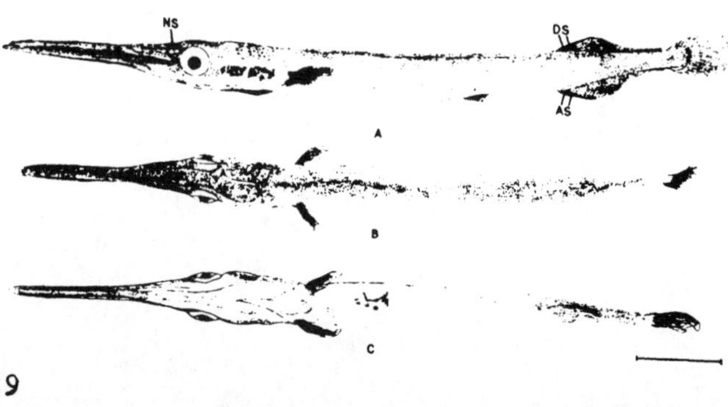

29

References

Ahmed, Nazir (1948). On the spawning habits and early development of the copper Mahseer, *Barbus (Lissochilus) hexagonolepis* Meclleland. Proc. Natn. Inst. Sci. India. 14, 21-28.

Balakrishnan, V. and K.V. Naryana Rao (1967). Some post larval and juvenile stages of the Indian Mackeral, *Rastrelliger Kanagurta* (cuvier) with notes on the changes in body form. Indian J. Fish. Vol. 14 : 97-114.

Banerji, S.R. (1974). Hypophysation and life history of *Channa punctatus* (Bloch) J. Inland Fish Soc. India. Vol. VI: 62-73.

Bensam, P. (1965). On the eggs and larval stages of the Malabar sole, *Cynoglossus semifasciatus* Day. Indian J. Fish. Section A. Vol. XII, No. 1 : 90-98.

—(1966). The eggs and early development of a Muraenid eel. J. Mar. Biol. Ass. India : 8(1) : 181-187.

—(1967a). On a few post larval stages of *Anodontostoma Chacunda*. Indian J. Fish. Vol. 14 : 48-53.

—(1967b). The embryonic and early larval development of the long finned herring *Opisthopterus tardore* (Cuvier) J. Mar. Biol. Ass. India 9(1) : 76-83.

—(1969). Notes on the post-larval stages of the white Sardine, *Kowala coval* (cuvier) J. Mar. Biol. Ass. India Vol. XI, No. 1 & 2; Ps: 251-254.

—(1970). Notes on the eggs, larval and juveniles of the Indian sprat, *Sardinells jussieu* (Lacepede). Indian J. Fish Vol. XIII, No. 1 pp: 219-231.

—(1973). On a few post-larval stages and juveniles of Sardine, *Sardinella dayi* Regan. Indian J. Fish Vol. 20; No : 1 : pp: 148-156.

Chakrabarty, R.D. and D.S. Murty (1972). Life history of Indian Major carps, *Cirrhinus mrigala* (Ham.), *Catla catla* (Ham) and *Labeo rohita* (Ham). J. Inland Fish Soc. India ; Vol. IV: 132-161.

David, A. (1961a). Early life history of the Cobitid fish, *Botia dayi* Hora. Proc. Zool. Soc. Vol. 11 : No. 1: 39-45.

David A. (1961b). Notes on the embryonic and larval development of the "Goonch", *Bagarius bagarius* (Ham). J. Zool. Soc. India, Vol. 13: No: 2, 194-205.

Desai, V.R. (1972). Notes on the early larval stages of *Tor. putitora* (Hamilton) from Narmada River. J. Zool. Soc. India. 24(1): 47-51.

Desai, V.R. (1973). Studies on the Fishery and Biology of *Tor tor* (Hamilton) from River Narmada. 2, Maturity, fecundity & larval development. Proc. Nat. Acad. Sci. India. Vol. 39 Part-B, No: 2, 228-248.

Desai, V.R. and S.J. Karamchandani (1967). On the larval development and Spawning seasons of two species of*Barilius* from Narbada River. J. Zool. Soc. India. 19 (1 and 2): 27-41.

Ganapati, P.N. and N.S. Raju (1960). On the eggs and early development of Eels of Waltair coast. J. Zool. Sco. India. Vol. 12; No: 2, 229-238.

Grover, S.P. (1971). Some biological notes on *Barilius benedelisis* (Hamilton). Indian J. Fish. 18(122): 182-83.

Gupta, S.C. (1982). Studies on Some Fish Life Cycles from Jammu, with observations on Larval Ecology and Juvenile Mortality. Ph.D. thesis, Department of Biosciences, University of Jammu, Jammu.

John, M.A. (1951). Pelagic fish eggs and larvae of the Madras coast. J. Zool. Soc. India. Vol. 3, No. 1. 41-69.

Jones, S; and P.M.G. Menon (1951). Notes on the Bionomics and Developmental stages of Some Indian Flat-fishes. J. Zool. Soc. India. Vol. 3, No: 1, 71-83.

Kulkarni, C.V. (1971). Spawning habits, eggs and early development of Deccan Mahseer, *Tor Khudree* (Sykes). J. Bomb. Nat. Hist. Soc. Vol. 67(3), 510-521.

Kuthalingam, M.D.K. (1957). Life history and feeding habits of *Cynoglossus lingua* (Ham. Buch). J. Zool. Soc. India. Vol. 9. No: 2, 208-215.

Kuthalingam, M.D.K. (1961). Observations on the life history and feeding habits of a Plankton-feeder, the Rainbow serdine *Dussumieria acuta* Cuv. and Val.) with a note on the method of securing stomach contents of live fish. J. Zool. Soc. India. Vol. 13. No: 1. 24-33.

Malhotra Y.R.; M.K. Jyoti and Subash Chander Gupta (1980). Periodization of larvae of *Aspidoparia maror* (Bleeker). with special reference to melanophore occurence and their role in identification of larvae. Matsya, No: 6, 58-70.

Natarajan, A.V; and S. Patnaik (1972). Embryonic and larval development of Chilka Mullet, *Liza macrolepis* (Smith). J. Inland. Fish. Soc. India. Vol. IV; pp. 14-19.

Parameswaran, S.C. Selvaraj and S. Radhakrishnan (1969). Observations on the life history and Bionomics of the carp Minnow, *Oxygaster bacaila* (Ham.). J. Bomb. Nat. Hist. Soc. Vol. 66(3), 443-454.

Parameswaran, S.C. Selvaraj and S. Radhakrishnan (1971a). Notes on the life history and biology of the catfish, *Ompok pabda* (Hamilton). J. Zool. Soc. India 23(2) : 137-150.

Parameswaran, S; S. Radha Krishnan and C. Selvaraj (1971b). Some observations on the life biology and life history of *Nandus* (Ham.). The proceedings of the Indian Acad. of Sci.; Vol. LXXII *nandus* II; No: 3, Section B; 132-147.

Parameswaran, S.C. Selvaraj and S. Radhakrishnan (1971c). Notes of the life history and biology of the catfish, *Mystus tengara* (Ham.). J. Zool. Soc. India. 23(2): 151-162.

Rao, S.N.R. (1919). Notes of the life history of two fishes, *Callichrous bimaculatus* (Bloch) and *Notopterus notopterus* (Pallas). Proc. Indian Sci; Congress. 6. 269.

Selvaraj, C; S. Radhakrishnan and S. Parameswaran (1972). Notes of the Breeding Season, Fecundity and life history of Minor Carp, *Labeo* (Sykes.) J. Inland Fish Soc. India. Vol. IV : 87-97.

Thakur, N.K.; R.N. Pal and H.A. Khan (1974). Embryonic and larval development of *Heteropneustes fossilis* (Bloch). J. Inland Fish. Soc. India. Vol. VI: 33-44.

Chapter 14

Rotifer Fauna of Devikoppa Tank : Dharwad (Karnataka, India)

*K. Vijay Kumar**

Introduction

Systematic study of the rotifers of the Indian water bodies has been carried out by several workers (Anderson, 1889); Edmondson and Hutchinson, 1934; Sewell, 1935; George, 1961; Gouder and Joseph, 1961; Arora, 1962, 1963, 1965; Vasist and Batiish, 1971a; Sharma, 1976, 1978a, 1979a, b; Choudhary *et. al.*, 1977; Tiwari and Sharma, 1977). Most of these studies pertain to the rotifers of North India region, the climate and topography of which differs widely from those of the Southern part. To mention few works from South India are those recorded from Ootacmund (Edmondson and Hutchinson, 1934), Madras (Pasha, 1961; Naidu, 1967), Kerala (Nayar and Nair, 1969) and there is a mention of 10 species in a lone study on the ecology of freshwater zooplankton of Karnataka (Gouder and Joseph, 1961 and Gouder and Patil, 1989 and Vijay Kumar, 1991).

The present report describes the rotifers collected from Devikoppa water body; Kalaghatgi taluk, Dharwad Dist, Karnataka State, India.

Material and Methods

Collections were made by filtering the water through a silk (crave) conical net (900 meshes/cm) with a vial attached to its cone. The animals were killed by dropwise addition of formalin in the concentrate and finally preserved in 10% formalin for feature quantitative determination. Later specimens were mounted in a mountant prepared by mixing equal parts of Lactophenol (Lactic acid 100 ml + Phenol, 100 gr + glycerol, 100 ml + distilled water, 100 ml) in which a drop of blue black ink per 10 ml was added and RENY'S mountant (Chloralhydrate 50 gm + Gum arabica, 30 gm + glycerol, 12.5 ml + distilled water 50 ml). For taxonomic confirmations, a thorough study of the living specimens were also made and camera lucida tracings were made. Identification was made according to Ahlstrom (1940), Arora (1962); Edmondson (1963); Nayar and Nair (1969); Pennak (1953); Pasha (1961);

* Environmental Biology Division, Department of Zoology, Gulbarga University, Gulbarga-585106 Karnataka.

Sharma (1976, 1978a, 1979b, 1979c) and Gouder and Patil (1989). Koste's (1978) scheme of classification was adopted in the present study.

List of Species Described

1. *Anuraeopsis fissa* (Gosse, 1851)—D.V.
2. *Brachionus angularis* (Gosse, 1851)—D.V.
3. *Brachionus bidentata* (Anderson, 1889)—V.V.
4. *Brachionus caudatus var personatus* (Ahlstrom, 1940)—D.V.
5. *Brachionus forficula* (Wierzejski, 1891)—D.V.
6. *Brachionus urceolaris* (Muller, 1773)—D.V.
7. *Keratella tropica* (Apstein, 1907)—D.V.
8. *Lepadella ovalis* (Muller, 1786)—D.V.
9. *Trichocera longiseta* (Schrank, 1802)—V.V.
10. *Asplanchna priodonta* (Gosse, 1850)—L.V.
11. *Filina longiseta* (Ehrenberg, 1832)—L.V.
12. *Rotifer tardus* (Ehrenberg, 1832)—L.V.

Observations and Discussion

Family : Brachionidae

Anureopsis fissa (Gosse, 1851) Fig. 1.

The lorica is cylindrical and flexible with heavily stipped dorsa and ventral plates. The trophi is malleate.

Measurements : LL = 91, *um*, WL = 46 *um*

Recorded from Punjab (Edmondson and Hutchinson, 1934; Vasist and Battish, 1971b); Baroda (Wulfert, 1966); Rajasthan (Nayar, 1968); Chandigarh (Vasist and Gupta, 1967); Kerala (Nayar and Nair, 1969) and in West Bengal (Sharma, 1979).

Brachionus angularis (Gosse, 1851) Fig. 2.

The lorica is firm and heavily stippled. Antero-dorsal margin has two small median spines. The front openings is large with V-shaped sinus on the ventral plate of lorica. It is widely distributed in India.

Measurements : LL = 110 - 125 *um*, WL = 87 *um*

Recorded from Punjab and Kashmir (Edmondson and Hutchinson, 1934), Rajasthan (Nayar, 1968); Kerala (Nayar and Nair, 1969); North-West India (Vasist and Battish, 1971 a: Sharma, 1976); Tamil Nadu (Michael, 1973); Andhra Pradesh (Dhanapathi, 1974); Karnataka (Gouder and Patil, 1989 and Vijay Kumar, 1991).

Brachionus bidentata (Anderson, 1889) Fig. 3

The lorica is barrel shaped. It has 6 anterior spines, the laterals are longer than medians and intermedians. There are two unequal posterior spines. The foot opening is wide.

Measurements : TL = 192 *um*, WL = 100-105 *um*

Distribution: Calcutta (Anderson, 1889); Kashmir (Edmondson and Hutchinson, 1934); North-West India (Vasist and Gupta, 1967; Sharma, 1976) and in Andhra Pradesh (Dhanapathi, 1974).

Brachionus caudatus var personatus (Ahlstrom, 1940) Fig. 4

The lorica is firm and stippled. The anterior spines are 2. The posterior spines are long, stout and tapering towards end.

Measurements : LL = 160 um, WL = 70-70 *um* & PLS =67 *um*.

Brachionus forficula (Wierzejski, 1891) Fig. 5

The lorica has 4 anterior spines, the laterals are longer than medians with slight swellings at their inner side. The posterior spines are long and stout with swelling of innerside near base.

Measurements : LL = 181 *um*, WL = 100 *um*, ASP = 68, 28*um* & PSP = 78-81.

The species is widely distributed and reports of its occurrence are available from every part of the country. From Maharashtra (Arora, 1963), Gujarat (Wilfert, 1966), Punjab (Vasist & Battish, 1971) Kerala (Nayar & Nair, 1969), Tamil Nadu (Michael, 1973) Andhra Pradesh (Dhanapathi, 1974), Karnataka (Gouder & Patil, 1989).

Brachionus urcelaris (Muller, 1773) Fig. 6.

The lorica has 6 anterior spines, the lateral and intermedians are almost equal in length. The inner spines have thickened ridges.

Measurements : TL = 172 um, WL = 138 *um*.

Recorded from West Bengal (Anderson, 1889, Sharma, 1979) Maharashtra (Arora, 1963), North West India (Sharma, 1976), Punjab (Sharma 1981), Karnataka (Gouder & Patil, 1989).

Keratella tropica (Apstein, 1907). Fig. 7.

Lorica with arched dorsal and flat ventral surface, narrow markedly at the base of posterior spines, anterior dorsal margin with six spines medians stoutest, longest and curved ventrally, laterals directed outwards and longer than intermediates, posterior spines two and unequal, right longer than left, dorsal surface of lorica with 3 hexagons and a small four sided figure between posterior border of lorica and last hexagon.

Measurements : TL = 215 *um*, WL = 78 *um*, ASP = 26 *um*, PSP = 21 *um*.

Recorded from Punjab and Kashmir (Edmondson and Hutchinson, 1934), West Bengal (Sewell, 1935, Vasist and Gupta, 1967, Sharma, 1976, 1979), Delhi (George, 1961) Kerala (Nayar and Nair, 1969) Punjab (Vasist and Battish, 1971b, Sharma 1981), Tamilnadu Michael, 1973), Andhra Pradesh (Dhanapathi, 1974), Karnataka (Gouder and Patil, 1989, Vijaykumar, 1991).

Family : Colurellidae

Lepadella ovalis (Muller, 1786). Fig. 8.

The body form is circular. The lorica is thin with dorsal and ventral plates, rigidly united at the edges having an anterior opening. The ventral plate has V shaped groove for foot opening.

Measurements : LL = 98 *um*, WL = 104 *um*, LT = 24 *um*.

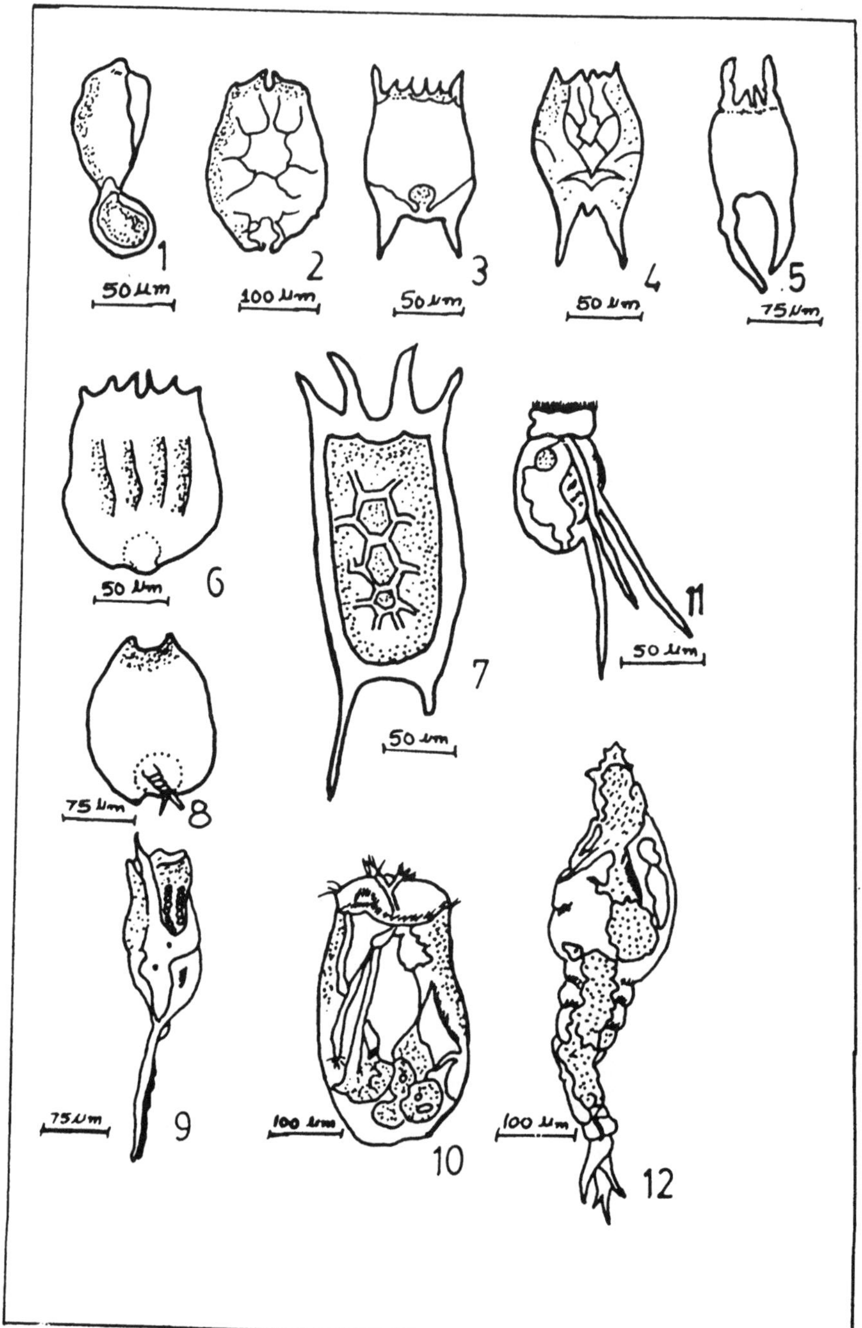
1
50 µm
2
100 µm
3
50 µm
4
50 µm
5
75 µm
6
50 µm
7
50 µm
11
50 µm
8
75 µm
9
75 µm
10
100 µm
12
100 µm

Figures 1–12

Recorded earlier from West Bengal (Anderson, 1889); Kashmir and Ladhak (Edmondson and Hutchinson, 1934); Punjab (Vasist and Gupta, 1967); North West India (Sharma, 1978).

Family : Trichoceridae (Harring and Myers, 1926)

Trichocera Longiseta (Schrank, 1802). Fig. 9.

The body is cylindrical and the lorica is of single piece with longitudinal grooves. The trophi is virgate.

Measurements : LL = 212 *um*, WL = 86-89 *um*, LT = 156 *um*.

Reported from Jammu (Choudhary, *et al.*, 1977) Karnataka (Gouder and Patil, 1989).

Family : Asplanchnidae (Harring and Myers, 1926)

Asplanchna priodonta (Gosse, 1850) Fig. 10.

This illoricate is conical in form. The intestine and anus are absent. The vitellarium is rounded. The trophi is incudate.

Measurements : TL = 516 *um*, UB = 251 *um*.

Recorded earlier from Kashmir (Edmondson and Hutchinson, 1934); Karnataka (Gouder and Joseph, 1961); Gujarat (Wulfert, 1966); Jammu (Choudhary, *et al.*, 1977); West Bengal (Sharma, 1979); Karnataka (Gouder and Patil, 1989 and Vijay Kumar, 1991).

Family : Filinidae (Bartos, 1959)

Filina longiseta (Ehrenberg, 1832). Fig. 11.

The body is oval and without foot. It has two lateral and one posterior long spine-like appendages.

Measurements : LL = 165-200 *um*, LRSP = 245 *um*, LLSP = 291 *um*.

Recorded from Karnataka (Gouder and Joseph), 1961); Maharashtra (Arora, 1962); Gujarat (Wulfert, 1966); Punjab (Vasist and Gupta, 1967); Rajasthan (Nayar, 1968); Tamil Nadu (Michael, 1973); West Bengal (Sharma, 1979).

Family : Philodinidae

Rotifer tardus (Ehrenberg, 1832) Fig. 12.

This species is conical and dull brown coloured. Corona is large. The body tapers posteriorly and its anterior part is cylindrical, truncate having a small proboscis at its tip. The retractile foot ends in 3 toes. Trophi is ramate, without fulcrum and with two distinct teeth in each ramus.

Recorded newly from the Indian Subcontinent.

Summary

In the present report described 12 species of the rotifers belonging to 6 families out of which *Rotifer tardus* (Ehrenberg, 1832) is recorded for the first time from India.

References

Ahlstrom, E.H., 1940. A revision of the rotatorian genera *Brachionus* and *Platycas* with description of one new species and two new varieties—*Bull. Am. Mus. Nat. Hist.* 77, 143-184.

Anderson, H.H., 1889. Notes on Indian rotifers—*J. Asiatic. Soc. Bengal. Calcutta*, 58, 435-458.

Arora H.C., 1962. Studies on Indian rotifers Part-I—*J. Zool. Soc. India. 14., 33-44.*

Arora, H.C., 1963. Studies on Indian rotifera Part—II. *J. Zool. Soc. India., 15, 112-121.*

Arora H.C., 1965. Studies on Indian rotifera, Part V. On collection of rotifera from Nagpur, India with four new species and a new variety. Hydrobiologia, 26 (3/4). 444-456.

Choudhary, S.K., Sharma, J.P. and Srivastava, J.B., 1978. On the rotifer fauna of Jammu ponds. *Abs. Indian Sci. congr. Sect. Zoology, Entomology & Fisheries.*

Dhanapathi, M.V.S.S.S., 1974. Rotifers from Andhra Pradesh. India—I, *Hydrobiologia, 45 (4). 357-372.*

Edmondson, W.T., 1963. Rotifer in Freshwater Biology, John Wiley & Sons. Ins. New York.

Edmondson, W.T. and Hutchinson, C.E., 1934. Report on Rotatoria.Article IX Yale North India. *Expedition Mem. Conn. Acad. Sci.* 30. 266-269.

George, M.G., 1961. Observations on the rotifers from shallow ponds in Delhi. *Curr. Sci.* 30. 266-269.

Gouder, B.Y.M. and Joseph, K.J., 1961. On the correlation between the natural population of freshwater zooplankton (Cladocera Copepodaand Rotifera) and some ecological factors. J. Karnatak Univ. Sci, 6. 19-96.

Gouder, B.Y.M. and Patil, C.S., 1989. Freshwater invertebrates of Dharwad. Karnatak University. Prasaranga Publications, *Dharwad.*

Koste, W., 1978: Rotatoria. Die Raderuers mitteleuropas (Uberordunung monogononta) in bestimmugg swerf. bergrundert von may augut *Text-4 faff (234-taf). Gebrudev. Bornfraeger. Berlin, Stuttgart.*

Michael, R.G., 1973. Rotatoria In : A Guide to the freshwater organisms. *J. Madurai Univ. Suppl.* 7. 23-36.

Naidu, K.U., 1967. A contribution to the rotatotrian fauna of South India. J. Bomb Nat. Hist. Soc. 64-66.

Nayar, C.K.G., 1968. Rotifer fauna of Rajasthan. *Hydrobiologia*, 31 (2). 168-185.

Nayar, C.K.G. and Nair, K.K.N., 1969. A collection of Brachionid rotifers from Kerala. *Proc. Ind. Acad>Sci. 69, 223-233.*

Pasha, S.M.K. 1961. On a collection of freshwater rotifers from Madras. *J. Zool.Soc.India*, 13 (1). 50-53.

Pennak, R.W., 1953. Rotatoria in freshwater Invertebrates of the United States. The Ronald Press Company, New York.

Sewell, R.B.S., 1935. Studies on bionomics of freshwater in India. II on the fauna of the tank in the Indian Museum compound and seasonal changes observed. Int. Rev. Ges. Hydrobiol. 31. 203-236.

Sharma, B.K. 1976. Rotifers collected from North West India. *New. Lett. Zool. Sur. India. 2 (6)* 255-256.

Sharma, B.K., 1978a. Contributions to the rotifer fauna of West Bengal II. Genus *Lepadella bory* dest Vincent 1826. Hydrobiologia. 58 (1) : 83-88.

Sharma B.K., 1979a . Rotifers from West Bengal III. Further contribution to the *Eurotatoria.* Hydrobiologia 64 (3) : 239-250.

Sharma B.K., 1979b. Rotifers from West Bengal IV. Further contribution to the *Eurotatoria.* Hydrobiologia 65 (1) : 39-47.

Sharma B.K. 1981. Contributions to the rotifer fauna of Punjab State, India I. Family. Brachionidae. Hydrobiologia. 76 : 249-253.

Tiwari, K.K. and Sharma, B.K., 1977. Rotifers in the Indian museum tank. Calcutta. Sci & Cult. 43 (6) 280-282.

Vasist, H.S. and Battish, S.K. 1971a. The rotifer fauna of North India. Brachionus. *Res. Bull (N.S.) Punjab Univ.*, 22 (1) 179-188.

Vasist, H.S. and Gupta, C.L., 1967. The rotifer fauna of Chandigarh. *Res. Bull (N.S.)., Punjab Univ.* 18 (3-4): 231-237.

Vijaykumar, K., 1992. Limnological studies on the perenneal and standing water bodies of Gulbarga area. Ph.D. thesis Gulbarga University, Gulbarga.

Wulfert, K., 1966. Rotatorien aus dem stausse ajwa and der trinkwasser Aufbereitung der stadt Baroda (India.) Limnologica, 4(I) : 53-93.

Chapter 15

Terminology of Various Developmental Stages of Fish Larvae Inhabiting Jammu Freshwaters

*Subash Chander Gupta & Arun Kumar Gupta**

Abstract

The present paper deals with the terminology used in defining various stages in the life history of fishes, with specific meanings of the terms used. The terminology has been proposed in relation to those used earlier. The terms used in the present context are : (i) Embryo and (ii) Larva. The embryonic development is sub-divided into : (a) cleavage stages, and (b) post-cleavage stages. The larval development is sub-divided into : (i) fry stages, (ii) post-fry stages, and (iii) fingerling stage.

Introduction

Perusal of literature reveals that there exists lot of confusion and errors in the use of various terms denoting various developmental stages of the fishes. However, now much data is available so as to trace a plan of intervals in development from fertilization to the death of the organism. In order to avoid confusions, and for comparative studies of development, it is proposed to unify the terminology and characteristics of each period of development.

The early stages in the life cycles of fishes are markedly different from the adult, and the degree of difference varies in different groups. In the study of larval fish ecology, biology and development, and in the preparation of formal descriptions and keys to larval identity, it is most convenient and useful to divide the larval period into a number of smaller intervals or distinct phases. The literature provides us with an amazing variety of terms and definitions applied to intervals of early development following hatching or birth. Unfortunately, many of these terms have several different definitions; others have not. Most of the terms are insufficiently defined and so vague that the precise characters used to distinguish one interval from another are not known. A few are more precisely defined but applied only to certain groups of bony fishes, and require intricate time consuming techniques. Some are presented because of their more common use by various groups of fisheries workers; others are less well known but are of such a nature that they should certainly be considered in selecting or developing a terminology for standard use. Thus, in order to facilitate comparability of descriptions, the terms and definitions used with respect to the larval period and its stages of development are standardized in this paper.

* Department of Bio-Sciences, University of Jammu, Jammu-180001, J and K State, India.

Terminology used by various workers

Stage 1	*Definition* 2	*Source* 3
Larva	Developmental stages well differentiated from the Juvenile and intervening between the times of hatching and of transformation (i.e.; completion of formation of adult complements of rays in all flus).	Hubbs (1943), May & Gasaway, (1967), Berry & Richards (1973), Scotton *et. al.* (1973), Snyder & Snyder (1976) (c.f. Snyder, 1976).
	Begins with the disappearance of yolk-sac, although vestiges of the yolk may remain in the gut region during the early parts of the stage (termination of larval stage with acquisition of the adult-fin ray complement. These counts are frequently attained before the rays are fully ossified).	Mansueti and Hardy (1967)
	Period during which the fish feeds on external food, and external appearance & internal structures have not yet taken on the characteristics of the adult organism.	Kryzhanovsky (1956).
	Commences when the transition to exogenous nutrition takes place & usually lasts unit the formation of ossification of the axial skeleton and until the undifferentiated (embryonic) median fin-fold is differentiated or no longer apparent.	Balon (1975).
Fry	Fish upto the time the yolk is absorbed and feeding begins.	Titcomb (1910), Rodd (1947).
Sac-fry	—	Jones (1950).
Yolk-sac larva	—	Mansueti & Hardy (1967), Ahlstrom (1968), Berry & Richards (1973), Scotton *et. al.* (1973).
Pre-larva	—	Kryzhanovsky (1956).

Contd....

1	2	3
Pro-larva	—	Hubbs (1943), Winn & Miller (1954), May and Gasaway (1967).
Eleutero–embryonic phase.	—	Balon (1975).
Stage 32.	The assimilation of yolk has proceeded so far that the liver reaches the mid-ventral line. The cleithrum is ossified.	Balinsky (1948).
Fry	Post-larvae and juveniles upto a length of about 1.5 inches.	Jones (1950).
Advanced fry.	Fish from the end of the fry period until they have reached a length of 1 inch.	Titcomb. (1910).
	Fish for a period of two weeks from complete absorption of yolk-sac.	Rodd (1947).
Stage 33	The assimilation of yolk is accomplished, there is as yet no trace of fin-rays in the caudal fin.	Balinsky (1948).
Early-post-larvae.	Larva after the time of absorption of yolk to a stage when all the soft-rays in the ventral fins have developed.	May and Gasaway (1967).
Post-larva.	Larva following the time of absorption of yolk; applied only when the structure and form continues to be strikingly unlike that of the juvenile.	Hubbs, (1943, Scotton *et. al.* (1973).
	Period after the absorption of yolk to a stage when fish resembles the juvenile. It seems the prolarva when (i) a continuous fin-fold is present (ii) there are pigment characters different from those of juvenile, or (iii) the fins are absent, unossified or developed to various degrees.	Winn and Miller (1954).
Late Post-larva.	Fishes whose soft-vertical fin-rays developed to a point where they are identical in count with those of the adult, and with any combination or degree of development of juveniles or larval characters and extending until the time that all rays are developed and the spines, or in some cases rays are ossified.	May and Gasaway (1967).

Contd...

1	2	3
Pre-juvenile.	Pre-juvenile stage is the intermediate stage between larval and juvenile forms which begins with acquisition of the adult fin-rays complement and continues, with rapid metamorphosis until the more adult-like form of juvenile is attained.	Mansueti & Hardy (1967)
	Refers to forms that are not appropriately termed larval (or post-larval) or juvenile; it is commonly characterized by strong spines, boney plates, and peculiar body form. Transformation into the juvenile stage is usually very rapid and often associated with movement from a pelagic to a benthic habitat.	Berry and Richards (1973), Hubbs (1958).
Juvenile	Young essentially similar to adult.	Hubbs (1943).
Fingerling	Fish between the length of 1 inch and the yearling.	Titcomb (1910).
	Fish from two weeks after complete absorption of yolk-sac until they are one year old from date of hatch.	Rodd (1947).
	May be used when the length of specimens ranges from about 1.5-5.0 inches, should be applied only to young of those species that grow to a fairly large size and not to those whose maximum size falls within or near the range given above.	Jones (1950)
Alevin Phase	Exists in species that do not have a larval period and lasts until the development of complete scalation or ossified spines (the yolk-bearing larva transforms directly into the juvenile).	Balon (1975), Hubbs (1943).
Transitional (or metamorphic-larva).	The period in which the larva is acquiring the juvenile shape, structure and pigmentation; does not occur in the life history of many fish species, i.e., development may proceed directly from the larva to the juvenile.	Scotton *et. al.* (1973), Ahlstrom (1968). Bertalson (1951), Moser, & Ahlstrom (1970).
Protopterygiolarval Phase.	Encompasses the interval between the transition of exogenous feeding and the commencement of differentiation of embryonic median finfold-the first appearance of mesenchymatic buds of lepidotrichia and, dorsal and anal fin outlines within the fin fold.	Balon (1975).

Contd...

1	2	3
Preflexion phase.	Characterized by a straight notochord.	Ahlstrom (1968).
Straight-Noto-chord Stage.	Begins at hatching (breaking of the chorion). The duration of the stage is characterized by a straight notochord extending out into the caudal region of the finfold.	Faber (1963).
Proto-lar-va.	Larval phase in which distinct median fin-elements (dorsal, anal and caudal spines or rays) are not yet apparent.	Snyder & Snyder (1976), (e.f. Snyder (1976)
Pterygio-larval Phase.	Lasts from the beginning of the differentiation of unpaired fins until the embryonic median finfold is entirely differentiated or no longer apparent.	Balon (1975).
Flexion P-hase.	Begins upon the formation of the candal rays.	Ahlstrom(1968)
Caudal-fin stage	The transition into the caudal fin stage occurs when two rays of caudal fin appear. Shortly before the ventral portion of the caudal fin-fold becomes opaque, actinotrichia radiate from the notochord towards the edge of the finfold. Caudal fin stage larvae are characterized by numerous rays developing in the caudal fin and a dorsal bending of the notochords.	Faber (1963).
Post-Flex-ion phase	Begins with the completion of notochord flexion and caudal formation.	Ahlstrom (1968).
Pelvic bud stage	The appearance of pelvic buds heralds the next stage. The local reduction of the embryonic finfold and the gradual appearance of fins occuring in this stage constitute a great transformation.	Faber (1963).
Stage-39	Pelvic fins are in the form of paddles. No trace of fin-rays is visible in them.	Balinsky (1948).
M e s o - Larva.	The larval phase in which atleast one but not the full complement of distinct principal rays in the median fin is apparent; or if the full complement is present and the adult possesses pelvic fins, the pelvic buds or fins are not yet apparent.	Snyder and Snyder (1976) (c.f. Snyder, 1976).

Contd...

1	2	3
Pelvic ray stage.	The appearance of rays in the pelvic bud and the accompanying loss of bluntness mark the start of pelvic ray stage. The larvae begin to resemble the adult in fins and in morphometric feature marking the end of pelvic-ray stage could not be found, so the end of this stage is based entirely on the total length of the larva (a pre-determined length).	Faber (1963).
Meta larva.	The larval phase in which the full complement of distinct principal rays in the median fins and of the adult possesses pelvic fins, the pelvic buds or fins are apparent.	Snyder & Snyder (1976), (Snyder, 1976).
Stage-40	In the pelvic fins rudiments of rays, not yet ossified, have appeared.	Balinsky (1948).

Present Terminologies

On the basis of present investigations, five terms to denote the development of fish have been worked out. These terms are based on the morphological criteria:

1. **Embryo :** The interval of development within the egg membrane from fertilization until hatching. It is covered in two stages:

   (a) Cleavage Stages :

   The interval of development within the egg membrane from fertilization until the beginning of organogenesis.

   (b) Post-cleavage stages :

   The interval of development within the egg membrane from organogenesis until hatching.

2. **Larva :** The interval of development from hatching upto a time when all the complements of an adult fish are formed. It is covered in three stages:

   (a) Fry Stages :

   The interval of development from hatching until absorption of yolk-sac.

   (b) Post-Fry Stages :

   The interval of development from complete absorption of yolk-sac and the initiation of exogenous feeding upto the time when all the complements of an adult fish are formed.

(c) Fingerling Stages:

> The fish larva in which all the complements of an adult fish are formed, but the larva is sexually immature. It may (Titcomb, 1910; Jones, 1950), as in *Tor tor* & *Xenentodon cancila* (of present investigations) or may not, as in *Puntius conchonius* & *Barilius bendelisis* (of present investigations) exceed 1 inch in length; and should be applied only to young of those species that grow to a fairly large size and not to those whose maximum size falls within or near the range given above (Jones, 1950).

Discussion

The present terminology agrees with the terminology of embryonic development proposed by Balon (1971). He divided the embryonic period into cleavage phase, Embryonic phase and Eleuterembryonic phase (from greek-eleuteros=free embryo); but the usage of eleuterembryonic phase appears to be incorrect because (according to Balon, *op. cit.*) it includes the interval from hatching to the beginning of independent feeding; but according to present investigations, the embryonic period lasts within the egg membrane until hatching; and, once the hatching comes out, it is denoted as fry. So there appears to be incongruity in including eleuterembryonic phase within the embryonic period.

Titcomb (1910) and Rodd (1947) used the identical terms but entirely different definitions for "advanced fry" and "fingerling" (one based on length and other on time). Jones (1950) adds the term "hatching" and substitutes "Sac fry" for "fry" and "fry" for "advanced fry", and defines these intervals in terms of Hubb's (1943) terminology as "pro-larva" and "post-larva" etc. Snyder (1976) suggested that the use of length or time since hatching is certainly the most precise and easiest expression for a phase or a period boundary. He further states that the boundaries between each developmental period or phase must be very precise and easy to determine & the terminology should provide intervals which are indicative of relative age & state of development. Unfortunately, there is no direct correlation with the development of morphological character that is consistent, when comparing various species. Some fishes like *Xenentodon cancila* may appear very adult like well before the end of advanced fry phase while others may not have all their fins developed until well into the fingerling interval which is often considered synonymous with young of the year juvenile. In addition, use of time to determine interval boundaries is usually not possible with field collected specimens since time of hatching is usually not known.

The present investigator divided the embryo into (i) cleavage stages, and (ii) post-cleavage stages, and fry, to substitute eleuterembryonic phase wherein the larva hatches out of the egg membrane. The present terminology agrees with Titcomb (1910) and Rodd (1947) in the description of "fry" but disagrees with Jones (1950) who described "fry" as the post-larvae and juveniles upto a length of about 1.5 inches. Although Hubbs (1943) used the term "pro-larva" to which the term "fry" is used in the present context, because the term "post-larva" should mean the fish after complete larval development (i.e., adult), the juvenile itself means the adult like fish. So the appropriate term may be to use "fry" in place of "advanced fry" (of Titcomb, 1910, and Rodd, 1947) or "Sac fry" (of Jones, 1950) and "post-fry" in place of "post larva" (of Hubbs, 1943). The present description of fingerling disagrees with that of Rodd (1947) because in the fishes of present investigations the larvae, two weeks after complete absorption of yolk-sac have produced very few characters (contrary to Rodd, 1947), and only one or two fins are formed but none of the fins have got the full complements within such a short span of time. These larvae after complete absorption of yolk-sac are still termed as "post fry" in the present investigations.

The terms alevin phase (Balon, 1975 and Hubbs, 1943) & transitional or metamorphic larva (Scotton *et. al.* 1973; Ahlstrom, 1968; Bertelson, 1951; & Moser and Ahlstrom, 1970) have been used in case of fishes in which there is no larval period and the yolk-sac larva transforms directly into the juvenile, which is not the case in presently investigated fishes. Similarly the terminologies used by Balon (1975), Faber (1963), Snyder & Snyder (c.f. Snyder, 1976) based on the sequence of notochord and fin development, which basis does not seem to apply correctly in the present case. Balinsky (1948) gave so many developmental stages (for cyprinid fishes inhabiting European freshwaters) which is a long process and may confuse the workers in the field. In the present investigations, an attempt has been made to compress the stages logically, so as to make them appropriate & easy to pronounce.

Besides above terminologies, there are some of the terms which are current in Bengal for denoting early stages of fishes (as given by Jones, 1950 in his publication): (i) Dim: Eggs or spawn (ii) Dimpona : prolarvae of carps. In trade, post-larvae may also be usually associated with prolarvae. (iii) Dhanipona: carp fry from one-half to one inch. (iv) Charapona: young of major carps, known by shortened term Chara also. (v) Chalapona: young of major carps about 4-9 inches in length, known also by shortened term of Chala. (vi) Nala or Nalapona: young of major carps slightly bigger than chalapona, upto about 12 inches long. (vii) Kaira (West Bengal) and Jatka (East Bengal) : young hilsa *(Hilsha ilisha)* upto a length of about 6 inches. In some places, the species *Gadusia chapra* also is called by the above names. (viii) Chuna or Chunamach: A general term for small fishes whether young or adults.

The term Dhanipona (Jones, 1950) may correspond with the term fry of presently investigated fish species. The term *chalapona* can also be used for the young of *Barilius bendelisis* and *Tor tor* (from major carps) and *Xenentodon cancila* (from Belonids) of present investigations, because amongst presently investigated species only the young of these fishes grow to a size of 4-9 inches and more; and not those of *Puntius conchinius.* As the young of major carps upto about 12 inches long are termed as Nala or Nalapona; and this term can also be used in case of *T. tor* & *X. cancila,* as the young of these species grow upto this size. The term Chuna or Chunamach can also be used in case of *P. conchonius* and *B. bendelisis* of present investigations as they are small fishes.

In Jammu region, the eggs or spawn of fishes in general are called as "Bee", or "Seed" by the fishermen (also by the people of villages). The larvae of fishes in general are called "Seed" or "Buch" but the small sized larvae generally fry and first few post-fry stages are called as "Poang".

The term Dim, used in Bengal for eggs & spawn corresponds to "Bee", "Beej" or "Seed" used in Jammu. Dimpona of Bengal corresponds with the term "Seed" or "Buch" and in Jammu region for the fry of present proposals. Dhanipona of Bengal may correspond with the term "Poang" used in Jammu and expresses the characters of post-fry and fignerlings as presently recommended.

Summary

After brief historical reviews and discussions, different terms which are in current use, used in the characteristics of each period of development are proposed. To indicate the period of development, two terms viz; Embryo & Larvae are used. Embryonic development comprises of cleavage and post-cleavage stages where as larval development comprises of fry stages, post-fry stages and fingerling stages. The difficulty in specifying a particular time for calculating age-groups for fishes in India is explained and the suggestions regarding the usage of general terms, current in fishery literature and fish-seed trade are also given in the present communication.

Acknowledgements

The senior author is highly grateful to the CSIR and the University Grants Commission, New Delhi for financial assistance during the course of investigations, and the University of Jammu, Jammu for providing necessary facilities to carry on these investigations.

References

Ahlstrom, E.H. (1968). "Development of fishes of the Chesapeake Bay Region; an atlas of eggs, larval and juvenile stages, Part I, by Alice J. Mansueti and Jerry D. Hardy Jr." Copeia 3 : 648-651.

Balinsky, B.I. (1948). "On the development of specific characters in Cyprinid Fishes". Proc. Zool. Soc. Lond. 118(2): 335-344.

Balon, E.K. (1958a). Development of the Danube wild carp *(Cyprinus carpio carpio* L.) in time of prelarval phase and the larval period. Biologicke´ prace (Bratis larva) 4 : 5-54.

Balon, E.K. (1958b). "About development Intervals in the life of fishes." Bull. Zakl. Biol. Stawo´w PAN 7 : 5-15 (in Polish).

Balon, E.K. (1959). About life periods of fishes. Vesmi´r 38: 341-343.

Balon, E.K. (1971). The intervals of early fish development and their terminology (a review and proposals). Vest.Cesk. Spol. Zool. 35(1) : 1-8.

Balon, E.K. (1975) Terminology of intervals of fish development. J. Fish. Res. Board. Can. Vol. 32(9) : pp. 1663-1670.

Berry, F.H. and W.J. Richards (1973). Characters useful to the study of larval fishes. Pages 48-65 *in* A.L. Pacheco, ed. Proceedings of a workshop on egg, larval and juvenile stages of fish in Atlantic Coast estuaries, U.S. Dept. Comm., NMFS, Middle Atlantic Coastal Fisheries Centre, Tech. Publ. No. 1.

Bertelsen (1951). Cited from Snyder D.E. (1976). "Terminologies for intervals of larval fish development. Pages 41-58 *in* Boreman, J. (editor). Great lakes fish egg & larval identification. (Proceedings of a workshop). USFWS, OBS. National Power Plant Team (Ann. Arbor, Michigan) FWS/OBS-76/23, 220 p.

Faber, D.J. (1963). Larval fish from the pelagic region of two wisconsin lakes. Ph.D. Thesis, University of Wisconsin, 122 pp.

Hubbs, C.L. (1943). Terminology of early stages of fishes. Copeia No: 4: 260.

Jones, S. (1950). A note on the terminology of the early developmental stages of fishes. J. Zool. Soc. India, Vol. 2, No: 1, pp. 39-41.

Kryzhanovsky, S.G. (1956). Development of clupeid fishes. Tr. Insorfol. Zhivoto. Serercova 17: 1-254 (In Russian).

Mansueti, Alice, J. and Jerry D. Hardy Jr. (1967). "Development of the fishes of Chesapeake Bay Region, an atlas of egg, larval and juvenile stages. part I". Nat. Resource. Inst., Univ. Maryland Baltimore, Md. 202 p. (Copeia No: 3, pp. 648, 651, 1968).

Moser, H.G. and E.H. Ahlstrom (1970). Development of Lantern fishes (Family Mycotophidae) in the California current, part-I. Species with narrow eyed larvae". Los. Angeles Country Museum of Natural History Science, Bull. No. 7 ; 145 pp.

Rodd, J.A. (1947). Standard terms and methods of measurement in fish culture. Trans. Amar. Fish. Soc. 74: 19-20.

Scotton, L.N., R.E. Smith: N.S. Smith; K.S. Price & D.P. De Sylva (1973). Pictorial guide to fish larvae of Delaware Bay, with information & bibliographies useful for the study of fish larvae". College of Marine Studies; Univ. Delaware, Newark, Dal. 206p.

Snyder, D.E. (1976). Terminologies for intervals of larval fish development, pages 41-58 *in* Boreman, J. (editor). Great lakes fish egg & larval identification." (Proceedings of a workshop). USFWS, OBS. National Power Plant Team (Ann. Arbor, Michigan) FWS/OBS—76/23, 220 p.

Titcomb, J.W. (1910). Fish cultural practices in the United States Bureau of fisheries. Bull. U.S. Bur. Fish. 28(2): 607-657.

Winn, H.E. and R.R. Miller (1954). Native post larval fishes of the lower Colorado River basin, with a key to their identification. Calif. Fish. Came. 40(3). 273-285.

Chapter 16

Inter-specific Competition in a Mixed Culture of Cladocera

*Y.R. Malhotra & Seema Langer**

Abstract

Cladoceran zooplankton *(Daphnia similis, Moina macrocopa, Simocephalus vetulus, Ceriodaphnia cornuta and Alona intermedia)* were successfully raised in the Laboratory (mono-species culture). A mixed culture of these species was also practised so as to determine their vitality for composite culture. Of the various spp. Moina was found to be the most resistant and less affected (in terms of inter-specific density) in the presence of other cladocerans, whereas the species viz. *Simocephalus, Daphnia, Ceriodaphnia* and *Alona* were less resistant in similar set up.

Competition ability (survival rate and natality) is directly related to body size. But *D. Similis* (of present studies) inspite of being a large sized species, suffers competitive exclusion in the presence of smaller sized species. Inter-se prey predatory relationships have also been investigated to further strengthen the idea highlighted in present communication.

Introduction

The purpose of this study is to identify suitable species of zooplankton which are successful not only in mono-culture but can equally be exploited in mixed/ composite culture. Competitive exclusion in polyculture is a subject of interest and is a field which defies scientific understanding at present. There is a considerable, though indirect, evidence (from field observations and experiments) that cladocerans may compete among themselves and with rotifers suppressing their population.[1] Although, it has generally been assumed that the mechanism involved in such a suppression is exploitative competition, but Burns and Gilbert [2] have recently shown that such a suppression in population growth occur even when food resources are not limiting.

In the present studies, an attempt has been made to analyse the interactions that operate with the various culturable cladoceran species viz. *D. similis, M. macrocopa, S.vetulus, C. cornuta* and *A. intermedia.*

---

* Dept. of Bio-Sciences, University of Jammu, Jammu—180 004, India.

Materials and Methods

(i) Collection of material.

Zooplankton progenitors were collected from nature (viz. Gadigarh nallah, Kunjawani pond and ponds located at Regional Research Laboratory, (Jammu). Healthy individuals were isolated as progenitors and acclimatized to produce progeny in laboratory, which was used as stock.

(ii) Experimental Design

Four identical experimental layouts were set (providing similar ecological conditions) in which the following combinations were introduced.

(a) monospecies culture of *Daphnia, Moina, Simocephalus, Ceriodaphnia* and *Alona.*

(b) mixed culture of all the five species.

(c) mixed culture of *Daphnia, Simocephalus, Ceriodaphnia* and *Alona* spp. (a combination of 4 spp. only).

(d) mixed culture of *Daphnia, Ceriodaphnia* and *Alona* (a combination of 3 spp. only).

Culture medium was prepared by thorough mixing of food nutrient (cow dung) in water at a concentration of 1g/1. Container troughs (filled with prepared nutrient medium) were inoculated with 5 organisms from laboratory raised progeny. Observations were continously made regarding the density variations, individual mortality and physical damage done to any of the cultured species.

Experiment was conducted for 80 days; with 1 litre culture volume at 20-25°C temperature. Each of the experiment set was accompanied with its replicate.

Results and Discussion

Single and mixed species population of *D. similis, M. macrocopa, S. vetulus, C. cornuta* and *A. intermedia* were observed in the laboratory for a period of 80 days (Tables 1, 2). When alone, all the five species usually persist successfully. *M. macrocopa* consistently persists and cause the elimination of *D. similis* and reduces the production of other species viz. *C. cornuta, S. vetulus* and *A. intermedia* in mixed cultures (Table 1). *D. pulicaria* in mixed culture is already reported to cause elimination of *D. magna,* the other companion species in the culture.[3] Results of present experiments indicate that larger cladocera do not necessarily exclude smaller ones (although) it is on record that growth of smaller species is inhibited by the presence of large sized individuals in competition experiments). Such findings appear in consistent with those obtained by Frank,[3] who predicted that competition ability increases with increase in body size but as is clear from present studies, it is not so in *Daphnia similis,* which inspite of being larger in size approaches extinction.

To start with, when density is low, production of individuals of different species in mixed culture is similar to that obtained in monospecies culture, but with progress of time, production rate of Moina becomes quite high (followed by Simocephalus

Table 1 : **Inter - specific Competition Among Various Cladoceran Species in Composite Culture (Temp. 25°C). M - *Moina*, D - *Daphnia*, S - *Simocephalus*, C - *Ceriodaphnia*, A - *Alona*.**

Days after Inoculation	*Density (Inds./ml)*									
	Composite Culture					*Mono-species Culture*				
	M	*D*	*S*	*C*	*A*	*M*	*D*	*S*	*C*	*A*
Initial	.01	.01	.01	.01	.01	.01	.01	.01	.01	.01
5	.013	.01	.012	.013	.012	.19	.014	.19	.016	.017
10	.21	.02	.09	.06	.03	.26	.035	.26	.036	.028
15	.59	.046	.07	.08	.06	.72	.052	.72	.093	.08
20	1.3	.016	1.14	.098	.03	1.5	.89	1.5	.2	.16
30	2.0	.005	.43	.13	.19	2.1	.12	2.1	.26	.37
40	.9	.02	.78	.29	.24	3.5	.30	3.5	.49	.5
50	1.2	.012	.9	.43	.33	4.9	.37	4.9	.53	.57
60	3.2	.014	1.2	.38	.41	6.1	.45	6.1	.5	.49
70	4.1	.02	1.1	.44	.44	4.8	.67	4.8	.66	.56
80	5.0	.029	1.6	.53	.46	5.8	.73	5.2	.78	.62

Table 2 : Inter-specific Competition Among Various Cladoceran Species in Composite Culture Experiments in Different Combinations (Temp. 25°C). M - *Moina*, D - *Daphnia*, S - *Simocephalus*, C - *Ceriodaphnia*, A - *Alona*.

Days after Inoculation	*Density (Inds./ml)*											
	M	*D*	*S*	*C*	*A*	*D*	*S*	*C*	*A*	*D*	*C*	*A*
Initial	.01	.01	.01	.01	.01	.01	.01	.01	.01	.01	.01	.01
5	.013	.01	.012	.013	.012	.007	.015	.018	.013	.013	.015	.014
10	.21	.02	.09	.06	.03	.03	.085	.048	.022	.032	.03	.48
15	.59	.046	.07	.08	.06	.036	.12	.078	.064	.046	.08	.089
20	1.3	.016	1.14	.098	.08	.049	.19	.12	.087	.078	.15	.1
30	2.0	.005	.43	.13	.19	.03	.52	.31	.092	.099	.34	.16
40	.9	.02	.78	.29	.24	.038	.84	.47	.078	.14	.51	.18
50	1.2	.012	.9	.43	.33	.052	1.24	.49	.082	.32	.42	.21
60	3.2	0.14	1.2	.38	.41	.04	1.97	.33	.045	.24	.40	.28
70	4.1	.02	1.1	.44	.44	.062	.88	.36	.078	.62	.61	.34
80	5.0	.029	1.6	.53	.46	.054	2.4	.30	.093	.40	.65	.38

and other species) (Table 1), thus resulting in significant increase in overall population density. This increase in population density exerts a competitive inhibitory effect on all the population in the polyculture as indicated by the population decline of various species (Table 1) In the absence of dominant species *(Moina)*, however, smaller species (*Alona* and *Ceriodaphina*) show better success with low food densities and thus compete better than larger animals *(Daphnia* sp.) and increase in number (Table 2). Gilbert and Mac Isaac[4] also observed an exceptional case where they observed that natural densities of *Daphnia*, *D. ambigua* do not exert much direct effect on the rotifer population. Whereas another very small cladoceran *Scapholebris kingi* strongly interfered with the rotifer, *Synchaeta oblonga*.[5]

Tactics exhibited by different species may lead to successful co-existence of various competing species, and even to a condition in which the competitively inferior species (small species) may become numerically abundant. Over exploitation of food resources by the competitively dominant population of large species can provide an opportunity for a less successful competitor of smaller species to co-exist and even become numerically dominant.[4]

Similar results have been reported by Mitchell and Williams[6] in which case both *D. carinata* and *S. expinosus* are food limited at high densities. As long as *M. macrocopa* and *S. vetulus* are in phase of density decline before the exclusion of *Alona* and *Ceriodaphnia* populations, then latter (especially *Ceriodaphnia*) can reproduce and become more abundant, at least temporarily (Table 1). These results are further strengthened by the evidence that competitive ability of co-existence of species (that exploit similar resources) not only depends upon survival in a limited food environment, but also upon the age structure of the population.[5]

As indicated in Table I, production of *Daphnia* is inhibited by the increased densities of *Moina* and *Simocephalus;* such a reduction of *Daphnia* has also been recorded by Allan[6] in competition of *Daphnia* against *Holopedium*. Allan[7] observed that *Daphnia* is rare and declines throughout the study whereas *Holopedium* exhibits a pronounced burst of growth.

In culture, where both *Simocephalus* and *Alona* are simultaneously present, high rate of mortality of *Alona* has been observed (Table 2). Such a high rate of mortality of *Alona* (observed during present studies), however, can be attributed to a large extent to the non-feeding damage done on *Alona* by *Simocephalus;* the latter generally tends to feed along the sides and bottom, where *Alona* usually resides, thus causing physical damage to the small species, *Alona*. Similar sort of physical damage (to *Keratella* species) has been observed by Burns and Gilbert[8] while studying co-existence of *Daphnia* and *Keratella* in a culture. The frequency of negative interaction is reported to increase with increasing body size of *Daphnia*. As regard the mechanisms of damage, although Williamson[9] recorded that physical contact with prey elicits an instantaneous attack response, attack without physical contact are also common when the prey is at close range. However, in the present case of *S. vetulus,* a contact response may be established through thoracic appendages of the size dominant species.

One of the probable reason that speaks of failure of *D. similis* in composite culture can be the temperature of surrounding medium during experimental period (25°C) which otherwise also is not much favourable for *Daphnia;* indicated by its small average brood size (8-10) eggs/brood) as compared to the brood size of dominant *M. macrocopa* (20-25 eggs/brood) and *S. vetulus* (18-20 eggs/brood), which of course is accompanied by density stress and ensuing food competition.

On the basis of present investigation, however, it is not advisable to visualize a complete picture of inter-specific competition along a thermal gradient, as the studies have not been conducted at different temperatures and the temperature is known to influence the food uptake, brood size, growth and other biological aspects of cladocerans as has also been recorded by Tessier[10]. Tessier (op. cit.) reported that the population density and food stress vary seasonally, and it is this fluctuation of regulating factors that produce seasonal pattern of abundance exhibited by each species.

Aknowledgements

The author is thankful to the University Grants Commission for providing the financial assistance.

References

1. Neill, W.E. *Oecologia* (Berlin)., 1984, **61**, 175-181.
2. Burns, C.W. and Gilbert, J.J. *Limnol. Oceanogr.*, 1986, **31**, 859-866.
3. Frank, P.W. *Physiol. Zool.*, 1957, 178-204.
4. Ibid.
5. Ibid.
6. Allan, J.D. *Ecology*, 1973, **54(3)** 485–498.
7. Ibid.
8. Burns C.W. and Gilbert J. J. Limnol. Oceanogr., 1986, **31**, 859–866.
9. Williamson, C.E. *Hydrobiologia*, 1983, **104**, 385-396.
10. Tessier, A.J. *Ecology*., 1986, **67**, 285-301.

* Key words : Zooplankton/ mixed culture/ inter-specific competition.

Chapter 17

Limnology of Farooq Nagar Pond, Jammu Part-II—Rotifera

*S.P.S. Dutta and Jyoti Sharma**

Abstract

The composition and seasonal fluctuations of rotifers in relation to various abiotic and biotic parameters in a perennial pond in Jammu are described. Rotifers recorded annual maximum diversity and density in the month of August. Analysis of coefficient of correlation (r) of rotifers with various abiotic parameters is insignificant. Among biological factors, rotifers have shown significant correlation with Euglenophyceae (r=0.6024), Bacillariophyceae (r=0.5656), Cladocera (r=0.6047) and Copepoda (r = 0.7748). Species diversity index analysis has indicated stress in the sewage-fed station.

Introduction

Rotifers are an important link in the energy transfer to a secondary level and form a good bulk of food for the fishes and other aquatic animals. Their importance as bioindicators is also well known. Jyoti and Sehgal (1979), Gupta and Sudan (1985), Puri (1989), Kumar (1990), and Kumar *et. al.* (1991) have described ecology of rotifers from this part of the country. In order to add to the existing information, one year study was undertaken in Farooq Nagar Pond, Jammu.

Study Area

The pond investigated (Fig. 1) is situated near the river Tawi bridge (30°40'15" N Lat; 74° 50' 30" E long, at 300 msl), Jammu. It is elliptical in shape with surface area of about 5400²m. The pond is fed by overflow from pumping station, rainwater and sewage nullahs from the catchment area. Bottom of the pond is soft and rich in macrophytic detritus. Macrophytic vegetation is represented by floating *Lemna* sp. and *Wolffia* sp. For the present work, four experimental stations were selected at different spots of the pond (Fig. 1).

Materials and Methods

Monthly planktonic samples, at five spots of each station, were collected by filtering five litres of water through a silk bolting cloth planktonic net (3,136 meshes/ cm²) and the collected material was fixed at the spot in 5% formaldehyde. In the

* Department of Biosciences, University of Jammu, Jammu-180004.

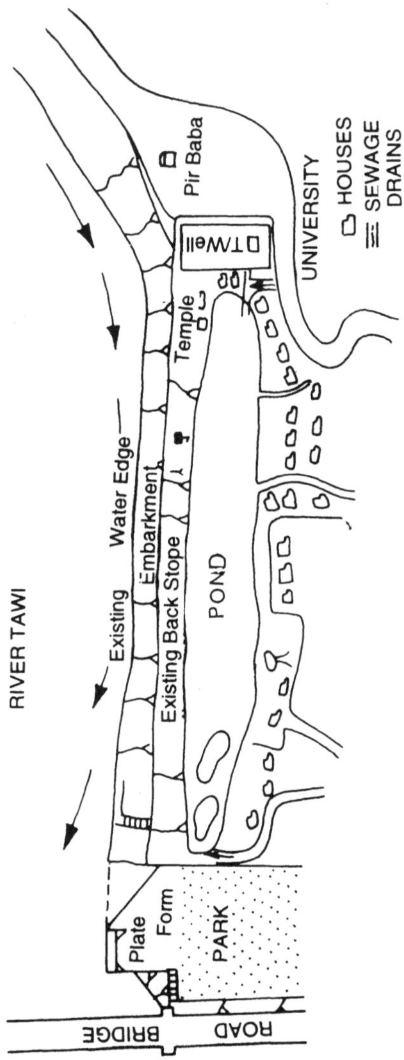

Fig. 1: Map of the Farooq Nagar Pond, Jammu.

Table 1 : Monthly Mean Variations in Physico-chemical Parameters at Station I-IV, Farooq Nagar Pond, Jammu (June, 1991—May 1992)

	Months											
	1991							*1992*				
Parameters	*June*	*July*	*Aug.*	*Sept.*	*Oct.*	*Nov.*	*Dec.*	*Jan.*	*Feb.*	*March*	*April*	*May*
Depth (cm)	40.00	51.25	49.75	44.25	41.50	40.50	40.70	53.00	52.50	51.75	48.25	50.00
Transparency (Cm)	27.02	21.25	21.75	25.00	34.13	28.00	28.75	30.75	32.50	30.13	38.00	23.75
Air temp. (^oC)	34.94	32.80	27.75	31.13	25.63	19.00	18.00	16.13	23.50	24.00	35.50	36.63
Water temp (^oC)	26.69	28.75	27.75	28.75	19.83	14.50	13.43	12.43	19.25	22.13	29.25	30.75
pH	7.36	7.31	7.11	7.80	7.86	7.48	7.40	7.30	7.47	7.57	7.84	8.37
Electric cond. (μ mhos)	166.25	140.24	118.15	123.85	181.22	127.06	131.22	136.08	120.53	136.56	178.56	156.15
DO (mg/1)	4.19	4.72	1.09	8.94	6.27	3.7	3.18	1.17	3.12	2.46	5.02	14.78
FCO_2 (mg/1)	6.89	7.72	7.27	2.37	5.45	5.15	9.71	18.18	14.37	9.83	8.73	-
CO_3^- (mg/1)	-	-	-	-	-	-	-	-	-	-	-	18.49
HCO_3^- (mg/1)	260.24	219.02	194.13	220.28	260.30	285.12	299.79	306.14	269.04	256.50	271.48	226.20
Cl^- (mg/1)	23.39	26.47	21.09	26.67	29.85	30.24	31.33	37.08	26.27	25.49	28.42	30.31
Ca++ (mg/1)	47.41	63.68	37.59	43.08	47.32	55.45	62.48	63.97	52.77	55.01	53.68	51.52
Mg++ (mg/1)	10.02	8.86	5.90	8.05	8.88	6.20	6.65	11.01	10.25	7.99	9.54	8.30
Total Hardness	159.51	195.33	119.25	140.10	154.58	167.97	186.90	201.15	173.84	170.11	173.10	160.12
PO_4–P (mg/1)	**	0.0046	0.015	0.014	0.014	0.006	0.027	0.013	0.016	0.019	0.022	0.02
SO_4— (mg/1)	**	0.760	0.520	1.200	0.730	2.015	0.970	1.425	1.480	0.970	0.571	0.75
NO_3^-N (mg/1)	**	4.20	6.12	3.74	1.36	1.15	6.66	0.47	0.60	1.33	2.41	0.50
S_iO_2 (mg/1)	**	0.177	0.275	0.299	0.27	0.099	0.159	0.146	0.149	0.284	0.138	0.15

** = Data not analysed.

laboratory, the rotifers were identified by consulting Pennak (1953), Edmondson (1959), Arora (1965), Nayar (1968), Vasisht and Battish (1970, 1971, 1972 a-c), Tonapi (1980), Jyoti and Sehgal (1980), Adoni (1985), and Battish (1992). Numerical counts of rotifers were made with the help of Sedgewick Rafter Counting Cell and the results are expressed as number 1.

At the time of planktonic sampling, water samples were collected and analysed for various physico-chemical parameters (Table 1) by standard methods (ISI, 1973; APHA, 1975 and Trivedy *et. al.* 1987).

The coefficient of correlation 'r' was calculated to study rotifer relationships with different physico-chemical and biological characteristics of water.

The numerical values of species diversity index were calculated by applying the following formula (Shannon and Weaver, 1963):

$$d = -(n_i/N) \text{Log}_2 (n_i/N)$$

where d = diversity

n_i = number of individuals of ith species

N = Total number of individuals.

Results and Discussion

Qualitatively, the rotifer fauna of Farooq Nagar pond includes forty-five species, belonging to twenty-eight genera (Table 2). Most of the species excepting two species of *Brachionus* viz. *B. quadridentatus,* and *B. bidentata,* remained seasonal (Table 2). Seasonal presence of rotifers has also been worked out earlier by Jyoti & Sehgal (1979), Gupta & Sudan (1985), Yousuf *et. al.* (1986), Puri (1989), Kumar (1990) and Kumar *et. al.* (1991) from this part of the country.

Qualitatively, rotifer fauna of this pond is more diversified when compared to the earlier findings of Jyoti and Sehgal (1979) from littoral (14 species) and limnetic zone (11 species) of lake Surinsar; Gupta and Sudan (1985) from Bantalab (12 species), Kherwanda (8 species), Bhalwal (15 species) and Nagbani (11 species) ponds and Kumar *et. al.* (29 species) from Kunjwani pond, Jammu. Rotifer's qualitative richness in Farooq Nagar pond may be due to absence of fish, except for *Gambusia* sp.

An important qualitative feature of rotifers observed in this pond has been the simultaneous presence of six species of *Brachionus (B. quadridentatus, B. bidentata, B. calyciflorus B. rubens, B. plicatilis* and *B. angularis)* in August and five species *(B. quadridentatus, B. bidentata, B. calyciflorus, B. plicatilis* and *B. angularis)* in April and May; three species of Lecane *(L. decipiens, L. luna* and *L. curvicornis)* in July and *(L. quadridentata, L. decipiens* and *L. bulla)* in May, four species *(L. ploenensis, L. luna, L. curvilinealis* and *L. bulla)* in August and four species *(L. luna, L. quadridentata, L. decipiens,* and *L. closterocerca)* in June. The earlier recording of simultaneous presence of six *Brachionus* species by Mandal (1979), four by George (1961), five by Puri (1989), nine by Mishra and Sharma (1990) and seven by Kumar *et. al.* (1991) supports our findings. However, contrary to these observations, Pennak (1957) and Jyoti and Sehgal (1979) documented that in a few genera of rotifers, no more than two

Table 2 : Monthly Mean Variations in Rotifers (n/l) at Station I-IV Farooq Nagar Pond, Jammu (June, 1991—May, 1992)

Months / *Organisms*	*1991* *Jun.*	*Jul.*	*Aug.*	*Sep.*	*Oct.*	*Nov.*	*Dec.*	*1992* *Jan.*	*Feb.*	*Mar.*	*Apr.*	*May*
Platyias quadricornis (Ehern.)	-	8	2	-	-	-	-	-	-	-	-	-
P. patulus (Muller)	-	6	210	83	-	-	-	-	-	29	105	4
Total *Platyias* spp.	-	14	212	83	-	-	-	-	-	29	105	4
Brachinous quadridentatus (Hermann)	4	610	2096	219	70	39	121	23	6	126	167	32
B. bidentata Anderson	2	607	265	11	434	411	646	1194	348	2633	669	7
B. calyciflorus Pallus	5	110	6	-	-	148	-	-	-	389	73	12
B. rubens Ehern	-	468	1027	62	8	-	110	501	34	-	14	-
B. plicatilis Muller	-	79	178	-	-	-	6	211	-	-	-	1
B. angularis Gosse	3	26	9	2	4	5	9	-	5	-	155	8
Total *Brachionus* spp.	14	1900	3581	294	516	603	892	1929	393	3148	1078	60
Anuraeopsis fissa Gosse	-	7		-	-	-	17	18	-	-	13	-
Keratella tropica Apstein	-	-	8	-	-	-	-	-	-	-	-	-
Euchlanis dilatata Myers	-	-	9	-	-	-	-	12	-	11	77	-
Mytilina ventralis (Enren)	-	35	19	-	-	-	141	-	27	-	-	-

Contd......

Months	*1991*							*1992*				
Organisms	*Jun.*	*Jul.*	*Aug.*	*Sep.*	*Oct.*	*Nov.*	*Dec.*	*Jan.*	*Feb.*	*Mar.*	*Apr.*	*May.*
Colurella obtusa Gosse	15	19	8	-	-	-	-	-	-	-	-	8
Lepadella ovalis Muller	15	13	7	1	-	-	-	-	-	11	-	8
L. patella (Muller)	-	-	8	-	-	11	-	-	-	-	3	-
Total *Lepadella* spp.	15	13	15	1	-	11	-	-	-	11	-	8
Lecane ploenensis (voight)	-	-	6	-	-	6	-	-	-	-	-	-
L. elasma Hauer	-	-	-	8	-	-	-	-	-	-	14	-
L. luna Muller	3	5	44	-	-	-	-	-	-	5	-	-
L. curvilinealis Harring	-	-	4	-	-	-	-	-	-	5	-	-
L. curvicornis Murray	-	4	-	-	-	-	-	-	-	-	-	-
L. quadridentata Ehren.	3	-	-	-	-	-	-	-	-	-	-	1
L. decipiens Murray	14	13	-	-	-	-	-	-	-	-	-	10
L. closterocerca Schmarda	2	-	-	14	-	5	17	-	-	11	-	9
L. bulla Gosse	-	4	4	-	-	-	-	-	-	-	-	-
Total *Lecane* spp.	22	26	58	22	-	11	17	-	-	21	14	20
Notholca sp.	-	10	4	3	-	5	-	5	-	-	5	3
Proales sp.	-	-	-	-	-	-	-	-	-	-	-	5
Ascomorpha sp.	-	-	-	-	-	-	-	-	-	-	-	15
Cephalodella auriculata (Dixon, Nutall and Freeman)	3	-	-	2	-	-	-	-	-	-	-	3

Contd......

Month	*1991*							*1992*				
Organisms	*Jun.*	*Jul.*	*Aug.*	*Sep.*	*Oct.*	*Nov.*	*Dec.*	*Jan.*	*Feb.*	*Mar.*	*Apr.*	*May*
Polyarthra vulgaris Carlin	-	-	-	8	-	-	-	-	-	-	-	-
Asplanchna sp.	29	35	144	-	-	5	-	12	-	37	13	3
Harringia sp.	-	-	-	-	-	-	-	-	-	-	-	4
Synchaeta sp.	-	3	-	-	-	-	-	5	-	-	6	-
Asplanchnopus sp.	-	-	-	7	-	5	-	5	-	-	4	-
Testudinella sp.	-	-	-	-	-	-	-	-	5	-	4	-
Conochilus sp.	-	24	-	-	-	-	-	-	-	-	-	16
Hexarthra sp.	19	24	-	-	-	-	-	-	-	-	5	-
Filinia longiseta (Ehren).	2	15	153	36	-	-	-	-	-	-	14	-
F. terminalis (Edmondson & Hutchinson)	-	-	29	-	-	-	-	-	-	-	9	-
Total *Filinia* spp.	2	-	182	36	-	-	-	-	-	-	23	-
Habrotorcha angusticollis Murray	-	-	-	-	-	-	4	-	-	-	-	-
Philodina sp.	178	51	149	125	-	16	-	11	6	177	174	126
Rotaria neptunia weber	8	19	6	57	-	6	-	22	16	236	75	1
R. citrinus weber	-	21	25	2	-	-	-	-	-	-	-	-
Total *Rotaria* spp.	8	40	31	59	-	6	-	22	16	236	75	1
Epiphanes sp.	-	-	-	-	-	-	-	-	80	-	-	-
Voronkowia sp.	4	-	4	-	-	-	-	-	-	-	-	-
Collotheca ornata Myers	2	-	-	-	-	-	-	-	-	5	9	-
Total Rotifera	311	2216	4424	640	516	662	1071	2019	527	3675	1608	276

species of the same genus appear occasionally in a season, no matter how many more species may be representing it within a particular habitat.

An overall study of rotifers has indicated August, November, February and April qualitative peaks (Table 2), with highest annual qualitative count in August and lowest in October.

Quantitatively, rotifers recorded a July, August, January and March increase. Rotifer count remained low in the months of June, September, February and May. These recorded annual lowest count (276/1) in May and the highest (4424/1) in August (Table 2). Monsoon highest count of rotifers has also earlier been worked out by Kumar *et. al.* (1991) from the area.

Earlier, George (1961) recorded an increase in rotifer fauna during April to July from Delhi, Vasisht and Dhir (1970) noticed winter and summer peaks from Chandigarh. Vasisht and Sharma (1975) documented September and January peaks from an urban pond in Ambala. Saha *et. al.* (1971) observed January peak in he first year and November peak in the second year from Cuttack pond, Orissa. Jyoti and Sehgal (1979) worked out four peaks (November, February, April and August) with annual maximum February and April peaks, in littoral zone and trimodal (November, March and August) peaks, with annual highest count in March, in limnetic zone of lake Surinsar. Gupta and Sudan (1980) documented three peaks of abundance in Bantalab (March, July and October), Kherwanda (February, June and November) and Bhalwal ponds (February, June and November) and bimodal peaks in Nagbani (March and July) pond, with annual highest count in March, February, June and July, respectively. Jindal and Vasisht (1985) reported bimodal viz winter and summer rotifer peaks from a pond at Nabha, Punjab. Puri (1989) noticed October, December, March and May quantitative peaks of rotifers, with annual highest count in the month of March, from some pools adjacent to the river Tawi, Jammu. Kumar (1990) found March, May and September peaks during first year and February, May, August and October peaks during second year from R.S. Pura pond, Jammu. Kumar *et. al.* (1991) recorded June, August and January peaks, with annual highest August count from Kunjwani pond, Jammu.

Present record of variable seasonal behaviour of rotifers, as compared to the findings of the earlier workers (op. cit.), may be attributed to the changed environmental conditions in different water bodies.

Analysis of coefficient of correlation (r) of total rotifers with various physico-chemical parameters has indicated insignificant positive correlation with temperature (r=0. 5128), transparency (r=0.0641), DO (r=0.5469), free CO_2 (r=0.2865), NO_3'–N (r=0 .4345) and SiO_2 (r=0 .3486) and negative with pH r = 0.4593), electrical conductivity (r=– 0. 2266), H CO_3, (r = – 0.2809), HCO_3' (r= – 0.3409), Cl^- (r= – 0.3469), Ca^{++} (r = – 0.0862), Mg^{++} (r = – 0.3093), total hardness (r= – 0.1469), PO_4'' (r = – 0.3162) and SO_4' (r = – 0.0256). This indicates that no one physico-chemical parameter appears to be a strong determinant for rotifers in Farooq Nagar pond.

Contrary to the present observations, Banik *et. al.* (1994) reported significant coefficient of correlation 'r' of abundance of total rotifers with water temperature

Table 3: Density Species (Shannon - Index) Relationship of Rotifers at Station I-IV of the Farooq Nagar Pond, Jammu (June 1991-May 1992)

Months	*Station I*		*Station Ii*		*Station III*		*Station IV*	
	Density	*Diversity*	*Density*	*Diversity*	*Density*	*Diversity*	*Density*	*Diversity*
June 1991	359	1.7722	254	2.8487	192	2.8247	471	0.1729
July	3267	0.7624	1551	0.9116	1955	0.2709	3173	1.0500
August	7220	0.6709	4311	1.1830	3502	1.3595	2560	1.7290
September	675	1.5147	513	1.4047	934	2.0070	299	1.4622
October	576	0.0	972	0.0	322	0.3712	191	0.0
November	799	0.7613	1034	0.6056	253	0.0	577	0.5850
December	990	1.4716	796	0.0	0	0.0	2226	0.1843
January 1992	2093	0.0872	3142	0.4743	661	0.4048	2179	0.3737
February	621	0.2285	357	0.7918	673	1.5430	456	0.5816
March	3078	1.2516	1653	0.3859	6032	0.6928	3955	0.8736
April	1457	1.4496	500	1.1153	2307	1.7992	1532	1.3919
May	320	2.3159	101	1.8210	315	2.7863	400	1.8868
Total	21455	12.2859	15184	11.5419	17146	14.0593	18019	10.291
Mean		1.0238		0.961872		1.2781		0.9545

Correlation coefficient was 661, at 12 degrees of freedom, at station I, II and IV and .684, at 11 degrees of freedom, at station III.

(r=0.9795, p < 0.01), pH (r = 0.6954, p < 0.05) dissolved oxygen (r = 0.8686m p < 0.01), bicarbonate alkalinity (r=0.6501, p < 0.05) and phosphate-phosphorus (r=0.9831, p < 0.01). Among biological factors, rotifers in Farooq Nagar pond have shown a significant correlation with Euglenophyceae (r = 0.6024), Bacillariophyceae (r=0.5626) and an insignificant correlation with Chlorophyceae (r = –0.2550), Cyanophyceae (r=0.1119), Xanthophyceae (r=0.5278) and total phytoplankton (r=0.1936). Among various groups of zooplankton, rotifers have shown significant correlation with Cladocera (r=0.6047), Copepoda (r=0.7748) and total Zooplankton (r=0.9372) and an insignificant correlation with Protozoa (r=0.1017), Insecta (r=0.3308), Oligochaeta (r=0.1143), water nematodes (r=0.1924), Ostracoda (r=–0.5309) & Turbellaria (r=–0.0009).

Quantitatively, the order of dominance of various genera of rotifers has been seen as *Brachinous* (14/1–3581/1) > *Rotaria* (0–236/1) > *Platyias* (0–212/2) > *Filinia* (0–182/1) > *Philodina* (0–178/1) > *Asplanchna* (0–144/1) > *Mytilina* (0–141/1) > *Epiphanes* (0–80/1) > *Euchlanis* (0–77/1) > *Lecane* (0–58/1) > *Conochilus* (0–24) & *Hexarthra* (0–24/1) > *Colurella* (0–19/1) > *Anuraeopsis* (0–18/1) > *Lepadella* (0–15/1) > & *Ascomorpha* (0–15/1) > *Notholca* (0–10/1) > *Collotheca* (0–9/1) > *Keratella* (0–8/1) & *Polyarthra* (0–8/1) > *Asplanchnopus* (0–7/1) > *Synchaeta* (0–6/1) > *Proales* (0–5/1) & *Testudinella* (0–5/1) > *Harringia* (0–4/1) > *Habrotrocha* (0–4/1) > & *Voronkowia* (0–4/1) > and *Cephalodella* (0–3/1).

Among rotifers, quantitative predominance of *Brachinous*, noticed presently, is in agreement with the findings of George (1961), Gosh *et. al.* (1974), Vasisht and Sharma (1975), Jyoti and Sehgal (1979), Mandal (1979), Sandhu *et. al.* (1984), Gupta and Sudan (1985), Jindal and Vasisht (1985), Chourasia and Adoni (1987), Puri (1989), and Kumar *et. al.* (1991). This quantitative predominance of *Brachionus* genus may be due to its qualitative richness and perennial presence.

Annual mean analysis of Shannon and Weaver species diversity index (Table 3) indicates low records of species diversity index (0.0-1.88) at station IV, which may be due to the stress caused by sewage discharge at this station.

Acknowledgements

Thanks are due to H.O.D., Biosciences, for necessary laboratory facilities and Mr. Puran Chand for his help in field collections.

References

Adoni, A.D. (1985). Workbook on limnology. Pratibha Publishers, Sagar-470003, India.

APHA (1975). Standard methods for examination of water and waste water. Amer. Publ. Hlth. Assoc. Inc., New York.

Arora, H.C. (1965). Studies on Indian Rotifera-Part VI. On a collection of Rotifera from Nagpur, India, with four new species and a new variety. Hydrobiol., *26* (3-4) : 444-462.

Banik, S., Debnath, R., Debbarman, S., and Kar, S. (1994). Occurrence of Rotifers in a seasonal wetland in relation to some Limnological conditions. J. Freshwater Biol., *6* (3) : 221-224.

Battish, S.K. (1992). Freshwater zooplankton of India. Publ. by Oxford & IBH Publishing Co. Pvt. Ltd., New Delhi : 233 p.

Chourasia, S.K. and Adoni, A.D. (1987). Rotifers as indicators of eutrophication. In perspectives in Hydrobiology. Rao, K.S. and Shrivastava, S. (Ed.). Vikram University, Ujjain (M.P.), India, 4 (28): 147-149.

Edmondson, W.T. (1959). Freshwater biology. John Wiley and Sons Inc, New York, 1248p.

George, M.G. (1961). Observations on the rotifers of shallow ponds in Delhi. Curr. Sci., *30* : 268-269.

Gosh, A., Hanumantha, R.L. and Banerjee, S.C. (1974). Studies on hydrobiological conditions of a sewage fed pond with a note on their role in fish culture J. Inland Fish Soc., India, *6* : 51-61.

Gupta, V.K. and Sudan, A.V. (1985). Seasonal fluctuations in rotifer population from four freshwater fish ponds of Jammu, India. Bull. Env. Soc., *2* (2) : 7-14.

I.S.I. (1973). Indian Standard Methods for Sampling Test (Physical and chemical) for water used in industry. Indian Standard Institute, New Delhi.

Jindal, R. and Vasisht, H.S. (1985). Ecology of a freshwater pond at Nabha. Zoologica Orientalis, *2* (1-2): 74-83.

Jyoti, M.K. and Sehgal, H. (1979). Ecology of rotifers of Surinsar, A subtropical freshwater lake in Jammu (J & K), India. Hydrobiologia, *65* (1) : 23-32.

Jyoti, M.K. and Sehgal, H. (1980). Rotifera fauna of Jammu (J & K), India, Part I. Loricates, Limnologica (Berlin), *12* (1) : 121-126.

Kumar, N. (1990). Altitude related limnological variations in some fish ponds of Jammu province. Ph. D. Thesis, Unversity of Jammu, Jammu.

Kumar, S., Dutta, S.P.S., Malhotra, Y.R. and Kumari, V. (1991). An ecological study of rotifers in Kunjwani pond, Jammu. J. Hydrobiol., : 41-45.

Mandal, B.K. (1979). Limnological studies of freshwater fish pond at Burdwan, West Bengal, India. Limnol., *40* : 10-18.

Mishra, S.R. and Sharma, S.P. (1990). Studies on the zooplankton from Gwalior and Bhind, M.P.J. Hydrobiol., *6* (1) : 25-28.

Nayar, C.K.G. (1968). Rotifer fauna of Rajasthan, India. Hydrobiol., *31* : 168-185.

Pennak, R.W. (1953). Freshwater invertebrates of the United States, Ronald Press, New York. 769 p.

Pennak, R.W. (1957). Species composition of limnetic zooplankton communities. Limnol. and occanogr., 2 : 222-223.

Puri, A. (1989). Limnology of some pools along the river Tawi, near Nagrota, Jammu, with reference to the plankton and Macrophytes. M. Phil. Dissertation, University of Jammu, India : 1-154.

Saha, G.N., Sehgal, K.L., Mitra, E. and Nandy, A.C. (1971). Studies on the seasonal and diurnal variations in physico-chemical and biological conditions of a perennial freshwater pond at Cuttack, Orissa. J. Inland Fish Soc. India, 3 : 79-102.

Sandhu, J., Sehgal, H.S. and Toor, H.S. (1984). Rotifer species diversity in manured experimental fish tanks. Indian J. Ecol., *11* (2) : 303-308.

Shannon, G.E. & Weaver, W. (1963). The mathematical theory of communication. Univ. of Illionois Press, Urhana, U.S.A.

Trivedy, R.K., Goel, P.K. and Triscal, C.L. (1987). Practical methods in ecology and environmental science. Environmental Publications, Karad, India.

Tonapi, G.T. (1980). Freshwater animals of India. An Ecological approach. Oxford and IBH Publishing Co. Calcutta, 341 p.

Vasisht, H.S. (1968) and Battish, S.K. (1970). The Rotifer fauna of North India II. *Trichocerca porcellus* Res. Bull. Punjab, Univ., *21* (3 & 4) : 515.

Vasisht, H.S. (1968) and Battish, S.K. (1971). The Rotifer fauna of North India *Brachionus*. Res. Bull. (N.S.), Punjab Univ., 23 (1-2) : 179-188.

Vasisht, H.S. (1968) and Battish, S.K. (1972a). The Rotifer fauna of North India. *Asplanchna*. Res. Bull. Punjab Univ., 23 (1-2) : 105-106.

Vasisht, H.S. (1968) and Battish, S.K. (1972b). The Rotifer fauna of North India. *Filinia* and *Testudinella*. Res. Bull. (N.S.), Punjab Univ., 23 (1-2) : 93-95.

Vasisht, H.S. (1968) and Battish, S.K. (1972c). The Rotifer fauna of North India. *Polyarthra*. Res. Bull. (N.S.), Punjab Univ., 23 (1-2) : 107-108.

Vasisht, H.S. (1968) and Dhir, S.C. (1970). Seasonal distribution of freshwater zooplankton of four tanks, Chandigarh. Ichthyologica, *10* (1-2) : 44-56.

Vasisht, H.S. (1968) and Sharma B.K. (1975). Ecology of a tropical urban pond in Ambala city of Haryana State. Ind. J. Ecol., *2* (1) : 79-86.

Yousuf, A.R., Shah, G.M and Qadri, M.Y. (1986). Some limnological aspects of Mirgund Wetland. Geobios New Reports, *5* (1) : 27-30.

Chapter 18

Relative Population Abundance of Ichthyo-fauna of Lake Mansar.

Arun K. Gupta, Anil Khajuria, S. C. Gupta, Seema***

Abstract

A survey of the relative population abundance of fishes inhabiting lake Mansar, reveal the presence of ichthyofauna comprising of fish species like *Puntius conchonius, Rasbora rasbora, Channa gachua, Danio rero* and *Trichogaster fasciatus* belonging to five different genera. Of the various species *Puntius conchoinus* is the most abundant contributing 41.14% of the mean annual catch of total fish fauna of lake Mansar followed serially by *Rasbora rasbora* (Annual catch: 28.01%), *Channa gachua* (14.18%), *Danio rerio* (12.17%) and *Trichogaster fasciatus* (4.44%).

Introduction

In order to understand properly the biology of fishes, it is very essential to have some idea about the prevailing ecological condition and viz-a-viz their inter-relationships. Present study is an attempt in the direction of understanding competitive interaction of fishes among themselves and with other organisms in Lake Mansar, which is a perennial waterbody that spreads over an area of 3.29 sq kms, situated at an altitude of 666 meters MSH, between latitude 30° 45′ 5″ north to 33° 42' 36″ and longitude 75° 8' 32″ and 73° 9' 8″. The Lake is situated about 65 kms. away from Jammu Tawi, J & K State. In order to generate data on fishery biology of Lake Mansar, an attempt has been made for the first time during the course of present investigation to study relative population abundance of fishes from the present area of investigations based upon catch per unit effort.

Materials and Methods

For estimations of relative population abundance of ichthofauna of Lake Mansar, fishes were collected from six different stations, all the year round. Bi-monthly collection of different fish species of the Lake has been obtained every month and mean average is calculated.

* 1. Govt. Women College, Parade, Jammu.

** 2. Deptt. of Bio-Sciences, Jammu University, Jammu

Puntius conchonius

Puntius Conchonius is the most abundant fish species contributing **41.14%** of the mean annual catch of total fish fauna of lake Mansar as indicated in Table 1-3 and Figs. 1 and 2. The population of *Puntius conchonius* was recorded to be highest from November to August, maximum being in Feb. (81.95%) and minimum in September and October (25.25%). *P. Conchonius* has been regularly hauled round the year.

The abundance of *Puntius conchonius* is due to adoptability of the fish to : (i) low temperature (16.50°c), (ii) sufficient food supply, (iii) high fecundity & (iv) other abiotic factors governing the lake ecosystem. (Ber, 1854; Danilevaskii, 1862). The longest netted fish attained maximum size 85 mm and weight 7.600 gms.

Maximum percentage of species (81.95%) in Feb. may be due to complete absence of predactor which is temperature sensitive and less food supply, and thus fulfills the hypothesis of Manteifel (1961), .i.e. *Tritroph* according to which abundance of a member (occupying mid part of food chain viz, Puntius conchonius) is least effected due to complete absence of its predactor in the lake eco-system particularly in cold weather. Low population size of P. Conchonius, minimum in Sep-Oct because of abundance and increased competition and activity of predator Channa Gachua (Ber, 1854) in lake Mansar during above cited months (as also cited by Nikolsky, 1963) which also confirms the present findings i.e. predator is a limiting factor for the prey fish population.

Thus above findings are in confirmity with Nikolsky, 1963 who states that "Dynamics of abundance and bio-mass of population are the result of specific interaction of environment and adoptive properties of the population".

Rasbora rasbora

This fish comes second in order of relative abundance exhibiting an annual average catch of 28.07% of total fish stock, inhabiting lake Mansar. The studies conducted over the period of one year reveals quantitative composition (Table 1-3 & Figs. 1 & 2) of this fish and remains high from August to January maximum being 43.60% in June and minimum 11.20% in July.

Like *Puntius conchonius, Rasbora rasbora* appears in catches round the year thereby exhibiting invincivility against limiting factors for the species in the ecosystem which are identical with those discussed for *P. Conchonius.*

Low catch of the fish in July may be attributed to first monsoon showers which create turbid conditions (as also discussed by Nikolsky, 1963; for turbed condition). Low abundance in April is because of severe inter-specific competition for food during this time of year in lake Mansar when other competitors for *Puntius* food become active feeders.

Maximum annual population size of *R. rasbora* in December (Tables 1-3 and Figs. 1&2) is attributed to abundance of large number of Juveniles (of this very fish) which with adult fishes, have been enumerated.

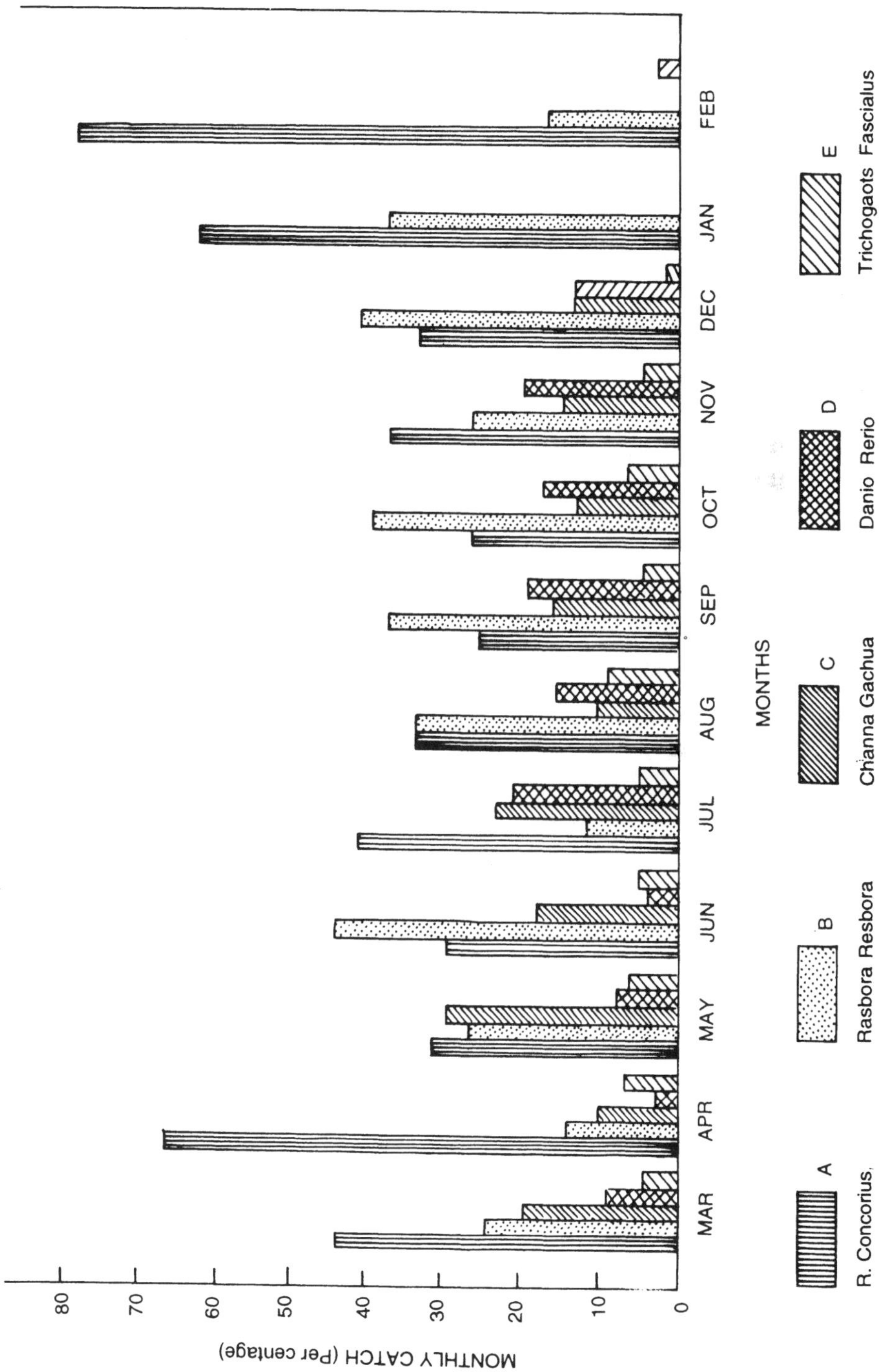

Fig. 1: Monthly Catch Composition of fishes of Lake Mansar 1989–90

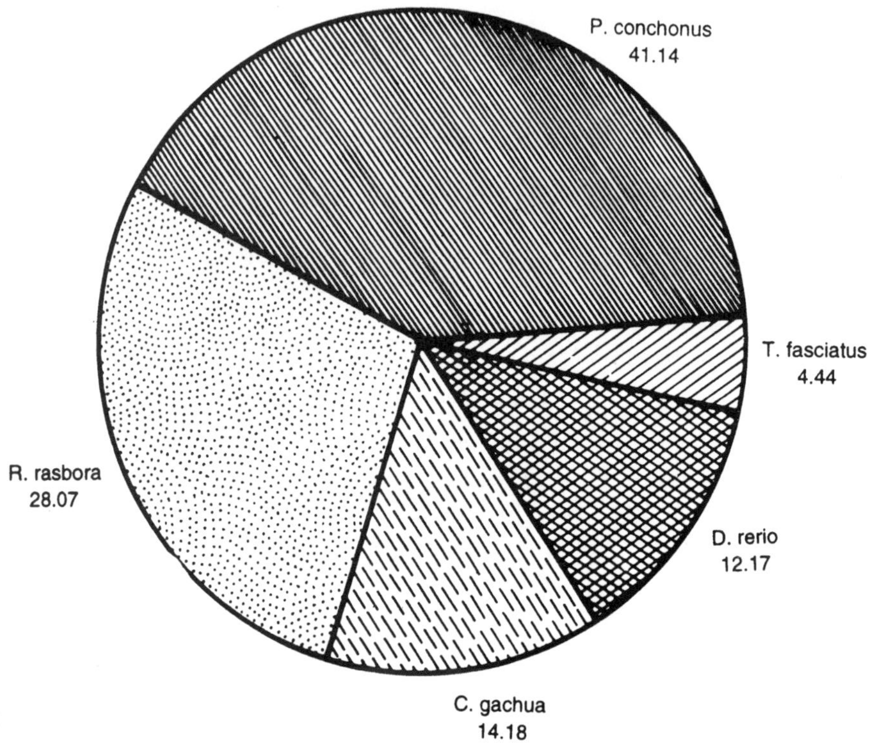

Fig. 2: Mean Catch Composition of Fishes of Lake Mansar (1989–90)

Table 1: Average Monthly Catch Composition of Fishes of Lake Mansar 1989-90

Name of the month	*Puntius Conchonius*	*Rasbora rasbora*	*Channa gachua*	*Danio rerio*	*Trichogaster fasciatus*
March	43.5%	24.10%	19.40%	8.80%	4.20%
April	67.2%	13.90%	9.80%	2.60%	6.50%
May	30.80%	26.3%	29.10%	7.50%	6.10%
June	29.30%	43.6%	17.70%	3.80%	4.90%
July	40.60%	11.20%	22.90%	20.60%	4.00%
August	33.20%	33.10%	9.90%	15.20%	8.60%
September	25.00%	36.50%	15.50%	18.80%	4.20%
October	25.90%	38.50%	12.46%	16.80%	6.34%
November	36.50%	25.78%	14.22%	19.30%	4.20%
December	32.70%	40.45%	12.90%	12.80%	1.65%
January	63.22%	36.78%	A	A	A
February	81.95%	16.35%	A	A	2.70%
Mean	41.14%	28.07%	14.18%	12.17%	4.44%

Table 2: Distribution of Fishes at Diffferent Stations of Lake Mansar 1989-90

S. No.	*Name of fishes*	*Station No. 1*	*Station No. 2*	*Station No. 3*	*Station No. 4*	*Station No. 5*	*Station No. 6*
1.	*Puntius Conchonius*	+	+	+	+	+	+
2.	*Channa gachua*	+	–	+	+	+	+
3.	*Rasbora rasbora*	+	+	+	+	+	+
4.	*Danio rerio*	+	–	–	+	+	+
5.	*T. fasciatus*	+	–	+	–	+	–

(+) = Present
(–) = Absent.

Table 3: Monthly Record of Different Fishes of Lake Mansar

S. No.	*Name of fishes/ species*	*Mar.*	*Apr.*	*May*	*June*	*July*	*Aug.*	*Sept*	*Oct.*	*Nov.*	*Dec.*	*Jan.*	*Feb*	*Remarks*
1.	*Puntius Conchonius*	+	+	+•	+•	+•	+	+	+	+	+	+	+	Abundant
2.	*Rasbora rasbora*	+	+	+	+	+	+	+	+.	+.	+	+	+	Abundant
3.	*Channa gachua*	+	+	+	+	+	+	+.	+.	+.	+	–	–	Common
4.	*Danio rerio*	+	+	+	+.	+.	+.	+.	+	+	+	–	–	Common
5.	*Trichogaster fasciatus*	+	+	+	+	+	+	+	+	+	+	–	+	Rare

(+) = Present
(–) = Absent
(.) = Juveniles present.

Channa gachua

Table 1-3 and Figs. 1&2 portrays annual catch (14.18%) of *Channa gachua* and is placed third in order of dominance with reference to fishes inhabiting lake Mansar.

The population of this fish show higher relative population abundance from March to July, maximum being in July (22.90%) and minimum in December (12.90%).

The high population from March to July may be due to favourable temperature, availability of food, better weather condition.

Lowest stock of this fish in April and December and their complete absence in January and Feb. is possibly due to very low temperature, reduced food supply and unfavourable weather condition. Over-wintering or hibernation in fishes has already been reported by Kalabucha, 1956, (C. f. Nikolsky, 1963) for common crucian carp and Black fish *Dallia pertoralis* Bean, in chutostsk peninsula, who observed that active mode of life is impossible during winter. Fishes dig themselves into mud fall into state of tor por and thus their metabolic activity almost ceases, unless their body fluids freeze. Qayumm and Qasim (1964) have reported aestivation in *Channa gachua* (extreme high temperature) from Aligarh. Dutta and Malhotra (1991) have also reported low catch of *C. gachua* in December (winter) due to cold water temperature from Gadigarh stream, Jammu.

Danio rerio

Stands fourth in order of dominance (Table 1-3 and Figs. 1 & 2) contributing 12.17% of annual catch, maximum being in July (20.60%) and minimum April (2.60%). Their total absence is recorded during colder months, Jan-Feb. in which respect they resemble Channa gachua. As the Tables 1-3 portrays that abundance of both these carnivorous fishes synchronous, being maximum in (22.90% & 20.60%) and minimum (12.90% and 2.80%) which may be due to different food items of the two predator species which are sumpatric and synchronous.

Lowest stock of this fish from December to June is possible because of low temperature and scarcity of food (Nikolsky 1963), and abundance of other fishes of the lake in March to June and intense competition for food during this period.

Highest population as recorded during the investigations from July to November may be attributed to greater availability of food of *D. rerio* and better conditions of other abiotic factors governing the lake ecosystem.

The increased population size is also attributed to presence of fingerlings of *D. rerio* during this time of study which also have been enumerated and thus contributes to the increase.

Trichogaster fasciatus

This fish contributes the minimum (4.44%) of total annual catch of the fish from lake Mansar as indicated in Tables 1-3 and Figs. 1 & 2 and stands fifth in order of dominance. The factor contributing to its lowest relative population is possibly due to greater competition with other herbivorous fishes viz., *Puntius conchonius* and

Rasbora rasbora. Besides, lower functionaries Channa gachua and *Danio rerio* also compete with this fish for algal and macrophyte food. Another factor for its low availability in the lake, is possibly due to its preference to abode in clean water with rich macrophytic vegetation as also observed by Das & Srivastava (1959). Another important factor responsible of its low catch is that it is a specific feeder (periphytonic).

This fish is present in lake Mansar from Feb. to December (absent in Jan).

Maximum of this fish is hauled in August (8.6%) because of disturbance of periphytonic flora and fauna (due to rains) on which fish feeds. Minimum of this fish has been caught in Dec. and winter (1.65%), due to low temperature, decline in marcrophytic vegetation which also perishes in lake (Kant and Vohra 1989). Being sensitive to temperature, and as it abodes among the vegetation its food supply is hindered and therefore, its low population in the lake results.

References

Ber, K.M. (1954). History of fishing in Russia and adjacent sea. Uah. Zap. Akad. Nahk. 2 No. 4.

Danilevsku, N. Ya (1962). Fisheries of white sea and Arabic Ocean. In research on the fisheries of Russia (In Russian). 7 : St. Pettersburg.

Dutta, S.P.S. and Malhotra, Y.R. (1991). Food and feeding habits of some fishes of Gadigarh Stream, Jammu, Proc. Nat. Acad. Sci. India 61. : (B) : 163-168.

Kalabucha, N.I (1956). "Hibernation o Animals'. Sov. Sci. Press.

Lagler, K.F., Bardach, J.E. and Miller, R.R. (1962). Ichthyology (The Study of Fishes). John Wiley & Sons, Inc. New York. London, 545 pp.

Manteifel, B.P. (1961). Vertical migration of marine organisms Part-III. Adaptive significance of the vertical migration of Plankton eating fish. Trudy Inst. Masphol. Zhirotz. Akad. Nauk SSR No. 39.

Nikolsky, G.V. (1963). Fish ecology (In Russian). Izd. Vyashaya Shkola.

Qayumm, A. and Qasim, S.Z. (1964). Studies on the biology of same fresh water fishes Part-I. *Ophiocephalus punctatus* (Block) J. Bom. Nat. Hist. Soc. India, *61* (1) : 74-98.

Vasentsov, V.V. (1953). Trends in the development and dynamics of fish numbers. Transactions of all Union Conference on fisheries (in Russia). Izd. Akad. Nauk SSR.

Chapter 19

On the Distribution and Ecology of Some Gastropod Molluscs of the Jammu Province of J & K State

*Anil K. Verma and P. L. Duda**

Abstract

Surveying of the Jammu province of J & K state for gastroped molluscs yielded bulk of gastroped molluscs constituting 19 species of the slugs and snails from different localities in the province. The collection proved very significant as 10 genera and 12 species turned out to be first records from the Jammu province of the state. Ecological make-up of the places from where the snails and slugs were collected has been described. Note on the geographical distribution of the collected species of the molluscs is made as complete as possible.

Introduction

Members of the phylum Mollusca, second only to Arthropoda in number of species, have colonised every possible habitat on the globe and are dominant in benthic communities of all aquatic ecosystems. The largest number of molluscs are found in marine environment and the species inhabiting the freshwater and terrestrial ecosystem are relatively fewer. However, our knowledge of the latter group is rather limited. Their participation in the way of life of many organisms as food for a number of aquatic animals and man; and in being instrumental in the transmission and spread of many diseases of live-stock and humans etc., has made them significant partners in the ecological communities.

Notwithstanding their role in aquatic ecosystem, molluscs have been little explored ecologically and biologically. A perusal of literature shows a marked paucity of information on the occurrence, distribution, ecology and population density of these molluscan elements from the state save for some reports on their taxonomy from within the state (Annandale, 1918; Theobald, 1878; Rajagopal & Subba Rao, 1971; & Duda and Verma, 1992). An attempt has, therefore, been made to study the distributional patterns and ecology of various gastropod molluscs occurring within the geographical range of the Jammu province of the Jammu and Kashmir state. The specimens collected have been identified through the Courtesy of Zoological Survey of India, Calcutta and preserved at the parasitology laboratory of Department of Biosciences, University of Jammu, Jammu.

* Department of Biosciences, University of Jammu, Jammu (J & K State) – 180004.

Materials and Methods

Collection of the gastropods were made mostly by the methods of Hauling, Hand - picking, Netting, Dredging and Sand-sieving rom submerged aquatic vegetation, edges of ponds, lakes, streams and paddy-fields. During brumation, the specimens were collected by Scrapping superficial layers of soft-mud from beds of streams, small road-side cess-pools, seasonal ponds and also from paddy-fields, and washing them repeatedly in water. As far as practicable, the gastropods were grossly studied at their collecting site itself and notes made on their habitat characteristics, vegetation, topography of the area and use of the water resource by inhabitants of that area. The snails were brought from the different collecting sites in moist vegetation and placed in glass-aquaria or separate labelled jars with perforated tops, in the laboratory and studied there.

Results and Discussion

Foraying of the 4 districts namely Jammu, Kathua, Udhampur and Doda of the Jammu province by the authors provided over 5,000 specimens of the gastropod molluscs. The various gastropod genera recorded so far under present investigation includes nineteen (19) species belonging to thirteen (13) genera representing nine (9) families from the province (Table 1). Of the nineteen species recovered during present study, twelve are collected for the first time from the region/province (Table 2) and eight species have been reported for the first time even from the J & K state having as many as five (5) major river systems and six (6) lakes draining the entire state.

Therefore, malaco-fauna (gastropod) of the province is rich both qualitatively as well as quantitatively in having both the freshwater and the land forms in a fairly good proportion.

Table 1: Showing various gastropod taxa collected so far from various collecting stations in the Jammu province of the J & K state.

S. No.	*Gastropod famly*	*Gastropod taxa*
1.	Viviparidae	*Bellamya bengalensis* (Lamarck) B. dissimilis (Mueller)
2.	Bithynidae	Digoniostoma pulchella (Benson)
		Bithyma tentaculata var. kashmirensis (Nevill)
3.	Thiaridae (= Melanoidae)	*Thiara* (=Melanoides) tuberculatus (Muller)
4.	Lymneaidae	*Lymnea* (=Pseudosuccinea) *acuminata Lamarck.*
		L. (P) *luteola* Lamarck
		L. (P) *lagotis* Kobelt.
		L. (=Radix) *auriculara* Linn.
5.	Planorbidae	*Gyraulus convexiusculus* (Hutton)
		G. pankongensis (Ven Martens) *Indoplanorbis exustus* (Deshayes)
6.	Enidae	Ena spp.
7.	Veronicellidae	*Laevicaulis alte* (Ferussac)
8.	Arionidae	*Anadenus altivagus* (Theobald)
9.	Ariophantidae	*Bensonia jammuensis* (Theobald)
		B. monticola (Hutton)
		Euansteniamonticola (Pfeiffer)
		Syama splendens (Hutton)

Table 2 : Showing List of various Gastropod Molluscs (Slugs and Snails) Recorded for the First Time from the J & K State

Gastropod orders		*Species of gastropod*
(A) Mesogastropoda		
Family:		
1.	Viviparidae	Bellama dissimilis (Muller)
2.	Bithyniidae	Digoniostoma pulchella (Benson)
		Bithynia tentaculata var. kashmirensis (Nevill)
(B) Basomatophora		
Family:		
3.	Lymnaeidae	Lymnea (=Radix) auricularia Linnaeus
4.	Planorbidae	Gyraulus pankongensis (Von Martens)
(C) Stylomatophora		
Family:		
5.	Enidae	Ena spp.
6.	Veronicellidae	Laevicaulis alte (Ferussac)
7.	Arionidae	Anadenus altivagus (Theobald)
8.	Ariophantidae	Bensonia jammuensis (Theobald)
		B. monticola (Hutton)
		Euanstenia monticola (Pfeiffer)
		Syama splendens (Hutton)

Ecological Spectrum of Gastropods

(A) Freshwater snails

1. *Bellamya* spp. lives in large flood-plains, standing waters (ponds, swamps and pools) and in river channels where it produces enormous population. It enjoys life when the rains flood the fields and it can live in comfort for sometime even when without water; and can also lie hidden in soft - mud.
2. *Bithynia* spp. inhabits various types of water - bodies viz., rivers, lakes and ponds i. e., lotic as well as lentic water bodies.
3. *Digoniostoma* spp. shares habitat with *Bithynia* spp., and *Lymnea* spp.
4. *Melanoides (= Thiara)* spp. inhabits running waters of rivers and springs where currents are mild and water clear, free-from organic matter and with a sandy-bed, it usually avoids the turbid and fouling waters.
5. *Lymnea* spp. lives in cold as well as hot springs; swamps, pools, lakes and rice/paddy fields. It occurs at considerable altitude of even 4000 m above sea-level. It also inhabits the marginal strip of sluggish water bodies and in the temporary water-bodies. Lymneaids also migrate from the freshwater, to the marshes and meadows, where they are often found on grass-stalks with their egg masses as the floating objects.
6. *Gyraulus* spp. occurs in standing and slowly running waters, springs, river

flood-plains, brooks, ponds, lakes and river margins on stones and aquatic vegetation.

7. *Indoplanorbis* spp., inhabits standing and the irrigated rice fields; sluggish waters (pools, swamps) and overgrown brooks; and can withstand temporary draught. *Indoplanorbis* spp., like *Lymnea* spp. have the habit of crawling upside down beneath the surface film of the pond, wherein they live, the foot gliding along as it were moving along a sheet of glass. In dry/hot weather when the ponds and the cess-pools etc. dry up, these snails estivate in mud after the fashion of *Vivipara (=Bellamya)*. Spawn of this genus is deposited in colourless gelatinous masses on stones and aquatic weeds.

(B) Land Snails and Slugs

1. **Land snails:** Among land snails *Ena* spp. *Bensonia* spp., *Syama* spp. and **Euanstenia** spp. have been collected as the pests on the vegetable crops and ornamental plants causing extensive damage. Of these snails, *Euanstenia* is endemic to the colder region of Bhaderwah area of distt. Doda.

2. **Slugs**

   (a) *Anadenus altivagus:* Occurs at higher elevations extending up to an altitude of 4,000 m in the Bhaderwah region of distt. Doda. The slug collected by the authors are straw coloured which afford protection to them according to their surroundings. This slug was collected by the authors only in pine wood where it harmonised with the general colouring of the ground. The straw-yellow colour of slug and the yellow-fallen-leaves may easily be mistaken for a piece of *Bark.*

   (b) *Laevicaulis alte:* The common garden slug is a widely distributed species of this region and causes extensive damage to vegetable crops and ornamental plants. This slug owing to its dark grey or blackish dorsal surface harmonises well with the prevailing tinge of the fields/soils, among which it finds a home. This slug, therefore, exhibits protective colouration (mimicry) which is not uncommon among the slugs. From stand point of ecological specialization and distribution, freshwater gastropods from the state have been categorised into following zoogeographical groups:

      Group – I Consists of wide- spread species (e.g. *Lymnea* spp) adapted to a broad range of environmental conditions. The molluscs of this group occurs in standing and slowly running waters and many of them can withstand a drop in water level and O_2.

      Group – II Having an ill-defined range and can survive in rivers and pure-waters of lakes e.g., *Gyraulus* spp.

      Group – III *Consists of largely of riverine species which* may visit slightly silted lake littorals e.g., *Bithynia* spp. and *Digoniostoma* spp.

Group – IV Comprises the species inhabiting soft-water in flood-plain, preferentially the slow-moving flood-plain waters etc., e. g., *Bellamya* spp.

Group – V Composed of species living in brooks or small limpid (waters) rivers, irrigation canals and ditches feeding rice paddies e. g. *Indoplanorbis* spp.

Group – VI Consists entirely of river and spring species e. g. *Melanoides (=Thiara)* spp.

A more thorough/extensive study of the areas adjacent to the Jammu province may show that some species occurring here have a wider distribution. Gastropods, therefore, comply with the well-known Zoological Law that "the more extensive the ecological range of a sp., the wider will be its geographical distribution". (Zahdin, 1965).

Estimation of Taxa

Out of a total of 37 families of Gastropods molluscs inhabiting Indian waters and land (Subba Rao, 1989) only 9 families are represented in the present collection (Table 3). A perusal of Table 3 reveals that against an estimated total of 12 families of freshwater gastropods in the phylum Mollusca represented in freshwater of India, about 5 families (41.66%) occur in the Jammu province of the state. Similarly of a total of 25 families of land snails and slugs, Jammu shares just 4 families (16%).

Table 3 : Showing various gastropod taxa recovered so far from Jammu province of the state in comparison to the existing knowledge of these elements from rest of the country (percentage of occurrence in parentheses)

Gastropod type	*Families*	*Genera*	*Species*
Freshwater forms:			
	Jammu 5	Jammu 7	Jammu 12
	Rest of India 12	Rest of India 32	Rest of India 156
	(41.66)	(21.87)	(7.69)
Land/terrestrial forms:			
	Jammu 4	Jammu 6	Jammu 7
	India 25	India 104	India 950
	(16.00)	(5.76)	(0.73)

At generic level Jammu shares (21.87%) of freshwater and (5.76%) of the land gastropods. The value becomes still lower at the species level wherein, just 12 species out of 156 species freshwater spp. from India (7.69%) and merely (0.73%) land gastropods (7 out of 950) fall in the present collection.

During present investigation, gastropod molluscs of freshwater ecosystems have been more thoroughly inventorised than their counterparts on land. As such, the estimation of number of families, genera and species of freshwater gastropods is reasonably accurate but with regards to the land gastropods it is only an inference drawn from the occasional and chance observations and availability of the collection of land gastropods.

As the province is still partially surveyed (depicted by striped portion of the districts, Figs. 2–5), some more forms are expected from the relatively lesser warmer and warm temperate regions of the districts Rajouri and Poonch of Jammu province and hence an accurate assessment of the gastropod fauna is bound to be a difficult proposition and therefore, our knowledge of the gastropod molluscs of the region is still far from complete at least for the present.

Acknowledgements

Authors are grateful to the Ministry of Environment and Forests, Govt. of India, New Delhi, for the financial assistance under the project No. 14/14–89 MAB/RE.

References

Annandale, N. 1918. Aquatic molluscs of the Inland lake and connected waters. *Rec. Indian Mus.*, 14: 103–182.

Duda, P. L. and Verma, A. K. 1992. Role of gastropods in the transmission of diseases among Herpetiles. Part I. Amphibia – A numerical Analysis. presented at the first International Congress of IUCN / ISRAG–SSG, at Bhubaneswar.

Rajagopal, A. S., and Subba Rao, N. V. 1971. Aquatic and Amphibious molluscs of the Kashmir valley, India. Proceedings of Symposium on Mollusca – Part I: 95–123.

Subba Rao, N. V. 1989. A Handbook of Freshwater Molluscs of India (Ed.).

Theobald, W. 1878. Notes on the land and freshwater shells of Kashmir, more particularly of the Jhelum valley, below Srinagar and hills of North Jammu. *J. Asiat. Soc. Beng.*, 47 (2): 141–149.

Zahdin, V. I., 1965. Molluscs of fresh and brackish waters of U. S. S. R. (Ed.) R. Finesilver Israel Program for Scientific Translations, Jeruselem: p. 1-368.

Chapter 20

Macrobenthic Fauna in Relation to Some Environmental Factors in Eutrophicating Lake Mansar, Jammu

*K. Gupta and Anil Khajuria**

Abstract

Taxonomic composition density, biomass of bottom fauna in lake Mansar were investigated from littoral to profundal zone. Density of macrobenthos ranged from 86 organisms/m^2 to 551 organisms/m^2 and biomass ranged from 0.83 gm/m^2 to 11.92 gm/m^2. Both parameters varied considerably along the lake profile. Distribution of bottom fauna in lake profile and species diversity were characteristics of eutrophicating lakes.

Key words: Macrobenthos/Littoral/Profundal/diversity/eutrophication.

Introduction

Macrobenthic organisms, occupying the bottom of water body, form an important link in detritus based food chain. This signifies the importance of benthic organisms in the economy of natural waters, and prompted many investigators to make studies on the qualitative and quantitative aspects of tis biota (Mandal & Moitra 1975, Vasisht & Bhandal 1979, Gupta 1979, Yadava *et. al.* (1984), Malhotra *et al.* (1990) and Kulshrestha *et al.* (1991). Macrozoobenthos at different zones of bottom shows marked variations Welch (1952) and Michael (1985). Contributions dealing with this aspect of zoobenthos are scanty and this aspect needs to be better attended. Keeping this in vew, present study has been undertaken to investigate the seasonal dynamics of zoobenthos in different zones of lake Mansar; a natural Siwalik Lake.

Material & Methods

Samples of Benthos were collected from all the three zones of the bottom viz. Littoral, sublittoral and profundal (Fig. 1) with the help of Eckman's dredge from March 89 to Feb. 90. The sediment samples were sieved through standard sieve. Bigger animals like, Molluscs, oligochaetes were sorted out manually using forceps. For the separation of smaller oganisms, solution of 750 mg/lit. sucrose was used. In each sample number of individuals of different species and groups was counted and

* Department of Biosciences, University of Jammu, Jammu.

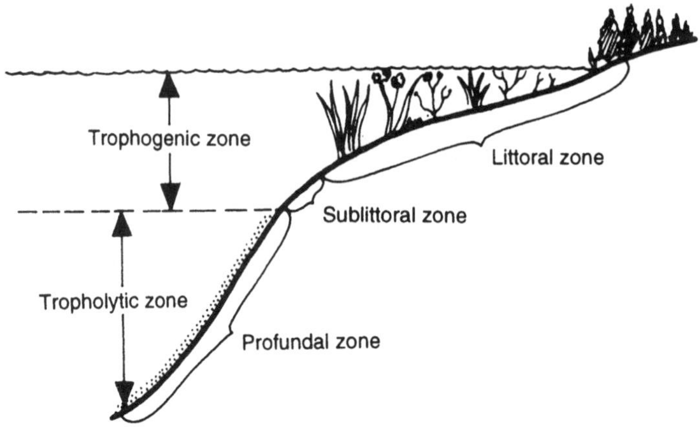

Fig. 1: Showing the Zonation of the Lake Bottom

their number was calculated per square metre of bottom area by following formula:

$$n = \frac{0}{as} \times 10,000$$

Where, n = number of organisms/m²

o = number of organism counted

a = area of Ekman's sampler in/m²

s = number of samples taken at each station.

Biomass of benthic organisms was calculated by Ash weight method after Adoni (1985).

To know the species diversity, following formula was used after Shannon & Weaver (1964):

$$\overline{H} = \sum_{i=1}^{S} Pi \ \log_2 Pi$$

Where, S = number of species

Pi = the importance probability for each sp (n_i/N)

N = total number of individuals in S species &

ni = number of individuals in ith species.

The samples of water overlying the bottom were analysed for pH, dissolved oxygen, free CO_2, carbonates, bicarbonates, calcium and Magnesium and Chlorides (APHA 1975).

Study Area

Lake Mansar 65 km east of Jammu city elevation of 710m above mean sea level and lies at an 32°42′ N and 75° 23′E.

The lake, 37 mts. deep has a circumference of 4.3 km and is primarily fed by surface run off and has some submerged spring sources.

The origin of lake is lost in antiquity and there are several myths about its origin. The common geological belief is that the lake owes its origin to the daming of the river which was flowing along the lower Siwalik ranges in main Siwalik belt. The observations gets support from the peat and sticky soil surrounding the lake which might have had even greater dimensions in the past.

Water temperature of lake Mansar remains always above 4°C at any depth. Further there is large vertical thermal gradient and single winter circulation period. These factors along with subtropical location place it as 'Subtropical monomictic lake'.

Results

(i) Physico-chemical characteristics of bottom water: Minimum and maximum average seasonal values of some selected physico-chemical parameters at different bottom zones have been shown in Table 1.

(ii) Composition and the occurrence of dominating groups: Table 2 reveals the occurrence of total number o 16 taxons of benthic organisms composed of molluscs, insects, oligochaetes and prawns. Out of these 15 taxons have been recorded from littoral bottom, 6 taxons rom sublittoral and 4 taxons from under profundal zone of the lake.

Table 1: Ranges of Seasonal Average Values o Physico-chemical Characteristics of Bottom Water at diferent zones of the lake

Parameters	*Zone*		
	Littoral	*Sublittoral*	*Profundal*
Water temp. (°C)	16.12–31.5	13.75–28.87	10–30.50
pH	7.97–9.23	7.61–8.70	6.62–7.87
Dissolved Oxygen (mg/lit.)	5.96–10.59	5.84–9.58	0.65–5.35
Free Carbondioxide (mg/lit.)	A	1.11–8.21	2.44–8.36
Carbonate (mg/lit.)	5.56–21.6	A	A
Bicarbonates (mg/lit.)	84.17–127.02	90.20–139.66	86.86–139.77
Chlorides (mg/lit.)	0.72–1.69	0.92–4.02	1.2–1.12
Calcium (mg/lit.)	15.92–36.87	17.82–31.88	15.40–41.94
Magnesium (mg/lit.)	4.61–18.17	4.91–11.34	4.97–11.19

A = Absent

Table 2 : Showing the Distribution of Different Benethic Organisms along depth profiles of the lake

Benthic organisms	Zones		
	Littoral	Sublittoral	Profundal
Annelida			
Oligochaeta			
Acolosomatidae			
Acolosoma sp.	+	–	–
Naididae			
Nais sp.	+	+	–
Arthropodes			
Decapoda			
Prawns	+	–	–
Insecta			
Ephemeroptera			
Baetis sp.	+	–	–
Caenis sp.	+	–	–
Diptera			
Tendipes sp.	+	+	+
Culicoides sp.	+	–	–
Chaoborus sp.	–	+	+
Tabanus sp.	+	–	–
Mollusca			
Gastaopoda			
Lymneacidae			
Lymnaea auriclaria	+	–	–
L. acuminata	+	+	+
Amnicolidae			
Digoniostoma pulchella	+	–	–
Planorbidae			
Gyarulus convexiusculus	+	–	–
G. pankongensis	+	+	+
Indoplanobis exustes	+	–	–
Pelecypoda			–
Sphaeriidae			
Pisidium sp.	+	+	–
Number of Taxons = 16	15	6	4

+ = Present

– = Absent

Of all the macrobenthic group molluscs were found to be dominant organism observed throughout the duration of present studies. Molluscs constitute about 47% of total macrozoobenthos at littoral bottom, 62.5% at sublittoral and 50.75% at profundal bottom of the lake. Among molluscs *Gyarulus convexiusculus* is the most abundant species in whole lake profile. Among insects Ephemeropteran & Dipteran are dominant in occurrence. Ephemeropterans, found only at the littoral zone, constitute about 4.5% of the total macrobenthic community. Chironomid larvae have been recorded at all the three zones of the lake but chaoborus larvae observed occasionally were restricted to the sublittoral and profundal bottom (even far out below profundal zone). The percentage occurrence of dipterans is 27.5% at littoral, 37.5% at sublittoral zone and 50.75% at profundal bottom of the lake. Oligochaetes, occurring in lower numbers, have been recorded at littoral zone only and constitute about 12% of total macrozoobenthos of littoral zone in lake Mansar.

(iii) Characteristics of Macrobenthos densities: Data on average monthly density of macrobenthos; reveals a considerable variation in their occurrence in the lake. Density of zoobenthos at littoral bottom registers a bimodal maxima, first in the month of March and second in the month of August (Fig. 2).

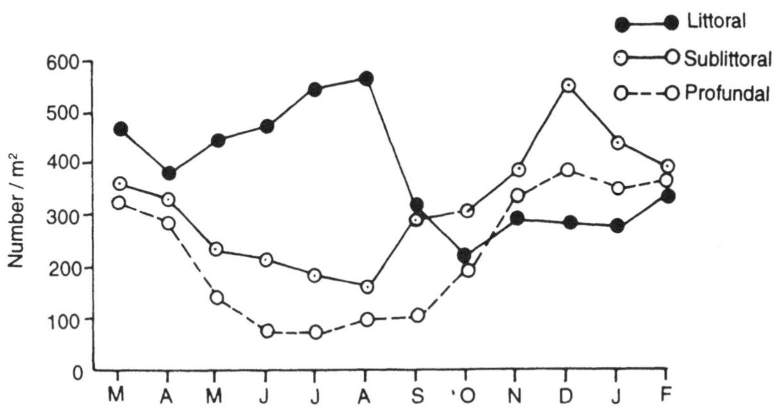

Fig. 2: Dynamics of Numbers of Macrobenthos at Littoral, Sublittoral and Profundal Zone of the Lake Bottom

Zoobenthic density at sublittoral zone, however, observed slight decrease with peak of population during winter (Fig. 2) and trough during summer and may be attributed to summer stagnation and decline in DO regime. At profundal zone density of benthic fauna decline significantly but maxima and minima of their population followed the pattern as recorded for sublittoral bottom (Fig. 2).

(iv) Characteristics of Macrobenthos biomass: Benthic biomass at littoral zone varied from 4.28 to 11.96 gm/m² at littoral zone, 3.39 to 11.92 gm/m² at sublittoral zone and from 0.83 to 5.75 gm/m^2 at profundal bottom. Biomass as observed presently undergoes similar seasonal fluctuations as observed for macrobenthic density (Fig. 3).

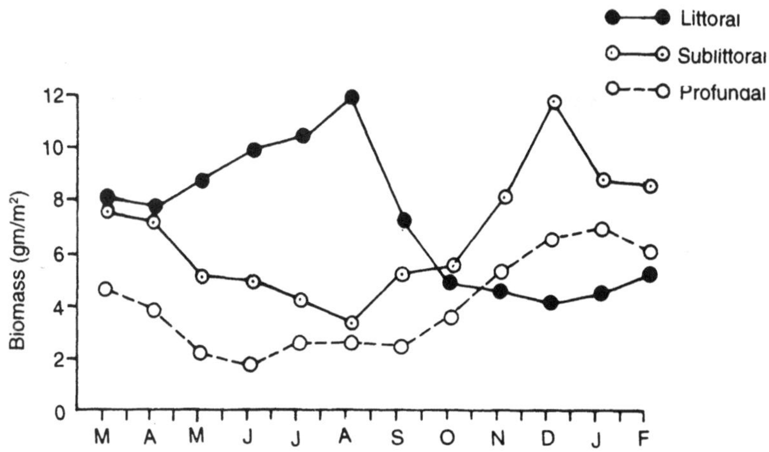

Fig. 3: Dynamics of Macrobenthos Biomass (Ash wt.) at Littoral, Sublittoral and Profundal Zone of the Lake

(v) Species diversity index: Shanon's diversity index (H) values of benthos have been shown in Table 3 for four sampling stations along the lake. Highest values (2.8747) were recorded from station 1 and lowest (2.6594) for station 2. Average species diversity index reveals decline from littoral through sublittoral to profundal bottom of lake Mansar.

Discussion

Present data along lake profile reveals that littoral zone has richest macrozoobenthic fauna. The predominance of benthic fauna at this zone can be attributed to the aquatic vegetation and high amount of DO. The direct relationship of aquatic vegetation with increase of benthic organisms has been projected by Lundbeck (1927), and Singh (1989). Present observations that DO plays an important role in distribution of benthic organisms in lake profile are in tune with those of Pennak (1989) who also advocated DO to be a limiting factor in the distribution of benthos in lake profile.

Table 3: Species diversity Index (H) Values of Macrobenthos In the Lake profile of Lake Mansar at four Sampling Stations

Zone	*Diversity Index Stations*			
	1	*2*	*3*	*4*
Littoral	2.8747	2.6594	2.6949	2.7302
Sublittoral	1.7880	1.3709	4.5848	1.4688
Profundal	1.5136	1.2796	1.4425	1.3597

On the basis of seasonal studies on benthic organisms at all the three zones of the lake, it has been observed that there is an increased depauperation of bottom fauna when studies are extended from littoral through sublittoral to profundal zone. Earlier Kajak & Dusoge (1975) also recorded similar trend from littoral to profundal bottoms in some polish lakes.

Present studies have not any fauna unique to profoundal zone, rather it is observed to be composed of few species which actually belong to littoral and sublittoral population and can easily tolerate the physico-chemical environmental attributes of this zone. Welch (1952) indicated that profundal population is the result of selective elimination where only few highly tolerant species can survive the specific environment.

As revealed by Fig. 2 highest macrobenthic density at littoral bottom exhibits a bimodal peak, first in the month of March and second during August. Earlier Michael (1968) documented maximum Zoobenthic density between January and April from a pond. Vasisht & Bhandal (1979) noticed summer and monsoon peaks. Yadava *et al.* (1984) noticed count of zoobenthos in the month of May. Kulshreshtha *et al.* (1991) observed the peak of benthos during postmonsoon. On the basis of comparison with available information from literature, it may be concluded that variable abiotic and biotic environmental factors prevalent in diferent areas are predominantly responsible for the difference in peak population of benthic organisms.

Benthos (all component species) at both sublittoral & profundal zones of the lake exhibit a unimodal peak in winter and trough in summer (Fig. 2). Maximum benthic volume during winter at these zones appears to be the outcome of winter overturn which results in the mixing of lake water. As a consequence of winter overturn appreciable quantity of DO is observed even at deeper zones of the lake which is favourably withstood resulting rich faunal composition. The summer minima, however, is possibly due to the summer stagnation period which causes depletion in DO in the benthic environment. Welch (1952) similarly also observed unimodal winter maxima and summer minima in benthic environment of third sister lake Michigan.

Similar to macrobenthic density, biomass of benthos shows direct correlation with peaks and trough recorded in the same months. Earlier Gupta and Pant (1983), Mortensen and Simonsen (1983) and Sunder & Subla (1986) reported the maxima and minima of density and biomass in the same months.

Olive (1971) stated that pollution status of water body is related to diversity index of benthic organisms. He suggests that when diversity value is > 3 the water is of high quality and harbours very few pollution tolerant species, and when diversity value is between 1–3, water is moderate water quality and such water harbours pollution tolerant species. He further suggests that water with species diversity < 1 is poor in quality and harbours 90% pollution tolerant species. Presently the diversity index in lake remain between 1.2796 to 2.8747 & thus water of this lake conforms to the 2nd category of Olive (*op. cit.*) viz. water is of moderate quality. However, Station No. 2 lowest values (Table 3) of species diversity index have been recorded which indicate benthic fauna to be under stress due to lot of human interference in the area being close to human habitation of village Mansar along the bank of the lake.

Sehgal (1980) and Wisniewski and Dusoge (1986) associated oxygen deficiency (at profundal bottom) and presence of Chaoborus larvae with the eutrophication of the water body. In the present case Chaoborus larvae have been recorded very occasionally in oxygen deficient (0.65 mg/lit.) environment. On this basis and species diversity index (Table 3), the present authors propose the lake under present investigation to be on way to eutrophication.

Acknowledgements

The junior author is highly thankful to Department of Environment and Forests, for financial assistance.

References

American Public Health Association (1975). *Standard Methods for the Examination of water*. APHA, AWWA, WPCF, 14th Ed. New York.

Adoni, A. D. (1985) . *Workbook on Limnology*. Indian MAB Committee. Govt. of India, New Delhi.

Gupta, P. K. (1979). Macrobenthos of freshwater fish pond, *Geobios* 6, 19–20.

Gupta P. K. and M. C. Pant (1983). Seasonal variations in the energy contents of benthic macro-invertebrates of lake Nainital, U. P. India *Hydrobiologia* 99, 19–22.

Kulshrestha, S. K., M. Srivasta, M. P. George, R. Saxena, A. Tiwari and M. Johri (1991). Seasonal variations in macrozoobenthic organism of Mansrover reservoir, Bhopal. *Proc. Nat. Acad. Sci. India*, 61 (B), 153–162.

Kajak, Z. and K. Dusoge (1975). Macrobenthos of lake Lal Lowisko. *Ikol. Pol.* 23, 295–316.

Lundbeck, J. (1927). Der "Fb–koeffcient" fur Tieche". *Zeitschr. f. Tischerei Bd. XXV. 4eft.* 4, 551–575.

Malhotra, Y. R., K. Gupta, and Anil Khajuria (1990). Seasonal variations in the population of Macrozoobenthos in relation to some physico-chemical parameters of lake Mansar, Jammu. *J. Freshwater Biol.* 2(2), 123–128.

Mandal, B. K. and S. K. Moitra (1975). Studies on bottom fauna of freshwater fish pond at Burdwan. *J. Inland Fish. Soc. India* 7, 43–48.

Michael, R. G. (1968). Studies on bottom fauna in tropical freshwater fish pond. *Hydrobiologia*, 31(2), 203–230.

Michael, P. (1985). *Ecological Methods for Field and Laboratory Investigations*. Tata McGraw–Hill Publishing Company Limited, New Delhi.

Mortansen, E. and J. L., Simonsen (1983). Production estimates of benthic invertebrate community in small Danish stream. *Hydrobiologia*, 102, 155–162.

Olive, J. H. (1971). Ohio Biol Serv. and water resource centre, *The Ohio State Univ. Res. Project Completion Report*.

Pennak, R. W. (1989). *Freshwater invertebrates of United States* 3rd Ed. A Wiley–Interscience Publication New York.

Shannon, C. E. and W. Weaver (1964). The mathematical theory of communication. Univ. Illincois Press.

Sehgal, H. S. (1980). Limnology of lake Surinsar, Jammu with special reference to Zooplanktons and fishery prospects. *Ph. D. Thesis, University of Jammu, Jammu*.

Singh, D. N. (1989). Studies on weed associated macrofauna of an ox-bow lake. *Proc. Nat. Acad. Sci., India*, 59(B), 271–278.

Sunder, S. and B. A. Subla (1986). Macrobenthic fauna of Himalaya river. *Indian J. Ecol.* 13(1), 127–132.

Vasisht, H. S. and R. S. Bhandal (1979). Seasonal variations of Benthic fauna in some Indian lakes and ponds, *Indian J. Ecol. 6(2)*, 33–37.

Welch P. S. (1952). *Limnological Methods*. McGraw-Hill Book Co. Ltd., New York.

Wisniew R. J., and K. Dusoge (1986). Macrobenthos of lake Zarnowieckie. Pol. Ec*ol. Stud. 12(3)*, 331–346.

Yadava, Y. S., V. Kolekar, R. K. Singh and M. Choudhary (1984). Studies on the macrobenthic fauna of Dighali Beel (Asam) *Proc. Nat. Acad. Sci. India, 54 (B)*, 177–184.

Chapter 21

Population Structure and Seasonal Succession of Zooplankton of Lake Surinsar, Jammu (India)

M. K. Jyoti and H. S. Sehgal***

Introduction

The sporadic rise in human population and the inability of most modernized and sophisticated agricultural techniques to meet the existing and future nutritional demands has impelled Scientists to look for viable alternate sources to meet the challenge of food insufficiency. Successful exploitation of aquatic resources (which cover a major part of earth's surface - 71%) seems to be a promising solution. The management of Inland Waterbodies towards maximizing the yield of a desired species is beneficial over marine fishery because of easy manipulation and proper returns with minimum inputs.

Jammu and Kashmir State and extensively hilly area, is bestowed with number of lentic and lotic waterbodies. Though the limnology of the waterbodies of Kashmir has been studied in some details, Jammu region remained almost virgin except the studies made by Anand (1977) and Dutta (1978) which too pertain to a stream. Keeping this in view, limnological studies on lake Surinsar were undertaken as sample studies for the lentic waterbodies of substropical areas so that management and development plans of culture fisheries could be successfully formulated.

The present paper, however, includes a part of the work i.e. the community structure and seasonal succession of zooplankton which forms an extremely important link in the food chain of the most of fishes and are largely responsible for determining the trophic status or the productivity potential of the water body. Seasonal changes in the abundance, migrations and movements of the zooplankton communities within the lake and the factors, both physico-chemical and biological, that are found controlling such changes have been discussed.

The lake—A Brief Description

The lake Surinsar (Fig. 1) is a freshwater substropical lake situated at 75° 2' 30"

* P.G. Department of Biosciences, University of Jammu, Jammu-180001.

** Regional Research Centre, Punjab Agricultural University, Ropar- 141004, India.

(E) longitude and 32° 46' 30" (N) latitude. It lies at an elevation of about 605 m above mean sea level about 40 km north-east of Jammu city. The maximum and the mean depths of the lake are 24.85 and 12.30 m respectively in the rainy season when it is fully swollen. During this period (monsoon) lake Surinsar has a storing capacity of 3,627, 170m^3 and covers a surface area of 29.4 hectares. The lake is warm monoholomictic i.e., it remains thermally stratified for about 9 months in a year (March to November) and shows single overturn during winter (December to February).

Material and Methods

1. The Study Stations

To have an overall idea of the zooplankton dynamics at the littoral and limnetic stations and also the physico-chemical factors various study stations were established. At surface, 9(1-9) stations were established at the littoral zone (8 along the lake margin and 1 close to the island) and 6 (10-15) at the limnetic zone. The selection of stations was based on seeming differences in topography, soil conditions and floral characteristics of littoral zone which could possibly lead to the variations in physico-chemical features and zooplankton communities. The limnetic stations were selected with an idea of their distance from the shore and mianly to cover the entire lake surface.

2. Sampling and physico-chemical conditions of water

For collecting water samples, the stations were approached with the help of an ordinary row boat. The boat was made stationary by tying it with a nylon rope, anchored across the lake margins or lake margins and the island. Water samples from surface were collected by dipping the sampling bottle in lake water, observing all precautions necessary for the estimation of a particular element.

Measurements of physico-chemical parameters like temperature, pH, dissolved oxygen, free carbondioxide, carbonates, bicarbonates and chlorides were done on the spot while the water samples for the analysis of sulphates, water soluble phosphate, nitrate nitrogen, hydrogen sulphide, calcium, magnesium and total iron were fixed in the field by adding 2-3 drops of Chloroform in every 500 ml of water sample and were brought to the feild laboratory for further analysis. All samples were estimated within 10-12 h of their collection. The sampling was done between 0600 and 1100 h and for every station the fixed sampling time was always observed. An average of three readings at one station was taken as the result.

The temperature at the surface layers was recorded with an ordinary celsius thermometer.

The chemical factors such as dissolved oxygen, free carbon dioxide, carbonates, bicarbonates, alkalinity, water soluble phosphates, nitrate-nitrogen were estimated by the mothods recommended by Rend et. al. (1976). The pH was estimated with a portable 'Elico' pH meter (Model L1-11).

Zooplankton

Zooplankton were collected by filtering 50 litres of lake water through a silk (crave) conical net (900 meshes/cm^2) with a vial attached at its cone in which the zooplankton become concentrated. Since medusae escaped plankton net, because of the narrow mouth of the vial and also due to their sticking with the net cloth, they were collected by filling plastic buckets (each of 16 l capacity) with lake water. However, those from below the surface were collected by the water sampler.

The zooplankters except medusae thus collected were killed by dropwise addition of formalin till its concentration reached to 5% in which they were preserved. The medusae were killed, relaxed and were fixed by sudden pouring of Dawar's solution (200 ml distilled water + 2 ml conc. HNO_3 + 4.5 ml of 36-40% formalin + 2.59 $MgSO_4$) at room temperature when they just started relaxing after the contracted posture.

Preserved samples, from the field, were brought into the laboratory and for purposes of taxonomy were washed several times in water. All zooplankters except medusae were mounted in a mountant prepared by mixing equal parts of acetophenol (lactic acid 100 ml + Phenol 100g + glycerol 100 ml + distilled water 100 ml in which a drop of blue-black ink was added) and Reny's mountant (chloral hydrate 50g + gun arabica 30g + glycerol 12.5 ml + distilled water 50 ml). The identifications were based on these whole mounts. The microdissections of crustacean appendages were made under stero binocular microscop. With the help of fine needless mounted on knitting rods. For taxonomic confirmation, thorough study of the living specimens was also made. Medusae were stained with aqueous alum carmine, dehydrated and mounted in DPX. The identifications of various genera and species of protozoa, cladocera, copepoda and ostracoda followed Ward and Whiple (1959). Medusa was identified after Malhotra *et. al.* (1976) while the taxonomy of rotifers followed mainly Arora (1963, 1964, 1965 and 1966), Nayar (1968), Fernando (1969), Chengalath and Fernando (1973), Chengalath *et. al.* (1973).

The quantitative studies were made by centrifuging the sample and concentrating these to 5ml. One ml of the concentrate was filled in the Sedgwick Rafter (S-R) cell. Allowed the S-R cell to stand for 15-20 minutes for the settling of plankton. The plankton was then counted by the random field count method (Rand *et. al.*, 1976). Individuals in fifty random fields were counted. The number of zooplankton per ml of the concentrate was calculated by the formula:

$$\text{No./ml} = \frac{C \times 1000\ \text{mm}^3}{A \times D \times F}$$

Where C is the number of organisms count; A is the area of a field; D is depth of the field in mm (S-R cell depth = 1 mm) and F is the number of fields counted. For relating these data back to the content of original lake water, the number obtained by the above formula was divided with a correction factor obtained by the following formula (Lind, 1974).

$$\frac{\text{Volume of lake water filtered}}{\text{Volume of concentrate (ml)}}$$

Results and Discussion

Following taxa belonging to different groups were recorded during the course of present investigations :—

Protozoa

1. *Centropyxis aculeata* (Ehrenberg, Stein)
2. *Centropyxis* sp.
3. *Difflugia lebes* Penard
4. *Difflugia* sp.

Coelenterata

5. *Mansariella lacustris* Malhotra *et. al.*

Rotifera

6. *Anuraeopsis fissa* (Gosse)
7. *Brachionus calycflorus* Pallas
8. *B. patulus* Mueller
9. *B. quadridentatus* Hermann
10. *B. angularis* Gosse
11. *Brachionus sessilis* Varga
12. *Kera-tella tropica* (Apstein)
13. *Platyias quadricornis* (Ehrenberg)
14. *Mytilina ventralis* (Ehrenberg)
15. *Euchlanis dilatata* (Ehrenberg)
16. *Tricholoria tetractis* (Ehrenberg)
17. *Colurella* sp.
18. *Lepadella ovalis*
19. *Lecane (Monostyla) bulla* (Gosse)
20. L. *(M) decipisna* (Murrey)
21. L. *(M) clasterocerca* Schmarda
22. L. *(M) ludwiqi*
23. L. *leontina*
24. *Trichocerca similis* Wierzeiski
25. *T. stylata* Gosse
26. *Hexarthra mira* Hudson
27. *Polyarthra vulgaris* Carlin
28. *Filinia opliensis* Zacharias
29. *Testudinella patins* Herman

Cladocera

30. *Daphnia* spp.
31. *Ceriodaphnia cornuta* Sars.
32. *C. reticulata* Jurina
33. *Simocephalus vetulus* Schodler
34. *Scaphelebris* sp.
35. *Alona* sp.
36. *Alonella globulosa* Daday
37. *Oxyarella tenuicaudis* (Sars.)
38. *Chydorus sphaericus* (Muller)
39. *C. eurvnotus*
40. *Graptelebris testudinaris* (Fischer)
41. *Pleuroxus hastetus* Sars.
42. *Macrothrix rosea* (Jurine)
43. *M. latiocornis* (Jurine)

Copepoda

44. *Mesocyclops leuckarti* (Claus)
45. *Cyclops (Diacyclops) navus*
46. *C. (microcyclops)* sp.
47. *Eucyclops aqilis* (Koch)
48. *F. speratus* (Lillijeberg)
49. *Tropocyclops prasinus* (Fischer)
50. *Tropocyclops* sp.
51. *Ectocyclops* sp.

Ostracoda

52. *Cypricercus* spp.
53. *Eucypris* sp.

Insecta

54. *Chaoborus* sp. (larval forms)

Seasonal Succession and Community Structure

Of various zooplankters that form the zooplankton community, rotifers dominate by contributing about 56.31% at littoral and 51.22% at the limnetic zone. Next to rotifers in abundance, are the copepods which contribute to the tone of 29.41% and 31.32% at the littoral and limnetic zones respectively. Cladocera contribute mainly at the littoral zone (10.02%), protozoa, coelenterata and Ostracoda represent poorly

(Tables 1 & 2 : Figs. 2 & 9). A similar observation for lake Mize where rotfiers dominated zooplankton has been made by Nordlie (1976). Contrary to this, however, Seenayya (1973) recorded cladoceran dominance in fresh water ponds of Hyderabad.

Although rotifers and copepodes contribute equally to the zooplankton both at the littoral and limnetic zones, the zooplankton community structure at the two zones is different as coelenterates are present only at the limnetic and ostracods only at the littoral zone. Further at littoral zone, the maximum contribution to rotifers is made by *Keratella* (36.21%) (Fig. 5-8) whereas at the limnetic zone, it is by *Brachionus* (37.36%) (Fig. 5-8). Similarly among copepode *Cyclops* contribute maxmally (34.15%) at the littoral (Fig. 8) and immature copepod stages (63.34%) at the limnetic zone of the lake (Fig. 7).

Quantitatively, there are statistically significant differences between total zooplankton (Table 3) at the littoral and limnetic zones (t = 3.76, p 0.01). These differences are due to significant differences in coelenterata (t = 4.52; p 0.01), rotifera (t = 2.18; p 0.55), Cladocera (t=3.93, p 0.01), Copepoda (t=2.91; p 0.01), and Ostracoda (t=3.72; p 0.01) between two zones of the lake. Protozoa, however, do not show significant differences between littoral and limnetic zones of the lake.

Total zooplankton observe three well marked maxim at the lttoral zone Table 1 (Fig. 4). The first maxima is achieved in winter (February) after a continuous rise in zooplankton from October onwards. The October high values are due to rotier dominance. In November, both rotifers and copepods contribute whereas in December cladocera replace rotifera. Although in January, the rotifers contribute least, the zooplankton rise is maintained by major contribution of copepoda; there is a decline in cladocerns too. The peak values of winter (February) as well as those of other two peaks (one observed in spring i.e., April and the other in monsoon i. e., August) are caused largely by the outburst of rotifier number.

Table 1 : Mean seasonal variations in % contribution of various groups to total zooplankton (TZ) at littoral zone of lake Surinsar

Month	*Protozoa*	*Rotifera*	*Cladocera*	*Copepdoda*	*Ostracoda*	*TZ*
January	0.0	0.53	20.64	74.86	2.82	98.85
February	0.0	64.25	8.69	26.32	0.43	99.26
March	0.15	23.45	29.62	45.93	0.73	0.61
April	1.78	90.93	3.39	3.30	0.00	0.31
May	0.77	50.06	6.30	33.34	9.44	0.64
June	14.38	42.26	2.16	34.80	3.74	0.86
July	13.98	65.74	0.55	19.89	0.00	0.46
August	3.14	92.60	0.06	4.17	0.00	0.27
September	5.38	31.64	6.2	45.60	10.48	0.80
October	1.11	83.67	0.6	14.40	0.12	0.50
November	0.00	56.38	9.24	33.74	0.22	0.44
December	0.00	11.60	42.95	44.90	1.05	0.54
Mean	3.99± 5.07	51.13± 29.11	10.86± 12.86	31.77± 19.32	2.42± 3.56	

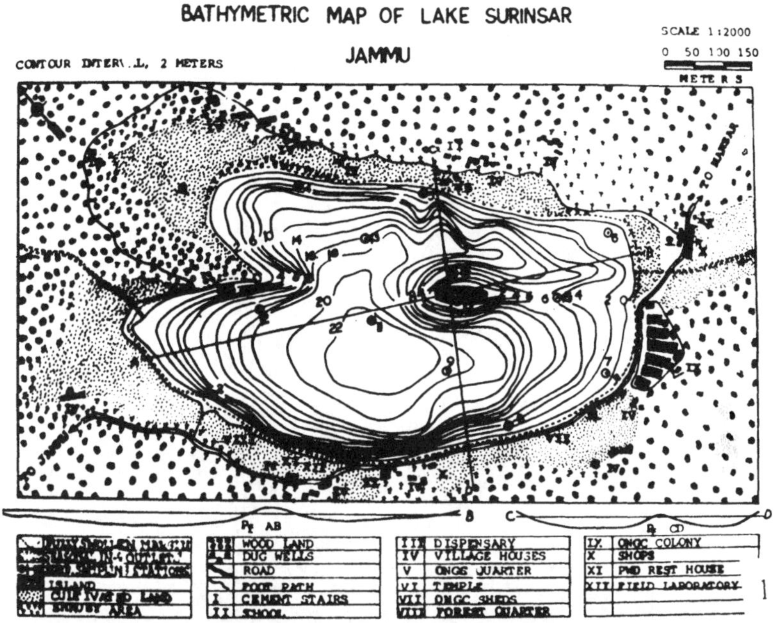

1

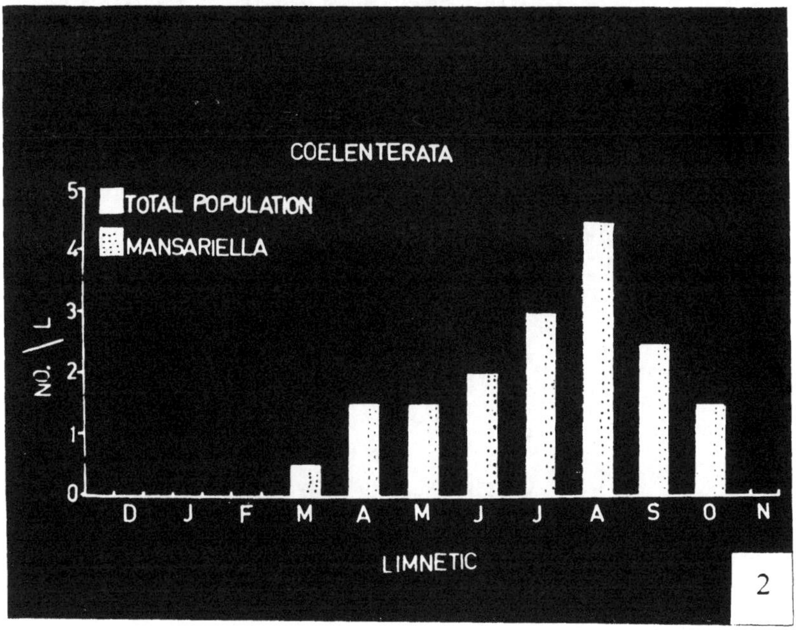

2

Fig. 1: Bathymetric map of lake Surinsar, Jammu.

Fig. 2: Seasonal variations of Coelenterates in the limnetic zone of lake Surinsar.

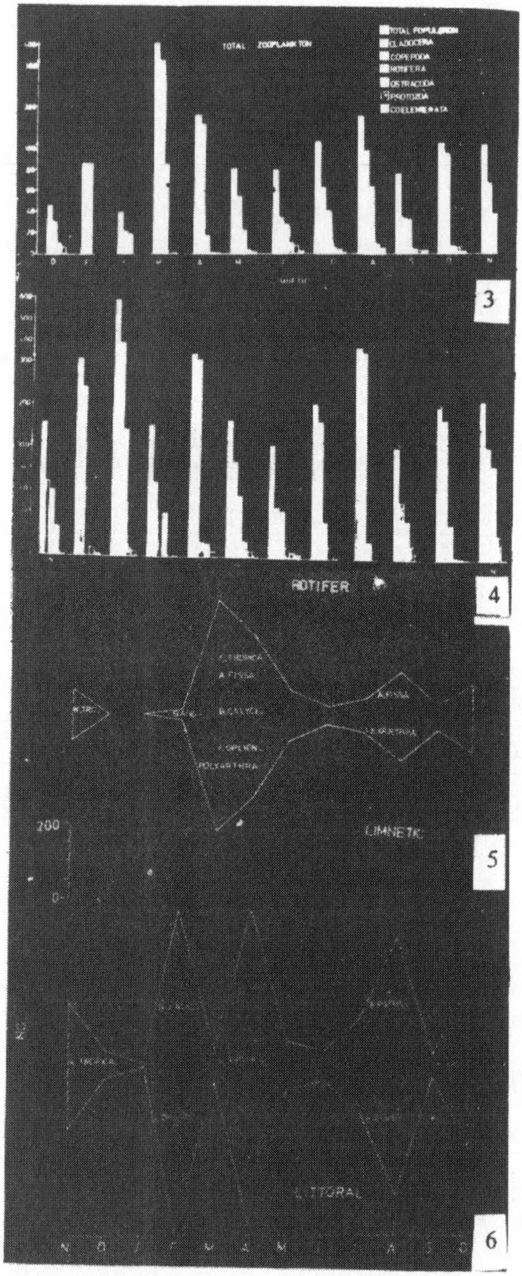

Fig. 3-4: Seasonal variations of total zooplanktons in limnetic (Fig. 3) and littoral (Fig. 4) zones of the lake.

Fig. 5-6: Seasonal variations of Rotifers in the limnetic (Fig. 5) and littoral (Fig. 6) zones of the lake.

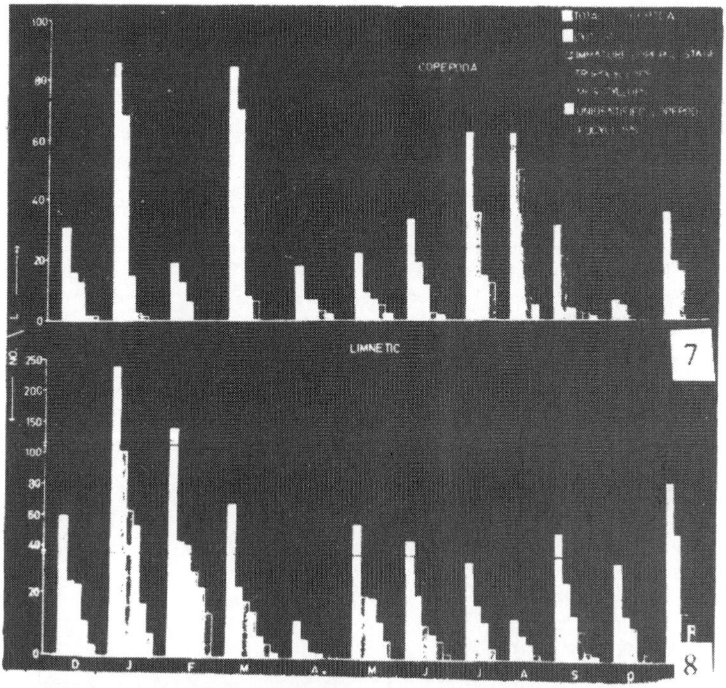

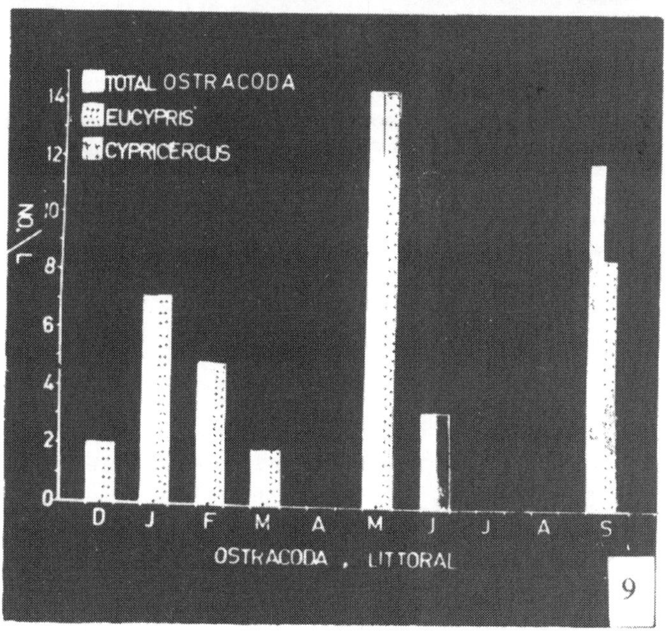

Fig. 7-8: Seasonal variations in copepoda population in the limnetic (Fig. 7) and littoral zones (Fig. 8) of lake Surinsar.

Fig. 9: Showing Ostracoda population in the littoral zones of lake Surinsar.

Limnetic zone, on the other hand, observe two zooplankton maxima (Table 2). The bigger one is observed in spring (March) and a comparatively smaller in monsoon (August). Although the spring peak is mainly due to rise in rotifier population, the monsoon peak is caused by the combined effects of Rotifera and Copepoda though Rotifera (61.9%) still represent double than the Copepoda (30.86%) (Fig. 3).

Table 2 : Mean seasonal variations in % total contribution of various groups to Total Zooplankton (TZ) at limnetic zone of lake Surinsar

Month	*Protozoa*	*Coelenterata*	*Rotifera*	*Cladocera*	*Copepoda*	*TZ*
January	0.00	0.00	115.95	0.48	83.42	0.97
February	0.00	0.00	45.57	3.55	50.59	2.45
March	1.02	0.07	79.91	0.00	19.73	0.31
April	1.22	1.52	89.75	0.00	8.61	0.58
May	0.00	2.20	63.25	4.98	29.04	1.26
June	12.9	2.68	34.90	11.05	38.59	1.34
July	23.71	1.75	32.90	0.31	40.88	0.62
August	4.07	2.56	61.94	0.00	30.86	0.54
September	5.44	3.57	44.65	4.89	39.97	1.30
October	0.00	0.83	86.19	4.77	7.88	1.11
November	0.61	0..00	65.25	0.27	26.40	0.87
December	0.00	0.00	19.57	6.31	73.57	2.25
Mean	4.08±	1.26±	54.12±	3.04±	37.67±	
	6.94	1.23	25.06	3.38	21.94	

Table 3: Table of significance of differences in zooplankton number between littoral and limentic zones of the lake (on the basis of student's t-test)

Zooplankton group	*'t'*	*'p'*
1. Total Zooplankton	3.76	0.01
2. Rotifera	2.18	0.05
3. Cladocera	3.93	0.01
4. Copepoda	2.91	0.01

As the seasonal succession of zooplankton of alke Surinsar is effected by the changes in the community structure of various groups mainly the Rotifera and Copepoda, it is imperative to look into these changes before assigning the ultimate physico-chemical and biological factors responsible for seasonal succession. As Protozoa, Coelenterata, Cladocera and Ostracoda contribute insignificantly to the total zooplankton of this lake (Table 1-2), the need of giving a detailed account of their seasonal succession is not felt essential here. An account of the seasonal succession of Rotifera and Copepoda is presented below:-

Rotifera

Of the various rotifers occurring in lake Surinsar only two (*Keratella* and

Anuraeopsis) and perennial. Some (*Brachionus, Lecane* and *Trichocerca*) show a considerably long existence while others (*Platyias, Mytilina, Euchlanis, Trichotoria, Lepadella, Hexarthra, Polyarthra, Filinia* and *Testudinella*) occur for short spells only. The present observation made for *Keratella* and *Brachionus* differs from that of Timms (1970a) who recorded their presence for short periods in some reservoirs of New South Wales Cowell *et. al.* (1975), albeit regarded *Polyarthra* to be perennial in Canadian and New Zealand waters but Carlin (1943) Schindler and Noven (1971) found it to be seasonal in ELA lakes as we found it true for lake Surinsar.

Although each individual rotifier genus undergoes seasonal changes and records considerable numerical fluctuations (Sehgal, 1980), yet total rotifer number (TRN) observes four distinct maxima at the littoral (Fig. 6) and three at the limnetic zone (Fig. 5). Increases in numerical values of one or at best two forms are found to be responsible for most of these maxima in lake Surinsar. This observation is clearly counter to one made by Cowell *et al.* (1975) that TRN maximum is invariably a combined effect of five or six forms, from amongst the many constituting the rotifer population in lake Thonolossasa.

At littoral zone (Fig. 6), the winter (February) and monsoon.(August) maxima are caused by the numerical increase in population of *Brachionus* and *Lecane* and *Brachionus* and *Anuraeopsis* respectively. The other two maxima, recorded in spring (April) and autumn (October to November) are largely contributed by *Keratella*.

Peaks at limnetic zone, on the other hand, are controlled by *Brachionus* in autumn (October to November) and *Anuraeopsis, Hexarthra* and *Polyarthra* in monsoon (August) (Fig. 5).

Summer rotifer peaks have been noted for waterbodies located both for distant (Lake thonolossasa, Cowell *et al.*, 1975 and Canadian Sheild Lakes, Schindler and Noven 1971) and near the site of present investigations (Ponds of Delhi, George, 1961) George's (*op. cit.*) view that summer Rotifer maxima is characteristic of tropical waters could not be confirmed for lake Surinsar. Here, instead, a minimum has been recorded in summer although the lake is subjected to almost near similar climatic conditions as his (George's 1961) site of study.

Copepoda

Among Copepoda inhabiting lake Surinsar, *Eucyclops* and *Ectocyclops* are seasonal whereas *Cyclops, Mesocyclops,* and *Tropocyclops* and the immature copepod stages are perennial. *Cyclops, Mesocyclops* and *Tropocyclops* are the abundant forms contributing 34.1, 20.6 and 18.3% respectively to the annual mean number of total copepod number (TCN) at the littoral zone of the lake (Fig. 8).

The TCN at the littoral zone reaches one well marked maxmum in winter (January) and two smaller and insignificant ones during summer (May) and post-monsoon (September) (Fig. 8) the winter copepod peak has also been noted by Plew and Robert (1949) in an Indian lake.

At the limnetic zone, TCN exhibit three distinct peaks in winter (January), spring (March) and summer (July-August) (Fig. 7).

It is quite interesting to note that all maximum (whether distinct or insignificant) noted at the littoral zone are contributed by the combined effects of all copepod forms living in this lake rather than a single genus or a group of genera as recorded by Welch (1952), Hutchinson (1967) and Wetzel (1975). This co-existence of all the copepod genera and the occurrence of their maxima probably suggests their selective needs and limited adaptability to the environmental conditions.

Causes of Seasonal Succession

Seasonal succession of total zooplankton (TZ) in lake Surinsar as in any other waterbody are controlled by the patterns of succession and community structure of various groups that contribute to it. The ultimate controlling factors are the physico-chemical and biological characteristics which control the seasonal dynamics at the species level. Since, the major contribution to TZ of lake Surinsar by Rotifera and Copepoda, the major ecological factors that govern the seasonal changes in the members of these two groups and hence the TZ are discussed here.

1. Temperature

Temperature is an important ecological factor which affects many viable biological phenomena and thus regulates seasonal numerical fluctuations of organisms. The population dynamics of almost every zooplanktonic group of lake Surinsar appears to be regulated by rhythmic thermal changes (Table 4). But here, we shall concentrate only on two groups (Rotifera and Copepoda) that contribute mainly to the TZ. Besides, the mention shall be made of other groups wherever required.

1.1 Rotifera

Many of the rotifera appear to be temperature sensitive and are stenothermal. Some (*Trichotoria, Testudinella, Euchlanis*) occur exclusively during low temperature (12.6–16.7°C) and others (*Hexarthra, Mytilina, Filinia*) show their presence during warmer months when temperature is high (Sehgal, 1980). However, still others (*Keratella, Brachionus, Anuraeopsis, Lecane*) which contribute considerably to TRN (Fig. 3.4) and also *Trichocerca, Lepadella* and *Polyarthra* are eurythermal. Contrary to present observations, Robinson and Robinson (1971) have recorded rotifers like *Keratella* as warm stenothermal and *Hexarthra* as cold stenothermal in lake Chad though temperature, tolerance range of *Keratella* besides above mentioned eurythermal forms has been recorded to be 12.6–32.4°C in lake Surinsar. This substantiate the experimental findings of Arora (1966) for *Brachionus* and *Keratella* which showed their temperature to tolerance range between 12 and 27°C. It appears, therefore, that rotifers as a group are temperature facultative and only marked temperature. Changes on either side of optima can influence their numerical abundance. That temperature does not govern the community structure of rotifers which are largely responsible for the community structure of TZ in lake Surinsar is evident by the absence of *Brachionus* and *Lecane* along with many other rotifers at the littoral zone during December to January and their outburst in February causing even a TZ maximum despite no significant difference between water temperature during December and February (Fig. 5,6). *Keratella*, although a eurythermal form which can withstand a wide range of temperature, variations, obeys temperature conditions as for as its numerical abundance is concerned. This is demonstrated by an outburst in its number in April leading to a spring maxima of TZ and also a considerable rise in October due to favourable temperature Zooplanktons Inhabiting conditions. The temperature almost same during April and October.

Table 4: Thermal Classification of Zooplanktons Inhabiting Lake Surinsar

Plankton	*Period of existence*	*(Temperature) tolerance range (°C)*	*Thermal status*	*Remarks*
PROTOZOA	July	8	Warm-stenothermal	
1. *Difflugia*	(March–Oct.)	(24–32)		
2. *Centropyxis*	Perennial	20 (12–32)	Eurythermal	
COELENTERATA		7		
3. *Mansariella*	(March–Oct.)	(24–31)	Warm-stenothermal	
ROTIFERA				
1. *Trichotoria*	February	4 (13–17)	Cold-stenothermal	1. Contrary to present observations, Robinson and Robinson (1971) have recorded. (i) *Keratella*–warm-stenothermal; (ii) *Hexarthra*–cold-stenothermal in lake Chad though *Filinia* has been regarded warm-stenothermal in both the lakes
2. *Testudinella*				
3. *Euchlanis*				
4. *Hexarthra*	June–Sept.	20 (12–32)	Warm-stenothermal	
5. *Mytilina*	June & Oct.			
6. *Filinia*	Aug. – Nov.			
7. *Brachionus*	Feb-Apr–July/Oct.	20 (12–32)	Eurythermal	2. Present observation in rotifers supports the experimental data of Arora (1966); for Eurythermal forms wherein he reported temperature tolerance limit from 12–32°C.
8. *Lecane*	Feb. – Nov.			
9. *Trichocerca*	March–Nov.			
10. *Lepadella*	Jan., Sept., Nov.			
11. *Polyarthra*	July–Aug.			
12. *Keratella*	Feb. – Dec.			
13. *Anuraeopsis*	Absent only in Jan.			
CIADOCERA				
14. *Graptolebris*	Dec.–March	(14–16)	Cold-stenothermal	Contrarily, Seenayya (1973) reported *Macrothrix* to exist at 40°C.
15. *Pleuroxus*				
16. *Chydorus*				
17. *Macrothrix*				
18. *Leydigia*				
19. *Daphnia*				

Contd...

Plankter	Period of existence	Temperature tolerance range (°C)	Thermal status	Remarks
20. *Alona*	Dec.–May & Sept.			
21. *Simocephalus*	Nov.–April	19		
22. *Alonella*	Feb.–May	(12–31)	Eurythermal	
23. *Oxyarella*	Nov.–June			
24. *Ceriodaphnia*	Absent in Aug.			
COPEPODA				Temperature facultative (Raylon 1948) regarded warm-stenothermal
25. *Eucyclops*				
26. *Cyclops*		(14–16)	Cold-stenothermal	– Peak in winter
27. *Mesocyclops*				– Peak in winter
28. *Tropocyclops*				– Peak in winter
OSTRACODA		15		
29. *Eucypris*	Dec.–May	(16–31)	Eurythermal	
30. *Cypricercus*	June–Sept.	(31–34)	Warm-stenothermal	

2. Copepoda

Inspite of their sensitivity to temperature demonstrated by Armitage *et al.* (1973) and Clady (1977a), copepoda are generally considered temperature facultative (Timms, 1970a, Saha *et al.*, 1971; Seenaya, 1973; Ghosh *et al.*, 1974; and Cowell *et al.* 1975) and show their peaks during acute summer and winter. That temperature has a limiting effect on TCN in lake Surinsar is evident by a single well mark copepod maximum in winter (January) at the littoral zone (Fig. 8) when the contribution to TZ is maximum both at the littoral and the limnetic zone (Fig. 7), and a fall in their number and migration to a zone of low temperature during warmer months (Sehgal, 1980).

The immature copepod stage do not show any response to the changing conditions of temperature and show their peaks in almost every season. Therefore, the increased number of maxima and minima in TCN at the limnetic (Fig. 7) then at the littoral zone is due to such changes in immature copepod stages as they contribute 63.34% to the annual mean TCN at the former zone.

Dissolved Gases

Since dissolved oxygen (DO) and free carbon dioxide (FCO_2) are two sisteril chemical elements that go in hand in hand during seasonal changes, their effects on biological components are interwoven. During winter (December and January) when DO (Fig. 5-6) and temperature reach lowest (Fig. 5-6) and FCO_2 reaches highest (Fig. 5-6) all rotifers except *Keratella, Polyarthra* and *Lepadella* disappear. The reappearance of *Brachionus* and *Lecane* (which lead to the rotifer and hence TZ maximum and *Anuraeopsis, Trichotorai, Euchlanis* and *Colurella* in February (Fig. 5-6) when temperature is almost the same as in December frees temperature as a factor responsible for ther disappearance in December and January. Now of the two (DO and FCO_2). DO seems to be of major value as FCO_2 in February 1978 (when TRN maximum was recorded (Fig. 5-6) acquired almost the same concentration as it had in December, 1977.

3. Alkalinity

Absence of lecane from rotifer collection made during October to January (Fig. 5-6) when HCO_3 alkalinity varies between 125.7 and 150 mg l^{-1}) (Fig. 5-6) speaks for the intimate relationship of the said rotifers with alkalinity. Statistically a strong positive correlation has been found between *Lecane* and HCO_3 alkalinity (r=0.53, p 0.01).

Multifactorial Influence

The relationships that exist between organisms and their abiotic environment are complex and intricate. Most often a combination of two or more factors is the cause of numerical abundance and seasonal dynamics of a species or a group of species. This feature is very well exhibited by *Euchlanis* and among rotifers *Trichotoria* and *Eucyclops* among Copepoda which are jointly governed by temperature and alkalinity. Besides the seasonal dynamics of Coelenterata (*Mansariella lacustris*), *Cladocera* (*Simocephalus*, *Cerciodaphia* and *Peuroxus*), and Ostracoda (*Cypericercus*) is governed by two or more than two physico-chemical factors (Sehgal, 1980). Since these groups (Cladocera, Ostracoda and Coelenterata) and the members of Rotifera and Copepoda do not effect seasonal dynamics of TZ, the details are not discussed here.

To sum up, the seasonal succession of TZ and their community structure in lake Surinsar are effected largely by Rotifera followed by Copepoda. The ultimate factors that control numerical abundance of important rotiers (*Keratella*, *Brachionus* and *Lecane*) and Copepods (*Cyclops*, *Mesocyclops* and *Tropocylops*) are water temperature, dissolved oxygen and HCO_3 alkalinity. The immature Copepod stages which have a great impact on the seasonal dynamics of Copepods at the limnetic zone, however, do not respond to the abotic environmental factors, supporting the earlier contention of Macan (1963) and Timms (1970b) that many copepods can tolerate a wide range of chemical factors and their populations are effected by the appearance of food, and predation.

References

Anand, V. K. 1977. Limnology of Gadigarh stream (Miran Sahib, Jammu) with special reference to producers inhabiting the stream. Ph. D. thesis submitted to University of Jammu.

A. P. H. A. 1976. Standard methods for the examination of water and waste water. Washington, DC: 1193 pp.

Armitage, Kenneth, B., Bhagwan Saxena and Ernst, E. Angino. 1973. Population dynamics of pond zooplankton I. *Diaptomus pallidus* Herrick Hydrobiol., *42* (2–3) : 295-333.

Arnemo, Rolf. 1965. Limnological studies in Hyttodammen, 3. Zooplankton Inst. Freshwat. Res. Drottningholm. Report. No: 46: 200-275.

Arora, H. C. 1963. Studies on Indian Rotifera – Part II. Some species of the genus *Brachionus* from Nagpur. J. Zool. Soc. India, *15* (2): 112-121.

_________. 1964. Studies on Indian Rotifera–Part III. On *Brachionus calyciflorus* and some varieties of the species. J. Zool. Soc. India, 16 (1-2) 1-6.

_________. 1965. Studies on Indian Rotifera-Part IV. On a collection of Indian Rotifera – from Nagpur, India, with four new species and a new variety. Hydrobiol. *26* (3-4): 444-456.

_________. 1966a. Responses of Rotifera to variations in some ecological factors. Proc. Ind Acad. Sci. (Sec. B), *63* (2): 57-66.

_________. 1966b. Studies on Indian Rotifera-Part V. On species of some genera to the family Brachionidae, subfamily Brachioninae, from India. Arch. Hydrobiol., *61* (4): 482-493.

Carlin, B. 1943. Die Plankton rotatorien des Motalestroms: Zurtaxonomie un Okalogie der Planktonrotatorien. Medd. Lunds. Univ. Limnol. Inst., *5* 255 pp.

Chengalath, R. and C. H. Fernando, 1973. Rotifera from Sri Lanka (Ceylon) 1. The genus *Lecane* with description of two new species. Bull. Fish. Res. Stn. Sri Lanka, 24(1 &2): 13-27.

_________, and W. Koste. 1973. Rotifera from Srilanka (Ceylon) 2. Further studies on Eurotatoria including new records. Bull. Fish. Res. Stn. Sri Lanka, *24* (1): 29-62.

Clady, Michael, D. 1977a. Crustacean zooplankton populations and concurrent survival of larval yellow perch in Oneida lake. New York Fish Game Jour. *24* (1) : 46-52.

Cowell, Bruce C., Craig, W. Dye and Robert, C. Adams, 1975. A synoptic study of the limnology of lake Thonotosassa, Florida Part I. Effects of Primary7 treated sewage and citrus wastes. Hydrobiol., *46* (2) : 301-345

Dutta, S. P. S. 1978. Limnology of Gadigarh stream (Miran Sahib, Jammu) with special reference to consumers inhabiting the stream. Ph. D. thesis submitted to University of Jammu: 305 pp.

Fernando, C. H. 1969. A guide to the freshwater fauna of Ceylon. Bull. Fish. Res. Stn. Ceylon, *20*: 15-25.

George, M. G. 1961. Observations on the rotifers from shallow ponds in Delhi. Curr. Sci., 30: 268-269.

Gosh, Apurba, L., Hanumantha Rao and S. C. Banerjee. 1974. Studies on hydrobiological conditions of a sewage fed pond with a note on their role in fish culture. J. Inland Fish. Soc. India, VI: 51-61.

Hutchinson, G. E. 1967. A treatise on limnology Vol. II Introduction to lake biology and the limnoplankton. John. Wiley and Sons, 1115 pp.

Jyoti, M. K. and Harjeet Sehgal. 1979. Ecology of rotifers of Surinsar, a subtropical freshwater lake in Jammu (J & K). India Hydrobiol., *65* (1): 23-32.

Lillijeborg, W. 1900. Cladocera suecae. Nov. Acta. Reg. Soc. Scient. Upsal, 19.

Lind, Owen, T. 1974. Hand book of common methods in limnology. The C. V. Mosby Company, Saint Louis, 154 pp.

Macan, T. T. 1963. Freshwater Ecology. Longman Group Ltd., London: 339 pp.

Malhotra, Y. R., P. L. Duda and M. K. Jyoti. 1976. *Mansariella lacustris* Gen. et. Sp. nov., a new freshwater medusa from Jammu, India. Curr. Sci., 45 (5): 190-91.

Nayar, C. K. G. 1968. Rotifer fauna of Rajasthan. Hydrobiol., *31*: 68-185.

Nordlie, F. G. 1976. Plankton communities of three central Florida Lakes. Hydrobiol., *48* (1): 65-78.

Patalas, K. 1972. Crustacean plankton and the eutrophication of St. Lawrence Great Lakes. J. Fish. Res. Bd. Canada, *29*: 1451-162.

Plew, Wayne, F. and Robert W. Pennak, 1949. A seasonal investigation of the vertical movements of zooplankters in an Indiana Lake. Ecology, *30* (1): 93-100.

Rand, M. C., A. E. Greenberg and M. J. Taras. 1976. Standard methods for the examination of water and waste water, 14th ed. American Public Health Association, American Water Works Association, Water Pollution Control Federation, Washington.

Robinson, A. H. and Patricia K. Robinson, 1971. Seasonal distribution of zooplankton in the northern basin of lake Chad. J. Zool., Lond. 163: 25-61.

Saha, G. N., K. L. Sehgal, Eva Mitra and A. C. Nandy. 1971. Studies on the seasonal and diurnal variations in physico-chemical and biological conditions of a perennial freshwater pond. J. Inland Fish Soc., India, *111*: 79-102.

Schindler, D. W. and B. Noven. 1971. Vertical distribution and seasonal abundance of zooplankton in two shallow lakes of the experimental lakes Area, northwestern Ontario. J. Fish. Res. Biol. Canada. 28(2): 245-256.

Seenayya, G. 1973. Ecological studies in the plankton of certain freshwater ponds of Hyderabad India, III. Zooplankton and Bacteria. Hydrobiol., *41* (4): 529-540.

Sehgal, H. S. 1980. Limnology of lake Surinsar, Jammu, with reference to zooplankton and fishery prospects. Ph. D. thesis submitted to University of Jammu, 367 pp.

Timms, B. S. 1970a. Aspects of the limnology of five small reservoirs in New South Wales. Proc. Linn. Soc. New South Wales, *94* (1): 105-112.

__________. 1970b. Chemical and zooplankton studies of lentic habitats in North-eastern New South Wales. Abst. J. Mar. Freshwat. Res., *21*: 11-33.

Ward, H. B. and G. C. Whipple. 1959. Freshwater Biology. 2nd Edition. John Wiley and Sons Inc. New York, London and Sydney: 1248 pp.

Watson, N. H. F. and G. F. Carpenter 1974. Seasonal abundance of crustacean zooplankton and net plankton biomass of lakes Huron. Erie and Ontario J. Fish. Res. Bd. Canada, *31*: 309-317.

Welch, P. S. 1952. Limnology. McGraw Hill Book Company, New York, Toronto, London: 538 pp.

Wetzel, R. G. 1975. Limnology "W. B. Saunders Company, Philadalphia. Pennsylvania, 743 pp.

Section II

Wildlife

Chapter 22

On the Habitat and Behaviour, Maturity-Size and Sexual Dimorphism in a Population of Freshwater Emydid Turtles of Jammu, J & K State

P. L. Duda, Anil K. Verma and D. N. Sahi

Introduction

Reptiles, the most primitive amniotes which successfully colonised the land in course of evolution are admirably well adapted for the aquatic as well as terrestrial mode of life. The Indian sub-continent shows extreme climatic variations from the temperate Himalayas in the North down to the tropical peninsula in the South and from the desert plains in West and North-West to deciduous thick forests in the North-east. Reproduction of vertebrates under such varying environmental conditions in Indian sub-continent should thus become considerbaly influenced and result in some interesting patterns.

The Indian sub-continent is richly endowed with a reptilian fauna comprising about 2,500 species, out of a total tally of 6,000 species of living reptiles (Das, 1985) which are comprised of all major groups like Lizards, Snakes, Turtles and Crocodiles.

Turtles which embrace the entire order Chelonia (Class: Reptilia) and evolved some 200 million years ago, have remained virtually unchanged in their basic morphology eversince. Within the group, however, the diversity is remarkable, and their distribution widespread; turtles being found in all the six zoogeographical realms of the world.

Everyone recognizes a turtle, of whatever size, shape, colour or origin. Would that the names applied to them were straight forward – "Turtles", "Tortoises", "Terrapins" — they mean one thing to one person, something else to other. Biologically, all members of the order: Testudines are correctly called 'turtles'. Terrapins and tortoises are certainly turtles, regardless of whatever else they may be; Tortoise is usually applied to the terrestrial turtles particularly the larger ones; Terrapins is usually applied to edible, more or less aquatic, hard-shelled turtles (Ernst and Barbour, 1972).

From man's view-point turtles are slow, harmless creatures of considerable

* Department of Biosciences, University of Jammu, Jammu. J & K State. 180001.

importance in several respects. But, although turtles have always been the object of much interest and speculation, surprisingly little is known about their biology; indeed, some species are known only as to their appearance. As of today, the turtles comprise as many as 13– families.

India represents one of Asia's most diverse Chelonian fauna comprising of 20 genera and 31 species. (Table 1) belonging to 5 families (Smith, 1931; Moll, 1983, 1984, and Das 1985).

Table 1: Showing family wise numerical (%) strength of turtles in India

	Family	Common Name	Genera	Species
1.	Emydidae	Hard-shell turtles	10(50.00)	16(51.63)
2.	Trionychidae	Soft-shell turtles	4(20.00)	6(19.31)
3.	Cheloniidae	Marine turtles	4(20.00)	4(12.91)
4.	Testudinidae	Tortoises	1(5.00)	4(12.91)
5.	Dermochelyidae	Leather back turtles	1(5.00)	1(3.24)
	Total		20 genera	31 species

Smith's "Fauna of British India" (1931) volume I, undoubtedly, a landmark in providing information on turtles of India, was followed up by serious studies on the fresh water turtles only a decade or two back, whereas a lot more attention has been paid to study of different biological aspects of lizards and snakes subsequent to his publication on lizards and snakes (Smith, 1935; 1943). Moreover, our knowledge about the biology of the turtles particularly on reproduction in males is significantly poor both qualitatively and quantitatively (Sud, 1956; Singh, 1977; Yadav, 1982; and Rao and Shaad, 1985).

The emydid turtles constitue the largest family of living turtles and is represented by 25 genera and more than 80 species world over (Bellairs, 1969). In India, the family is represented by 16 species in 10 genera (Das 1985), of which 6 species belong to the genus *Kachuga*. Of the six species namely *Kachuga tectum* (in 2 subspecies. *K. t. tectum* and *K. tentoria), K. smithi; K .sylhetensis; K. dhongoka; K. kachuga* and *K. trivittata*, two *K. tectum tectum* and *K. smithi*—are the most abundant of freshwater turtles in Jammu but have an extremely restricted distribution in being confined only to the plains of Jammu province (Sahi and Duda 1985). Both these turtle are the little known, secretive forms and no information on their male reproductive biology is available.

The present work was undertaken for generating information on sexual-dimorphism, habitat; minimum size at maturity; behaviour; and reproductive strategies of these turtles.

The present work on the *K. t. tectum* and *K. smithi* (family Emdidae) apart from being the pioneer work on these emydid species from northern region of India, could hopefully provide considerable material for comparative biological study with several other emydid turtles from elsewhere, so that the data generated in this behalf would be useful to conservationists in future.

The Study Area—Physiography & Topography

A stretch of Gho-Manhasan stream 1 km. in length, 5–8 m wide and about 0.5–0.8 m deep was selected as the principal site of collection and observations (Fig. 1). *Kachuga tectum tectum* and *K. smithi*, the emydids under investigation whose occurrence was reported from the state by Duda and Sahi (1979), are the most abundantly encountered turtles in the marshes, irrigation canals and riffles at Gho-Manhasan 74° 49' E longitude and 32° 38' N latitude and at an elevation of 302.6 m above sea level and a relatively wet area situated at about 18 km. in the north-west of Jammu city. The area is a poorly drained plain area and combines woodland, wetland and croplands.

The stream infact, is an anastomosing network of rivulets, shallow pools and spring-seepages characterized by a substrate of deep mud and constantly flowing clear, well lighted and relatively cool water, bordered on either side by cultivated fields planted with annual and/or perennial crops primarily the Wheat, Rice, Corn, Mustard and other seasonal vegetables from time to time (Fig. 2). Water is usually clear save for the monsoon period, (July – August) when water gets turbid after heavy run-off following the rains and the stream often flooded. A slow to fast current in almost all waterways (rivulets, pools and streams) is a regular feature. Tremendous quantity of cold water from river Chenab enters the stream, thereby maintaining the water temperature always below the air temperature throughout the year. However, because of the larger surface area of the waterbody, normal surface temperature (Air 15.5 – 23.3°C and water 16.2 – 20.5°C in winter and Air 27.1 – 39.8°C and water 24.5 – 35.8°C in summer) was recorded in most parts of the stream throughout the year.

The phenological features of the area Gho–Manhasan site from where the material was collected, during the present investigation are presented in Table 2. The warmest months as usual were June and July (Table 2) and the coldest December and January.

Table 2: Showing average monthly variations of some environmental parameters recorded at Gho-Mahasan collection site during present investigation

Month	*Temperature regime*		*Relative*	*Photoperiod*	*Rainfall*
	Air (°C)	*Water* (°C)	*Humidity* (%)	*(Sunshine)* (hrs./day)	(cms.)
January	17.1	17.0	89.8	10.5	3.60
February	18.5	16.6	85.5	11.5	14.70
March	23.3	20.5	78.2	12.0	6.50
April	27.1	24.5	64.9	13.5	1.35
May	32.8	30.7	68.0	14.0	0.70
June	39.8	35.8	72.6	15.0	2.63
July	36.1	35.0	85.9	14.5	39.12
August	32.8	31.2	89.3	14.0	21.18
September	33.2	31.3	90.2	13.0	14.05
October	32.2	30.0	75.8	12.0	0.60
November	23.5	21.4	82.0	11.0	0.70
December	15.5	16.2	88.6	10.5	13.17

Turtles were generally found inthe heavily vegetated regions along this stream away from the area of human interference. The population of emydid turtles *K. t. tectum* and *K. smithi* in the Gho-Manhasan stream at Jammu, was selected for the present study for following reasons:

(i) Readily availability of these turtles in this geographic region;

(ii) This study area is a unique natural habitat and display the conditions most conducive for the chelonians inhabiting therein;

(iii) Occurrence of extraordinarily large populations of shrimps, fishes, and amphibians which constitute the main food items of the turtles under investigation; and

(iv) Protection of the study habitat by the least interference of the public inhabiting that locality.

The site/area was usually visited once or twice a month and on each visit a stretch of about 500 m. was surveyed and the activity of the turtles recorded. Collecting of the turtles was, therefore, confined to this area alone and collections either personally or through the help of local fishermen, were done during the study period March 1989 to November 1991.

Materials and Methods

Various freshwater lotic waterbodies in an around Jammu city enlisted as the preferred habitat of various freshwater turtles (Sahi and Duda 1985) were surveyed to assess their potential for regular supply of the emydids, *Kachuga tectum tectum* and *Kachuga smithi;* and finally, a slow-running stream fed by river Chenab at Gho-Manhasan, was selected as a site for regular collections owing to reasons described above. The specimens were collected by various techniques (described below). They were brought live to the Dry laboratory of the Department of Bio-sciences, University of Jammu, Jammu and were subjected to the needful observations as soon as possible, usually within 24 hours of capture. The males were segregated arbitrarily by visual examination of their morphic characters like relatively longer tail and bright coloration as compared with their female counterparts.

Specific Diagonosis of Emydid Turtles Under Investigation

Kachuga tectum tectum (Gray)

Carapace elevated, with a promient vertebral keel which is produced into a strong spinous process at the posterior margin of the third vertebral shield (Fig. 3). Nuchal shield usually the broadest behind, first vertebral variable in shape; second vertebral usually longer than the third, obtusely pointed behind; third vertebral pointed behind, its tip touching the pointed extremity of the fourth which is much longer than the other vertebral shields, fifth vertebral broader than long, broader than others.

Plastron nearly or quite as long as the carapace, angulate laterally, truncate anteriorly (Fig. 5), top of head black (Fig. 7) the temporal region yellow (Orange);

1
2
3
4

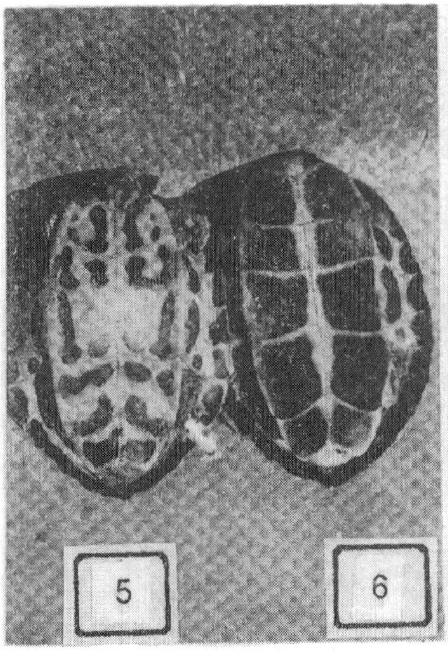
5
6

7

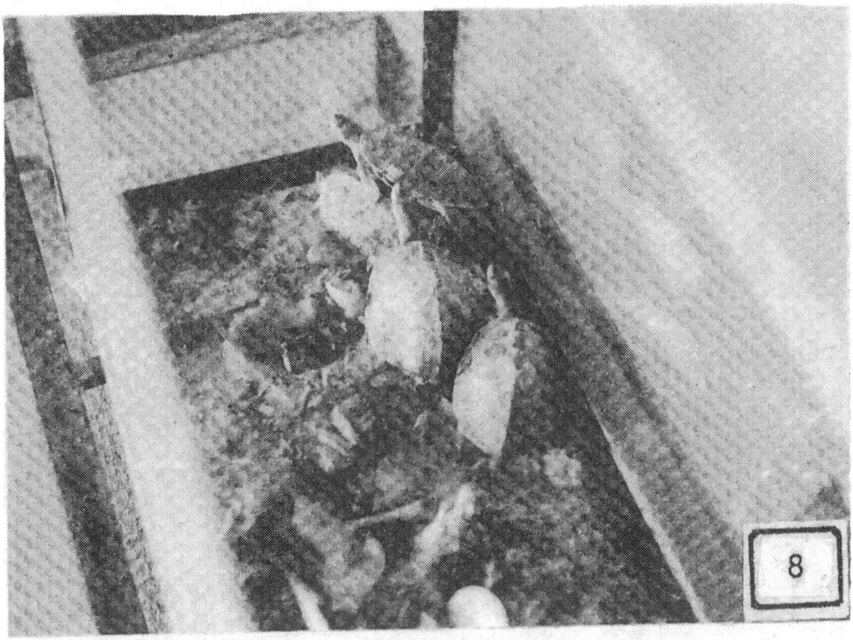
8

neck black, with fine yellow longitudinal lines (Fig. 3). Sheel olive in the young, with small black dots and an orange or red vertebral stripe, most conspicuous on the first three shields (Smith, 1931; pp. 126-128).

Kachuga smithi (Gray)

Carapace depressed, (Fig. 4) with obtuse interrupted, vertebral keel. Nuchal shield moderate, broadest behind; first vertebral usually longer than broad in the adult; second vertebral shortest, broader than long; third vertebral considerably longer than broad, pentagonal in the young, subquadriangular in the adult; fourth vertebral longest, tapering anteriorly and forming a narrow suture with the third; firth vertebral much broader than long, broader than the others.

Plastron nearly or quite as long as the carapace (Fig. 6) strongly angulate laterally in the young. Shell pale-olive, the vertebral keel usually blackish; yellow below, each shield with a large dark patch covering the greater part of it (Smith, 1931; pp. 125-126).

Collecting Techniques

Diverse collecting techniques employed to procure the turtles from their habitat during different seasons. Collecting of turtles was done once or twice a month and this schedule was adhered to, except during inclement of weather (e.g. flooding etc.) would make it impossible to reach the site or to collect the material from the waterbody. Specimens less than 56 mm. carapace length were considered immature and were excluded. Two or three matured males were sacrificed in each month, the tally of sacrifice deliberately kept low owing to the endangered status of the turtles under investigation. The turtles were collected either by Hands or Skin-diving or Netting or Muddling by the methods described by Mosimann and Bider (1960); Sahi (1978); Gupta (1979). The turtles, so caught were transferred into a water-filled bucket and were brought alive to the laboratory.

Maintenance of Turtles in the Laboratory

Within 2-3 hours of their catch, the turtles were washed thoroughly with tap water in the laboratory to remove debris/filth attached to their shell and limbs. Thereafter, turtles were segregated species wise and were kept in glass aquaria (4' × 5/4' × 5/4') size, (Fig. 8) half-filled with water and provided with a small wooden float to rest upon, whenever needed.

A pinch of salt was added to the water frequently to check fungal growth and the water was frequently changed to avoid putrefaction and the consequent skin infection of the animals. In captivity (Fig. 8) the turtles were fed on a mixed diet of aquatic plants like *Eicchorrnia* sp.; *Hydrilla* sp.; *Salvinia* sp.; *Myriophyllum* sp.; *Vallisinaria* sp.; and animals like small fishes, fingerlings; tadpoles of frogs; oligochaetes and molluscs etc.

Gross-Measurements

All the measurements, weighings and processing relevant to the present study were done in the laboratory after anaesthesizing the animals either with Ether or Chloroform, withing 24-48 hours of the catch. The body weights (in gms.) were taken with the help of a simple 2-pan balance. The measurements used in the present study are:

Carapace-length (CL) – is the median length of the carapace along the curve of the back.

Carapace-width (CW) – is the maximum curved line width of the carapace at rigth angle to the long axis.

Plastron-length (PL) – is the median length of the plastron along a straight line.

Plastron-width (PW) – is the maximum straight line width of the plastron.

Shell-height (SH) – is the maximum straight line height of the shell, measured as perpendicular to the plastron.

Tail-length (TL) – is the distance from the median posterior margin of the plastron to the tip of the tail.

Of all the measurements, shell height (= carapace height) was the most difficult to obtain and usually the least accurate. All linear measuremtns of the animal were taken with the help of a flexible steel-tape graduated in millimeters. Tail measurements were taken with the help of a divider. Measurements were recorded to the nearest millimeter.

Determination of Maturity Size

Because of lack of data pertaining to the *K. t. tectum* and *K. smithi* regarding the approximate snout-vent length or even the carapace length of reproductively mature males; the periodic dissections made by the author revealed that males were mostly mature at the carapace length ≥ 60 mm. in case of *K. t. tectum* and ≥ 56 mm. in *K. smithi.*

Consequently, turtles of both the species ≥ 56 mm. were, as far as possible, brought alive to the laboratory for subsequent processing. Those turtles which were regarded as juveniles or immature ≤ 56 mm. were let off back into the river. Visually, the turtles were considered sexually mature if they were found mating.

Observations

Habitat and Behaviour

Kachuga tectum tectum: Commonly known as 'Indian sawback turtle', prefers slow-moving shallow water like, Irrigation canals, streams and rivers with soft sandy/muddy bottom; abundant aquatic vegetation (Fig. 1), and suitable basking sites. It shares its habitat with the other freshwater turtles (viz. *k. smithi; Lissemys punctata punctata*) within its range.

The turtle spends much of the day time, basking on logs of wood, tree branches and banks of the waterbody. It is shy in nature and difficult to approach as it dives at the slightest alarm. Adults are omnivorous in nature, feeding on aquatic vegetation, snails oligochaetes and other organic matter of animal origin. It is a poorer swimmer than the brown roofed turtle *K. smithi* because of its greater elevated shell. It undergoes winter brumation from early December to late February.

Kachuga smithi: Commonly known as 'Brown roofed turtle' shares its habit and habitat with *K. t. tectum* with which it is found sympatric. It is, infact, a very active and fast swimming turtle, shy in nature; never attempting to bite.

It is carnivorous in habit, though vegetable food may be taken from time to time. It takes fingerlings, tadpole larvae and goat liver with relish.

Basking aggregations of the Brown roofed turtle may often be seen along river banks or logs of woods, or fallen branches of the trees growing nearby. During cold spells in winter, it undergoes winter brumation (early November to late February) and moves into the bottom under vegetation or burries itself inside mud at the bank and/or bottom of the stream, where they are sometimes trampled and fatally injured by cattle.

Size at Maturity: In order to assess the minimum size at which the species under investigation reach maturity in males, the dissections of the turtles of both the species pertaining to the different sizes, carapace length 56-140 mm (Fig. 9), were performed during the peak breeding-period i.e., July-September, whence, the females are maximally gravid (Gupta, 1979) and receptive for the advances of the male.

The minimum carapace length (measured as curved carapace length down the dorsal mid line) was used to determine the size at which the turtle species used in present study were mature, i.e. when on autepsy during the breeding season, the testes were found turgid, rounded, creamish coloured and from the epididymis oozed out a milky white fluid, when punctured with a fine needle. Furthermore, a portion of the testicular mass fixed in the Cornoy's fixative was squashed in Aceto-carmine stain to find if this tissue showed any spermatogenic activity. A scrutiny of the stained preparations of testes revealed that the testicular tissue of the specimens with a carapace length 60 mm and above in case of *K. t. tectum*, and 56 mm and above in *K. smithi*, had mature sperms in their germline. Therefore, the minimum carapace length of the turtle at which testes showed spermatogenic activity, was considered as the minimum maturity size of the turtle.

Accordingly, the minimum carapace length so determined for *K. t. tectum* was ≥ 60 mm. and males smaller than 60 mm in carapace length were, therefore, not included in the study and discarded as they may be immature. Of the 36 males of *K. t. tectum* used in the present study, all but one of the 60 mm carapace length were found mature and the specimens sacrificed during the present study had a carapace length 60-140 mm. (Fig. 9).

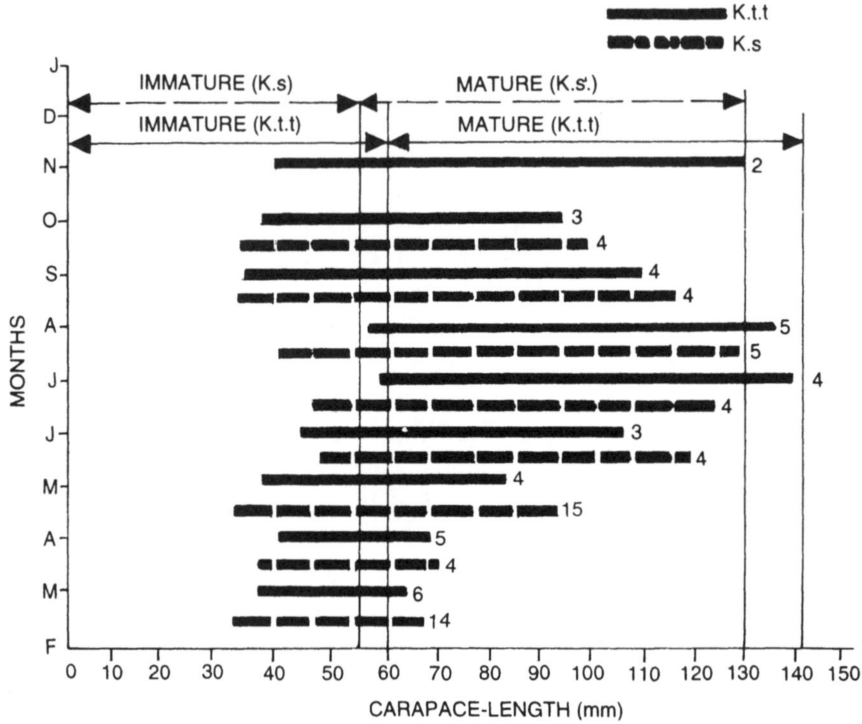

Fig. 9

Similarly, in case of *K. smithi*, periodical dissections of the turtles proceeded in the manner described above, revealed that at the minimum carapace length of 56 mm, the members of this species were found mature. Most turtles of this species of this size and larger sacrificed during present investigations were observed to be the potential breeders and mature. In all 34 males of *K. smithi* were assayed and ranged in carapace length between 56 mm and 130 mm. (Fig. 9).

Thus, turtles with a carapace length < 60 mm in *K. t. tectum* and < 56 mm in *K. smithi* were considered immature and therefore, released back in their natural habitat. The minimum size at which the females mature was found to be 142 mm in *K. t. tectum* and 160 mm in *K. smithi* (Gupta, 1979).

Sexual Dimorphism: In the emydid species, *K. t. tectum,* the males have a carapace, blanch-green in colour and the Plastron Eosin-red (Fig. 4), whereas, the females have a brown carapace and straw-yellow coloured plastron. Moreover, a mid-vertebral, deep, Eosin-red streak is more pronounced in case of males than in the females. The males have a concave plastron than the females which have a relatively flat plastron and there is much more pronounced humping of the dorsal sheilds in males (Table 3). Also, the females are bigger in size mature females having carapace length 100-180 mm (Gupta, 1970) as compared with the males which are of a relatively smaller size. Additionally, the tail in case of males (Table 4) is much longer than in the females. The ratio between tail length and plastron length in males was found to be 0.17 against 0.09 in females.

In case of *K. smithi*, the females are relatively much larger mature female having 102-219 mm carapace length than the males (Gupta, 1979). The colour of the carapace in males is light green and brownish in case of females (Table 3). Dorsal surface of the carapace in females is very much smooth to touch, while in males the surface of carapace give a decidedly gritty or sandy feel. The tail is much larger and thicker at the base in case of the males than in the female turtles. Moreover the ratio between the tail length (Table 4) and plastron length in males was 0.20 and its value was as low as 0.09 in case of the females.

Discussion

Size at maturity: A perusal of literature on the aspect of maturity in turtles reveals two divergent views, while one class of workers hold the view that the reproductive maturity in turtles is age related (Risley, 1938; in *Sternotherus odoratus*), the other view regards the attainment of maturity to be primarily size and not age related (Hildebrand, 1932; Cagle, 1948, 1954). The latter opinion has generated from the observations that the different populations of the same species of a turtle (*Pseudemys scripta troostii*) could show different rates of growth and hence at the same size could differ markedly in their ages.

The studies of Gupta (1979) on the female reproductive cycles in *Kachuga tectum tectum* and *Kachuga smithi* from Jammu, and Moll (1979) on further south populations of these turtles from India have shown that the turtles grow upto a certain size in accordance with their age (the age of the hatchlings being 35-40 mm in both), beyond which their size increases but only very slowly, adding barely a mm or so during each year of its growth; the latter growth rate being much slower in comparison to the earlier one. In view of such an observation, the possibility of their male counterparts, studied during the present investigation showing a similar pattern of growth cannot be ruled out.

In such an eventuality the males of *K. t. tectum* and *K. smithi* would attain a certain carapace or plastron length in a short period of time after their hatching, whereafter, the increments in their lengths would be much slower and smaller. This would therefore, lead one to the conclusion that in *K. t. tectum* and *K. smithi* in particular and in turtles in general (Cagle, 1954; Webb, 1961; Gibbons, 1968a, 1970a; Mahmoud, 1967; Moll and Legler, 1971; Ernst, 1971a, 1971b, 1972; Ernst and Barbour, 1972; Ernst and Ernst, 1972; Moll, 1973; Christiansen and Moll, 1973; Wilbur, 1975; and Gupta 1979) the attainment of maturity in the males is a function more related to the size than the age of the turtle. Although the minimum size at which the sexual maturity is attained could be indirectly a reflection of age, yet for want of a clear-cut criterion, the size or the age at which these turtles mature continues to be a matter of opinion and, therefore, arbitrary.

Table 3 : Showing characteristics size, general shell form, colour and Tail-plastron ratio observed in the two sexes of the turtles *K. t. tectum* and *K. smithi* investigated during present study

S. No.	*Characteristics*	*K. t. tectum*		*K. smithi*	
		Male	*Female*	*Male*	*Female*
1.	RELATIVE SIZE AND HABIT	Smaller and agile.	Generally larger and sluggish.	Smaller in size.	Mostly much larger than their male counterparts.
2.	SHELL-FORM	All the males irrespective of size had a concave plastron.	Females had a flat plastron.	Concave plastron in the males pronounced.	A relatively flat plastron.
		A mid vertebral streak on carapace more pronounced in males.	Vertebral streak lesser pronounced and humping is less pronounced.	...	...
		Dorsal sheilds conferring a pronounced humping of shell.		...	...
3..	COLOURATION	Males have a blanch-green carapace, and an Eosin-red plastron.	Brown carapace and straw-yellow coloured plastron.	Males have a light green carapace, plastron pale-yellow in colour.	Brownish carapace in females commonly observed. Plastron colour same as in males.
4.	TAIL LENGTH	Longer, thicker, prehensile and stouter in appearance.	Tail relatively smaller, as compared to the body size.	A longer, thicker and stouter tail in males is common.	A much smaller tail in the females is common.
5.	TAIL-PLASTRON LENGTH RATIO	0.17	0.09	0.20	0.09

Estimation of the age of turtles at which they mature from the counting of the abdomen scutes-annuli (growth rings) as suggested by Sexton (1959) became difficult because, these annuli tended to become smooth and difficult to count particularly, after their number increased beyond three. Assuming that each annulus represented the end of the previous year's growth, then with the fading out of the annuli above three signs per scute would make age determination extremely arbitrary.

Sexual dimorphism: A perusal of available literature reveals that a longer, stouter, more flexible, often prehensile and a cornified tail with a specialized scale on the tip; a concave plastron; elongated nails on the forefeet of some old males, an enlarged in-curved nail on the flippers of some marine males, and size dimorphism

Table. 4: Showing the relative size of plastron and tail of the male emydid turtles *K. t. tectum* and *K. smithi* investigated during present study

S. No.	Plastron length (P.L.) (mm)	*Kachuga tectum tectum* Mean (P.L.) (mm)	N	Tail length (T.L.) (mm)	Mean (T.L.) (mm)	N	T.L/P.L ratio	*Kachuga smithi* (P.L.) (mm)	Mean (P.L.) (mm)	N	(T.L.) (mm)	Mean (T.L.) (mm)	N	T./PL. ratio
1.	60–80	69.25	4	12.17	14.00	4	0.20	50–70	62.00	5	12.19	13.80	5	0.22
2.	80–100	88.18	27	18–23	19.74	27	0.22	70–90	82.44	9	13.21	12.55	9	0.15
3.	100–120	113.00	3	21–26	24.00	3	0.21	90–100	98.00	15	15–24	22.93	15	0.23
4.	120–140	137.50	2	28–30	29.00	2	0.21	110–130	120.00	5	22–25	23.20	5	0.19
	Total	118.38	36		19.97	36	0.17 ±0.007		91.82	34		18.87	34	0.20 ±0.032

where, N = Number of specimens studied.

in some forms, are some prominent characters possessed by males in species of turtles exhibiting sexual dimorphism (Graham, 1970). Male and female freshwater turtles usually differ in colouration, size, shell-proportions, and other features (Viosca, 1933; Carr, 1952; McCoy, 1968; Ernst and Barbour, 1972; Graham, 1970; Gupta, 1979).

A perusal of Table 3 reveals that the males of *K. t. tectum* and *K. smithi* differ from their females, in the respective species in having a relatively smallerbody size, a longer, thicker and stouter as well as prehensile tail; a brightly coloured shell, contrary to their female members in which the body size is bigger, tail is relatively smaller and the shell is of a duller colouration. Moreover, the shell has a concave plastron in males of both the species and a flat plastron in their females. Additionally, the ratio between the tail-length and plastron-length in the males of both the species is nearly twice the ratio observed in the females of *K. t. tectum* and *K. smithi* (Table 4) a finding which is consistent with that of Gupta (1979) in these turtle species.

Of the tabulated male-female differences (Table 3-4) the body size which is highly variable can't be regarded as a very dependable indicator of the sex of the turtle species, despite fact that the females have been often found and regarded as bigger of the two sexes in the turtles in general (Forbes, 1940; Cagle, 1944, 1954; Goin, Goin and Zug, 1978; Gupta, 1979). However, the bright colouration of the shell which together with the plastral concavity and a relatively larger tail (Table 3-4), all male characteristics emerge as a dependable group of features to help assort male from females in the two emydid turtles worked out here.

A larger, prehensile and stouter tail in the males of *K. t. tectum* and *K. smithi* presumably enables an easy access of cloaca of the male to that of the female during mating and, therefore, confers a mating advantage to the males, a view also held by Auffenberg (1964) in the land tortoise, *Geochelone travancortca*.

Acknowledgement

Authors are grateful to the Department of Environment and Forests, Govt. of India, New Delhi for the financial assistance provided under the project, "Herpeto Fauna of Kashmir Himalayas ... etc."

References

Auffunberg, W. 1964. Notes on the courtship of the land tortoise *Geochelone travancorica* (Boulenger). *J. Bombay Nat. Hist. Soc.*, 61: 247-253.

Bellairs, A. D'A 1969. *The Life of Reptiles*. II Universe Books, New York.

Cagle, F. R. 1944. Sexual maturity in the female of the turtle *Pseudemys scripta elegans*. *Copeia*, 1944: 149-152.

Cagle, F. R. 1948. Sexual maturity in the male turtle, *Pseudemys scripta troostii*. *Copeia*, 1948: 108-111.

Cagle, F. R. 1954. Observations on the life-cycles of painted turtles (genus *Chrysemys*) *Am. Midl. Nat.*, 52: 225-235.

Christiansen, J. L., and Moll, E. O. 1973. Latitudinal reproductive variation within a single subspecies of painted turtle, *Chrysemys picta bell*. *Herpetologica*, 29(2): 152-163.

Das, I. 1985. *Indian turtles: a field guide*. World Wild Life Fund India, Eastern region: 1-119. Ms. Bonani Kakkar (Publ.).

Duda, P. L. and Sahi, D. N. 1979. An uptodate checklist of herpetiles of Jammu and Kashmir. *Univ. Review.*, 10(6): 87-93.

Ernst, C. H. 1971a. Sexual cycle and maturity of the turtle, *Chrysemys picta*. *Biol. Bull. mar. biol. lab., Woods. Hole.*, 140(2): 191-200.

Ernst, C.H. 1971b. Population dynamics and activity cycles of *Chrysemys picta* in south eastern Pennsylvania. *J. Herpetol.*, 5: 151-160.

Ernst; C. H. 1972. Temperature activity relationship in the painted turtle, *Chrysemys picta*. *Copeia*, 1972: 217-222.

Ernst, C. H., and Barbour, R.W. 1972. *Turtles of the United States*: 1-347. University Kentucley Press, Levington.

Ernst, C. H., and Ernst, B. M. 1972. Biology of *Chrysemys picta bellii* in south western Minnesota. *Proc. Mim. Acad. Sci.*, 38: 77-80.

Forbes, T. R. 1940. A note on reptilian sex ratios. *Copeia*, 1940: 132.

Gibbons, J. W. 1968a. Growth rates of common snapping turtle, *Chelydra Serpentina* in a polluted river. *Herpetologica*, 24(3): 266-267.

Gibbons, J. W. 1970a. Reproductive characteristics of a Florida population of musk turtle. *(Sternotherus odoratus)*. *Herpetologica.*, 26: 268-270.

Goin, C. J., Goin, O. B., and Zug. G. R. 1978. *Introduction to Herpetology*. Third edition. W. H. Freeman and Co., San Francisco.

Graham, T. E. 1970. Growth rate of the spotted turtle, *Clemmys guttata*, in Southern Rhode Island. *J. Herpetol.*, 4: 87-88.

Gupta, V. K. 1979. Studies on the female reproductive cycle of some chelones from Jammu. Ph. D. Thesis, University of Jammu, Jammu.

Hildebrand, S. F. 1932. Growth of diamond-back terrapins. Size attained, sex ratio and longevity. *Zool.*, 9: 551-563.

Mahmoud. I. Y. 1967. Courtship behaviour and sexual maturity in four species of kinosternid turtles. *Copeia*, 1967: 314-319.

McCoy, C. J. 1968. The development of melanism in an Oklahoma population of *Chrysemys scripta elegans* (Reptilia: Testudinidae). *Proc. Okla. Acad. Sci.*, 47: 84-87.

Moll, E. O. 1973. Latitudinal and inter sub-specific variations in reproduction of the painted turtle, *Chrysemys picta*. *Hespetologica*, 29: 307-318.

Moll, E. O. 1979. *Reproductive cycles and adaptations*, In: *Turtles; Perspectives and Research*: 305-331. M. Harlers and H. Morlock (Eds.) John Wiley and Sons, New York.

Moll, E. O., and Legler, J. M. 1971. The life history of a neotropical slider turtle, *Pseudemys scripta* (Schoepff) in Panama. *Bull. Los. Angeles. County. Mus. Nat. Hist. Sci.*, 11: 1-102.

Mosimann, J. E., and Bider, J. R. 1960. Variations, sexual dimorphism, and maturity in a Quebec population of the common snapping turtle. *Chelydra serpentina*. Can. J. Zool. 38: 19-38.

Rao, R. J., and Shaad, F. U. 1985. Sexual cycle of male freshwater turtle *Trionyx gangeticus* (Cuvier). *Herpetologica*, *41*(4): 437-442.

Risley, P. L. 1938. Seasonal changes in the testes of musk turtle. *Sternotherus odoratus J. Morphol.*, 63: 307-317.

Sahi, D. N. 1978. A contribution to the Herpetiles of Jammu and Kashmir State. Ph. D. Thesis University of Jammu, Jammu.

Sahi, D. N., and Duda, P. L. 1985. Checklist and keys to the herptiles of Jammu and Kashmir State. *Cic. Herp. Soc.*, 20 (3-4): 86-97.

Sexton, O. J. 1959. A method of estimating the age of painted turtles for use in demograpic studies. *Ecology*, 40: 716-718.

Singh, D. P. 1977. Annual sexual rhythm in relation to environmental factors in a tropical pond turtle, *Lissemys punctata granosa*. *Herpetologica*, 33: 190-194.

Smith, M. A. 1931. *The Fauna of British India, including Ceylon and Burma*. Reptilia and amphibia. I. *Loricata, Testudines*, Taylor and Francis (Eds.), London.

Smith, M. A. 1935. *The Fauna of British India, including Ceylon and Burma*. Reptilia and amphibia. 2. *Sauria*. Taylor and Francis (Eds.), London.

Smith, M. A. 1943. *The Fauna of British India, including Ceylon and Burma*. Reptilia and amphibia. 3. *Serpentes*. Taylor and Francis (Eds.), London.

Sud, B. N. 1956. Studies on reptilian spermatogenesis II. Spermatogenesis of the freshwater turtle, *Lissemys punctata*, (Bonnaterre), with observation on the living materila under the phase contrast microscope. *Res. Bull. Punjab. Univ., 86: 31-47.*

Viosca, P. 1933. The *Pseudemys troostii elegans* complex, a case of sexual dimorphism. *Copeia*, 1933 (4): 208-210.

Webb, R. G. 1961. Observations on the life histories of the turtles (genus Pseudemys and Graptems) in Lake Texoma, Oklahoma. *Am. Midl. Nat.*, 65: 193-214.

Wilbur, H. M. 1975. A growth model for the turtle *Chrysemys picta*. *Copeia*, 1975: 337-343.

Yadav, M. R. 1982. Breeding biology of Indian tropical pond turtle, *Lissemys punctata granosa* (Schoepff). Ph. D. thesis, Banaras Hindu University, Banaras.

Chapter 23

Status and Management of Wildlife in Jammu and Kashmir State

*B. L. Kaul & Indu Kanwal**

Introduction

Among the ecological sub-regions of the Himalayan mountain complex are the North-Western Himalayas which are unique because the flora and fauna are typical of both the oriental and the palaearctic regions. The wildlife of this zone is markedly different from that noticed in the eastern, central and western Himalayas and change begins between the latitudes.

21,000 sq. Km of the forest land holds much of the diversified flora and fauna so that on the map of India, Jammu and Kashmir state which is a part of North-Western Himalayas is the hope of ecologists and conservationsits for a better future. Kashmir valley surrounded by high mountain-ranges has a generally distinct regional characteristics. Its salubrious climate and temperate and alpine vegetation, level alluvial plains and abundant water supply are some of the distinct features in which a wide variety of the wild-life flourishes. Although wildlife of the state is endangered yet it is regarded as one of the richest resources of the country. The three main constituent regions of the state of Jammu and Kashmir are the sub-Himalayan Jammu, the Himalayan Kashmir and the Trans-Himalayan Ladakh. These three regions can be likened to the three steps of a terrace, the lower step being Jammu region, the middle zone as the Kashmir valley and the upper one as Ladakh. In the Jammu region there are a range of hills known as *Siwaliks* and to the north of these are the middle mountains called Lesser Himalayas. The valley of Kashmir is surrounded by lesser Himalayas in the South and the greater Himalayas on the North. Ladakh is the land of mountain ranges which include the greater Himalayan range, the Zanskar range, the Ladakh range and the Korakoram range.

Status of Wild-Life

The wild fauna of Jammu and Kashmir state includes above 70 species of reptiles, 300 species of birds and about 90 species of mammals. This fact clearly

* Department of Zoology, Govt. College for Women, Parade, Jammu – 180001.

indicates the animal diversity of this crown of India. The fauna of the state has special place in the geographical realm of the world and some of the species of wild animals such as the Hangul (*Cervus elaphas hanglu*), the musk deer (*Moschus moschiferous*), the markhor (*Capra falconeri*), the wild Yak (*Bos grunniens*) the Tibetian antelope (*Pantholops hodgsoni*) the Tibetian wild ass (*Equus hemionus kiang*) are unique in the fauna.

Apart from having unique mammalian fauna the state also has a remarkable avifauna. One of the important reasons behind this wounderful collection of birds here may be numerous breeding and wintering sites that it provides.

The list of the resident and the migratory birds include the endemic, threatened as well as the protected species such as:

- Monal (*Lophophorus impejanus*)
- Kaleej (*Lophura leucomelans*)
- Chir pheasants (*Catreus wallichii*)
- Western tragopan (*Tragopan melanocephalus*)
- Himalayan snowcock (*Tetragallus himalyaensis*)
- Black necked crane (*Grus nigricolis*)
- Bar headed geese (*Anser indicus*)
- Golden eagle (*Aquila chrysaetos*)

Wild-life of Jammu

The Jammu region sustains many mammalian species and the most common being nilgai, barking deer, wild boar, rhesus monkey, leopard, goral, grey langur, cheetal in the foot hills and plains. Hangul, musk deer, markhor, Himalayan black bear, Himalayan brown bear etc. are found in the altitudinal range of 1,700 to 4,800 metres above M. S. L.

The avifauna of this region comprises of the Indian mynah, blue rock pigeon, pea fowl, red jungle fowl, jungle crow, golden oriole, white cheeked bulbul, Red vented bulbul, chir-pheasant and chukar pheasant.

Some areas of this range (Surinsar-Mansar wildlife sanctuary) may also serve as a refuge to a multitude of birds including black partridge, peafowl, grey partridges, rufous turtle dove etc. The jungle bush quail is a feature of the Jasrota wildlife sanctuary, about 70 kms North-West of Jammu.

A study of the climatic regions of India reveals that it is a tropical country and rainfall instead of temperature has been taken as the chief criterion for its division. Jammu accordingly falls in the region which can be named as the intermediate zone having a heavy rainfall between 100–200 cm. The chief vegetation is decidous. These factors play a major role in the rich bird diversity found in the Jammu region.

Wildlife of Kashmir

The Kashmir Valley is second component of the region and is called "Paradise on earth" for its breath taking beauty and rich wild fauna and flora. Here seeing is believing and so at the altitudinal range of 1,700 to 4000 M. S. L. apart from among 50 species of trees, 20 of shrubs and 500 of herbs we can also watch animals like Himalayan brown bear, Himalayan black bear, langur, Himalayan marmot, leopard etc.

On way to Ladakh through Lush Green Kashmir Valley
Photo: Indu Kanwal

This area also houses a variety of other high altitude mammals and avifauna including pheasants and partridges most of which are faced with a serious threat of extinction. At the altitudinal range of 3,050 to 5,425 m above M. S. L., the nature has gifted a multiple set of species ranging from rare to endangered and protected species. Main species are the hangul, musk deer, serow, brown bear, leopard, rhesus macaque, grey langur and Himalayan mouse hare etc. among the mammals and koklass, Himalayan snow cock, white crested kalij pheasant and western tragopan among the birds. Between the altitudinal range of 2,400 to 4,300 towards the south-west of Kashmir we come across a rich fauna which includes the Himalayan musk deer apart from other rare and protected species like the leopard, brown bear, black bear and the red fox.

Forest covers over 10,000 sq miles of Kashmir (depends on per year deforestation) only out of about 86,000 sq miles of the total area. This is the main reason why the vale of Kashmir (85 miles long and 20 miles wide) faces a great deal of threat from nature and man alike.

Wild-life of Ladakh

The most widespread component of the state is Ladakh. This highest inhabited territory of India is largely a barren wasteland of sand and stone, making up 70 per cent of the total area of the state.

The terrain is very difficult to access and although it is tree-less yet most of its land is bare of grassy vegetation. For miles and miles through this extremely arid, sun-beaten and wind swept territory one does not come across a single human being, bird or tree. Nevertheless nature has endowed its elevated plains and plateaus and its dizzy mountains heights and deep gorges with a rich and varied animal life which is wonder and delight of the naturalist and lure for a sportsman.

The wildlife of Ladakh has been decimated and many rare species brought to the very verge of extinction. The semi-desert tract of Ladakh which enjoyed quite a reputation of possessing a large variety of wildlife, has been brought very near to being a desert from the point of view of wildlife as well.

Although its lamas and gompas constitute a major attraction for the growing number of tourists, yet not the least among its manifold attractions are its peculiar wild animals.

Among the wildlife of Ladakh are Tibetean antelope wild sheep, wild goat, snow leopard, gazelle, lynx, donkh (wild yak) Kiang , Marmot etc.

The common characteristics of the wildlife residing at such a height are:

1. Inhabit high grounds
2. Thick hairy coats.
3. Colour blending is remarkable with their surroundings.
4. Highly developed powers of smell, sight and hearing.
5. They are fleet of foot and so they can run away instantly as they sense some danger.
6. Males of most species carry big spectacular horns which nature seems to have intended for protection.
7. Move about in herds and some keep a member of their family for sentry duty too.

Some of the wild-animals of Ladakh are as follows:

1. Wild Yak (*Bos grunniens*)

It is an enormous beast of 5 ft. at shoulder and a foot or two more in length weighing about 1000 lbs when grown-up. Black hair grows in length from top-downwards along sides and in front of the chest so that it appears that animal does not possess legs. Tail is long bushy but does not reach the ground. It has two big black but gracefully curved horns. It has a keen sense of smell though its hearing and sight are just average. They are by no means common in Ladakh and they never have been. They were found in the high plains and plateaus of Chang-Chenmo and Aksaichin.The animal is at its best above 14000 ft. and can not survive at lower heights.

2. Kiang (*Equus hemionus kiang*) or wild ass:

Found especially in Rupshu, Chang Thang and Chang chenmo. It is handsome, ruddy chestnut or brown above and white underneath the belly standing about 4 ft. at shoulders the animal has a stiff black mane and a dark brown stripe down back. Hair in tail tuft are black and transverse black stripes mark the legs.

3. Marmot (*Arctomys himalayanus*)

Found in Ladakh uplands with the size of small spaniel, it wears a thick coat of reddish brown or golden fur. Body measures about 24"–38" body and tail resembles that of a squirrel or mangoose. Common in mosquito infested Deosai plains in Baltistan in Ladakh. It is first met on the Zoji-la pass.

4. Hare (*Lepus* spp.)

It is called Ribong in Ladakhi due to the length of its ears. Common all over but in Rupshu and also in Chang chenmo they are numerous. It has soft thick wooly fur which is bluish in colour.

5. Snow Leopard or Ounce (*Panthera uncia*)

It is an animal with 90 cm tail and although a protected species yet hunted for its fur which fetches a very high price in the international market.

6. Ibex (*Capra ibex*)

It is the most handsome animal with deep notched sweeping horns found in western Himalayas in Kashmir. It is hunted for its meat, hide and fur. The finest Kashmiri shawls are made of Ibex wool and the Kashmiri long boots are made from its hide. The wool (Pasham) obtained from Chiru (*Panthlops hodgsoni*) is also used for preparation of soft and costly shawls (*Shah tush*) in Kashmir.

Game Animals: Nyan (*Ovis ammon hodgsoni*); Marcopolo's sheep (*Ovis poli*); Shapo (*Ovis vignei*) Bharal (*Ovis nahura*); Ibex (*Capra siberica*); Markhor (*Capra falconeri*) wild deer (*Gazella picticaudata*); Chiru (*Panthlops hodgsoni*).

Beasts of Prey: Snow–leopard (*Panthera uncia*); lynx (*Felix lynx isabellinus*); *Brown bear* (*Ursus arctus isabellinus*); Shanko (*Canis lupus*); Fox (*Vulpes vulpes*).

Avifauna of Kashmir and Ladakh

The avifauna of both Kashmir and Ladakh comprises of a rich as well as varied fauna as, at the altitudinal range of 1700 to 4,000 m of M. S. L. in Kashmir there are over 150 species of birds. Principal species are the Koklas, bearded vulture, griffon vulture, golden eagle, grey heron, starling, grey oriole, paradise flycatcher, western yellow-billed, blue magpie, kestrel, peregrine falcon, black bulbul etc.

In south east of Srinagar we find a rich assemblage of pheasants like the chukar, koklas, Himalyan snow cock with jungle crow, red-billed chough, giffon vulture, bearded vulture, Himalyan whistling thrush, rufous-backed shrike, Indian mynah etc. Kashmir valley also holds a good population of pheasants and the upland birds apart from other species both resident and migratory. Common birds are griffon vulture, Kashmir roller, European hoopoe etc.

White Crested Kalij Pheasant (*Gennacus hamiltoni*)

Western Tragopan (*Tragopan melanocephalus*)

Koklass (*Purasia macrolopha*)

Himalayan Snowcock (*Tetragallus himalyansis*)

Avifauna of Ladakh region consists of partridges, wild geese, black-necked crane, pheasants and the snow cocks. The catchments of Markha and Rumbak valleys and area opposite to the villages of Neimmu and Phyang has a rich and varied bird life, the principal species being chukar, Himalayan snow-cock, Magpie, Himalayan whistling thrush, red-billed chough and yellow billed chough etc.

Wild-life Management

For management of wild-life in Jammu, Kashmir and Ladakh, national parks, sanctuaries and biosphere reserves have been established. A short account of these is given hereunder:-

I. National Parks

(1) Dachigam N. P.

Location:	21 kms. from Srinagar in the north-east.
Area:	141 sq. kms.
Altitudinal Range:	1,700–4,000 m M. S. L.
Fauna:	20 mammal species as Hangul, Himalayan brown bear, Himalayan black bear, Langur, Himalayan marmot, Leopard etc.

(2) Kishtwar High Altitude N. P.

Location:	In the part of the Chenab catchment.
Area:	400 sq. kms.
Altitudinal Range:	1,700 – 4,800 m M. S. L.
Fauna:	Hangul, Musk deer, Markhor, Goral, Himalayan black bear, Himalayan brown bear, Langur, Leopard etc.

(3) Hemis High Altitude N. P.

Location:	Opposite the villages of Nimmu and Phyang and includes the catchments of Markha and Rumbak valleys.
Area:	600 sq. kms.
Altitudinal Range:	Ladakh zone.
Fauna:	Shapu, Bharal, Ovis ammon, Ibex, Himalayan marmot, Snow leopard etc.

II. Wild-Life Sanctuaries

(1) Overa W.L.S.

Location:	76 kms. from Srinagar in the South-east and located near Pahalgam.
Area:	32.27 sq. kms.

Altitudinal Range: Kashmir zone (Anantnag)

Fauna: Hangul, Musk deer, Serow, Rhesus macaque, Langur, Leopard and Leopard cat etc.

(2) Ramnagar W. L. S.

Location: In the catchement zone of the Kar-nalla of Tawi river.

Area: 31.50 acres (12 sq. kms.)

Altitudinal Range: 430–611 m above M. S. L.

Fauna: Nilgai, Barking deer, Wild boar, Rhesus monkey, Red Jungle fowl, etc.

(3) Nandini W. L. S.

Location: About 28 kms. from Jammu on Jammu/Srinagar Highway.

Area: 33.20 acres.

Altitudinal Range: Jammu zone.

Fauna: Leopard, Wild boar, Barking deer, Rhesus monkey, Goral, Grey langur etc.

(4) Trikuta W. L. S.

Location: Jammu zone (Distt. Udhampur)

Fauna: Leopard, Barking deer, Wild boar, Spiny anteater, Himalayan tahr, Hanuman langur etc.

(5) Surinsar-Mansar (Patwar) W. L. S.

Location: The above two lakes are situated around it.

Area: 97.82 acres.

Altitudinal Range: Jammu zone (Kathua)

Fauna: Goral, Wild boar, Barking deer, Leopard etc.

(6) Jasrota W. L. S.

Location: Right bank of the river Ujh, toward the north of the village Jasrota and about 70 kms. north-west of Jammu.

Area: 10.04 sq. kms.

Fauna: Cheetal, Barking deer, Wild boar, Rhesus monkey etc.

III. Biosphere Reserves

(1) Overa – Aru B. R.

Location: 76 kms. from Srinagar to its South-east.

Area: 400 acres.

Altitudinal Range:	3,050–5,425 above M. S. L.
Fauna:	Hangul, Musk deer, Serow, Brown bear, Leopard, Rhesus macque, Grey langur, Himalayan mouse hare etc.

(2) Gulmarg B. R.

Location:	South-west of Srinagar and located at a distance of 48 kms. from Srinagar.
Area:	180 sq. kms.
Altitudinal Range:	Srinagar zone.
Fauna:	Leopard, brown bear, black bear, Red fox etc.

References

Dar, Moti (1983). "Ladakh" in the Wildlife of Jammu and Kashmir Sanctuary Publications.

Dar, Moti (1995). "Wild-life of Ladakh" in Ecodegradation of Himalayas (Ed. B. L. Kaul) Yak Channel publishers, Jammu.

Ganhar J. N. (1979). The wild life of Ladakh. Harmukh Publications, Srinagar.

Gergan S. S. (1962). Big game of Jammu and Kashmir. A Government of Jammu and Kashmir Publication.

Kaul B. L. (1987). Ecology and Wildlife of Kashmir Himalayas, In Kashmir Ecology and Environment (Ed. S. K. Chadha). Mittal Publications, New Delhi.

Kaul B. L. (1989). "Ecology and Wildlife of Ladakh" in Ecological Hazards in the Himalayas (Ed. S. K. Chadha) Creative Publications, New Delhi.

Kaul B. L. (1992). "Habitat Ecodegradation in Kashmir Himalayas" In Conserving Wildlife in India (Ed. S. K. Chadha). Vinod Publishers, Jammu.

Kaul B. L. (1994). "Call of the Wild" In the Hindu, Madras (May 1, 1994).

Mir Innayatullah (1983). "The wildlife of Jammu and Kashmir". Sanctuary Publications.

Sehgal Bittu (1983). "The Grey Ghost" In the Wildlife of Jammu and Kashmir. Sanctuary Publications.

Chapter 24

Eco-geographical Distribution and Present Status of Herptiles in Kashmir Himalayas

*Deep N. Sahi and P. L. Duda**

Abstract

The present paper deals with eco-geographical distribution and ecological status of eleven species of amphibians and sixty three species of reptiles (7 turtles, 24 lizards and 32 snake species) in Kashmir Himalayas. Analysis of distribution of various herptilian elements available in Kashmir Himalayas has been done zone wise i.e. in Greater, Middle and Outer Kashmir Himalayas.

Data on the collection sites and their altitudinal ranges have been collated and zoo-geographical affinities of the forms enumerated discussed.

Known as 'the Heaven upon this earth' and blessed with diverse ecologies and climate, the state of Jammu and Kashmir is ingressed by Kashmir Himalayas and uniquely combines the Palearctic and Oriental faunas. While phenology ranges between subtropical to temperate, the topography of the state varies from plains to hilly and mountain areas. The state stretches between 32°.17 N and 36°.58 N latitude and 72°.6E to 80°.2E longitude.

The Kashmir Himalayas which form a part of North-West Himalayan chain are drawn into three longitudinal ranges, namely, Outer, Middle and Greater Himalayas, that divide the state into three ecologically distinct regions, namely; Jammu and Kashmir and Ladakh respectively.

Although largely arid subtropical, climatically the Jammu region tends to vary substantially with change in altitude and latitude of the place. Moist alpine to subalpine conditions prevail in higher mountains while summer temperature in excess of 45°C generally obtains in some low land areas. The foot hills exhibit a transition from semideciduous tropical forest to subtropical 'Chil' pine zone. The southern and western areas of Jammu region are subtropical and extremely hot during summer (May–June) with excessive monsoon precipitation (112–120 cm/ per annum) from July-September. The northern and eastern areas of the region are cold temperate in climate, simulating conditions similar to those of Kashmir region. In general, the three months rainy season is regarded as the dominant climatogical event of the year.

The Kashmir region is bounded on the north and northeast by the Greater

* Department of Bio–Sciences, University of Jammu-Jammu – 180001, India

Himalayas, in the south and southwest by the Pir-Panjal (Middle Himalayas) and in the west by a prominent spur of the Great Himalayas (Fig. 1). The predominant climate of Kashmir is, however, Sub-Mediterranean, with two dry spells, one in June and other from September to November. The temperature ranges from –9°C in winter to 35°C in summer and average rainfall is 92 mm per year.

The Ladakh region lies in the northern reaches of the Himalayas. It is uniformely cold and dry during most of the year, with very little precipitation in the form of rain. The temperature varies between –35°C in winter to around 27°C in summer. The region is a cold desert, having two distinct bioclimates – the cold and the cold temperate. The cold and arid conditions, along with nature of terrain, permits no development of forests, although cold temperature areas do show some vegetational growth.

The state of Jammu and Kashmir thus combines diverse climates with ecological diversity ranging from subtropical plains of Jammu through temperate valley of Kashmir to cold mountainous desert of Ladakh as one ascends north and northeast in altitude.

Knowledge about herpetofaunal elements of the state has steadily grown during the last few decades (Smith 1931, 1935, 1943; Das 1967; Sharma and Sharma 1975; Duda, Datta and Sahi 1977; Duda and Sahi 1978; Dubois 1978; Sahi and Duda 1978, 1983, 1985 and 1986), but no detailed account of distributional ranges, geographic location and ecology of the collections sites, which are important parameters for determining the status of a species, are available. The present communication makes available information on the present relative abundance, distributional ranges of the amphibians and reptiles recorded within the state of Jammu and Kashmir. Altitudinal distribution, zoogeographical analysis and zonal distribution of the forms collected within Kashmir Himalayas i. e. Greater, Middle and Outer Himalayas is recorded.

Amphibians

The amphibian elements of Kashmir Himalayas comprise of 11 species in 4 generas and 4 families (*Table 1*). The genus *Rana* is represented by 5 species of which 4 are Oriental. *Rana sternosignata* is a member of Central Asian mountane ranid complex, which is endemic to the Kashmir Himalayas, the North-west frontier Province (Pakistan) and Afghanistan. The genus *Bufo* is represented by 4 species. The *Bufo melanostictus* is only bufonid with Oriental affinities. Its distribution pattern in Jammu and Kashmir is very interesting. It has been collected at higher altitude in the Middle Himalayas in Poonch. This species is very common in peninsular India and it has not been collected from plains of outer foot hills of Jammu and its presence in Middle Himalayas near the border of Pakistan shows that it has taken mountainous route to reach Middle Himalaya from Pakistan from where it has been reported to occur (Minton, 1966). *B. latestei* is a Ladakh endemic at altitude above 2500 m. *B. Stomaticus* was collected at altitude below 1050 m and is apparently replaced by *B. viridis* at elevation between 1600–2700 m. The

Pelobtid, *Scutiger occidentalis* was a confirmed as a Ladakh endemic (Dubois, 1978) whereas *Microhyla ornata* is an Oriental species.

While among the ranids, *R. cyanophlyctis* is the most common species abunding the area, it is *B. stomaticus* among the bufonids. *R. Cyanophlyctis* is an eurythermal species and well spread in Outer and Middle Himalayas reaching upto an elevation of about 1900 m. *B. viridis* is a temperate species and well spread in Middle and Greater Himalayas in Kashmir. It is the only amphibian which is present in all the three regions within the state, i.e. Jammu, Kashmir and Ladakh. It is, however, present in Jammu at such places (Bhaderwah and Kishtwar) where the climatic conditions simulate those of Kashmir. All the amphibians are more or less aquatic in habitat some spending more time in water (ranids) then others (bufonids, Pelobatids and microhylids). Within ranids, the frequency of occupying aquatic life also varies; *R. tigrina* and *R. cyanophlyctis* being more aquatic than others.

Reptiles

Turtles are absent from Kashmir and Ladakh, and are known only from the low-lying areas and outer foot hills of southern Jammu Province. Chelonian fauna, which comprises of 7 species in 4 genera and 2 families, are all freshwater forms and Oriental in affinities.

The two trionychids, *Trionyx gangeticus* and *Lissemys punctata punctata* are present in good number in a subtropical Lake Mansar (710 m) in Outer Himalayas. The turtle population has been able to survive here from the cruel hands of man for the religious sentiments of local residents attached with the sacred lake. Otherwise, the overall picture of the state's Chelonian fauna is grim. All the Chelonian forms in the state are aquatic and found in or around water. The major cause of the decline of these turtle species has been the destruction of their natural habitat. However, soft shelled species have also greatly suffered in the hands of man for their high value of protein. Unless or untill some drastic steps are not taken by state government, their survival is in question.

The 24 species of saurians from Jammu and Kashmir represent 14 genera and 5 families. All the collected species of *Cyrtodactylus* (5), *Agama* (3) *Phrynocephalus* (2), *Scincella* (2), *Eumeces* (1), *Eublepharus* (1), *Acanthodactylus* (1) and *Ablepharus* (1) are Palearctic in their affinities whereas all the species of *Varanus* (2), *Calotes* (1), *Ophisops* (1), *Mabuya* (1) and *Sitana* (1) are Oriental in their zoogeographical affinities (Table 1). *Phrynocephalus* and *Ablepharus* are the two exclusive genera of Palearctic affinities which are represented in the state of Jammu and Kashmir (Sahi and Duda, 1983). Of the 17 Saurian genera endemic to the Indian subregion of the Oriental region (Jayram, 1974), only *Sitana* has been able to ingress the foot hills of Outer Himalayas in Jammu region.

The status of *Hemidactylus brooki, H. Claviviridis, Agama tuberculata, A. agronensis, Calotes versicolor* is satisfactorily as they are well spread from the plains Punjab in Jammu region to Middle Himalayan zone in the state. *Eumeces tacinolatus* and *Varanus bengalensis*, which also occupy the same zone, are relatively poorer in their abundance. Like *B. Viridis* (amphibian), *Scincella himalayanum* is a wide spread species inhabiting all the three zones of Himalayas in Kashmir. *C. stoliczkai, C. lawderanus P. reticulatus* though endemic to Ladakh, occupying nival habitat in Great Himalayas, are relatively fair in number while other Ladakh endemic (*C. montinum salsorum, S. ladacense* and *P. theobaldi*) which are so far known by only few specimens. *C. himalayanum* is very recently described from Kishtwar in Jammu region (Duda and Sahi, 1978) and seems to be endemic form. Among the plain and foothills species of sandy habitat (open country) *M. dissimilis* is fairly common unlike *S. ponticeriana* and *A. pannonicus* which seem to be on way to establishing a new home in this region as it forms the northern and eastern limit respectively. *Ophisops jerdoni, Acanthodactylus cantoris, Eublepharus macularius, Varanus flavescens* are the other species of open country (sandy habitat) which are poorly represented in the state. *C. scaber* occupying the rocky habitat is other less known species.

Of the 24 species of lizards collected from the various zones of Kashmir Himalayas, 17 species (70.8%) are Palearctic and only 7 species (29.2%) are Oriental in their affinities. Some of the Palearctic species are exclusively confined to plains (*A. cantoris cantoris* and *A. pannonicus*) and foot hills *E. macularisu* and *C. scaber* of Outer Himalayas in Kashmir. Interestingly no species of Lizard with Oriental affinity is able to cross Outer Himalayas to reach Middle Himalayan zone in Kashmir.

The ophidians are represented by 32 species in 21 genera and 5 families. The species of genera, *Typhlops, Ramphotyphlops Eryx, Python, Boiga, Sphalerosophis, Dendrelaphis, Lycodon., Oligodon, Ptyas, Sibynophis* and *Xenochrophis* are Oriental (Table 1) whereas those fo *Coluber, Echis* and *Agkistrodon* are Palearctic in their zoogeographical affinities. Of the two species of each genus *Amphiesma, Elaphe Naja* and *Vipera* one is Palearctic and the other is Oriental.

P. molurus molurus, D. tristis, E. Johni, P. leithi, S. saguittarius and *N. naja naja* are Oriental forms which are more or less confined to plains and foot hills of outer Himalayas. These are relatively rare in the state. Other Oriental species of snakes occupying the same zone such as *R. braminus, T. porrectus, E. conicus, O. arnensis, L. aulicus, E. helena* and *E. carinatus* are fairly common. The Oriental species of snake which have managed to reach upto Middle Himalayan zones from the plains of Punjab are *A. stolata, B. trigonata, B. multifasciata, E. helena, S. diadema-diadema, S. d. atriceps, S. arenarius, X. piscator, B. caeruleus* and *V. russelli*. These species are also well represented in the state of Jammu and Kashmir. The ubiquitous *P. mucosus* is the only Oriental snake which have managed to traverse the high mountainous region of Great Himalayas. Likewise, *C. rhodorachis* and *E. hodgsoni* are the only species among Palearctic snakes which are found in all the three zones of Kashmir Himalayas but they are relatively poorer in abundance and yet to be recovered from the plains of Jammu. The other mountain species, which are Himalayan endemic and more or less confined between Outer and Middle Himalayas, are *A. platyceps, C. ventromaculatus L. striatus, P. shokari, A. himalayanus,* and *V. lebetina*. These are relatively less known Palearctic species.

Table.1 : Status, Altitudinal Ranges and Places of Collection of Amphibians and Reptiles in Kashmir Himalayas

Name of the species	Places of collection (Station Nos.)	Altitudinal range in meters	Habitat	Status	Affinities
	1	2	3	4	5
A. AMPHIBIA					
Family : Ranidae					
1. *Rana tigrina*	1-6,9,12	245-1010	A	C	OR
2. *Rana Cyanophlyctis*	1-7,9,10,12,14,15	245-1020	A	VC	OR
3. *Rana limnocharis*	4,9,12	275-1010	A, Sf	UC	OR
4. *Rana breviceps*	4	275	A	VR	OR
5. *Rana sternosignata*	14	1650	A,Tf	VR	PE
Family : Bufonidae					
1. *Bufo viridis*	14,21,24,25	1650-2680	A,Af	C	PE
2. *Bufo stomaticus*	1-12	245-1010	A, S	VC	PE
3. *Bufo latestei*	25,28	2680-3500	N,A	UC	PE
4. *Bufo melanostictus*	12	1010	A,Sf	VR	OR
Family : Pelobatidae					
1. *Scutiger occidentalis*	25,27	2680-2750	A, Af	UC	PE
Family Microhylidae					
1. *Microhyla ornata*	12,20	1010-1890	Rd,A	R	OR
B. CHELONIA					
Family : Emydidae					
1. *Kachuga tectum tectum*	1,3,4	245-275	A	UC	OR
2. *Kachuga smithi*	1,3,4	245-275	A	UC	OR
3. *Geoclomys hamiltoni*	3	260	A	VR	OR
Family : Trionychidae					
1. *Lissemys pinctata pinctata*	1,3-5,9	245-710	A	UC	OR
2. *Trionyx gangetious*	1,3,5, 9	245-710	A	UC	OR
3. *Trionyx hurum*	3	260	A	VR	OR
4. *Chitra indica*	1	245	A	VR	OR
C. LACERTILIA					
Family : Gekkonidae					
1. *Hemidactylus brooki*	1-7,9,12	245-1010	S, Sf	VC	OR
2. *Hemidactylus flaviviridis*	1-7, 9-12	245-1010	S,Sf	VC	PE
3. *Eublepharis macularius*	6	390	Rk	VR	PE
4. *Cyrtodactylus himalayanus* Yanus	15	1670	Rd, Tf	UC	PE

Contd...

Name of the species		Places of collection (Station Nos.)	Altitudinal range in meters	Habitat	Status	Affinities
		1	2	3	4	5
5.	*Cyrtodactylus scaber*	8	800	Rk	VR	PE
6.	*Cyrtodactylus lawdernus*	25,26,28	2680-3520	N	C	PE
7.	*Cyrotodactylus stoliczkai*	25-28	2680-3520	N	C	PE
8.	*C. montium salsorum*	28	3520	N	VR	PE
Family : Agamidae						
1.	*Agama tuberculata*	6-17, 19, 21, 22	390-2680	Rk	VC	PE
2.	*Agama agronensis*	6,7,9,10	390-1010	Rk	VC	PE
3.	*Agama himalayana*	24-26, 28	2720-3520	N	C	PE
4.	*Calotes versicolor*	1-13, 16	245-1010	S, Sf, Rd	VC	OR
5.	*Sitana ponticeriana*	4,5	275-290	S	VR	OR
6.	*Phrynocephalus reticulatus*	28	3520	N	R	PE
7	*Phrynocephalus theobaldi*	25,28	2680-3520	N	R	PE
Family : Scincidae						
1.	*Scincella himalayanum*	14,15,18-20, 22-28	1650-3520	Tf N	C	PE
2.	*Scincella ladacense*	26	2700	N	VR	PE
3.	*Mabuya dissimilis*	4,6,7,9	275-810	S, Sf	C	OR
4.	*Eumeces taeniolatus*	4,14,15,20	275-1890	S, Tf	UC	PE
Family : Varanidae						
1.	*Varanus bengalensis*	1-7,9–12	245-1010	S, Sf	C	OR
2.	*Varanus flavescens*	3	260	S	R	OR
Family : Lacertidae						
1.	*Ophisops jerdoni*	4, 9	275-710	S	UC	OR
2.	*Acanthodactylus cantoris-cantoris*	5	290	S	VR	PE
3.	*Ablenpharus pannonicus*	4	275	S	VR	PE
D. OPHIDIA						
Family : Typhlopidae						
1.	*Ramphotyphlops bramina*	4,6,7,9	275-710	VC	Rd	OR
2.	*Typhlops porrectus*	1-4,6-8, 11, 14	245-1650	S	VC	OR

Contd...

Name of the species	Places of collection (Station Nos.)	Altitudinal range in meters	Habitat	Status	Affinities
	1	2	3	4	5
Family : Boidae					
1. *Eryx conicus*	1,2,4,5,7,11,12 14,17	245-1828	S	VC	OR
2. *Eryx johni johni*	1, 2	245-657	S	VR	OR
3. *Python millurus molurus*	2,4,6,7	275-657	Sf	VR	OR
Family : Colubridae					
1. *Amphiesma platyceps*	12,15,18, 19, 21	1010-2000	Tf	UC	PE
2. *Amphiesma solata*	1-12, 15, 17, 21	245-1890	S, Rd Sf	C	OR
3. *Boiga multifasciata*	4,7,13, 15, 17, 21	275-1890	S,Sf Tf	C	OR
4. *Boiga trigonata*	1,2,4,7,10-12, 16, 17	245-1828	S, Sf Rd	VC	OR
5. *Coluber rhodorachis*	7, 14,21,28	610-3520	Tf, Rd	R	PE
6. *Coluber ventromaculatus*	13, 21	736-1890	Tf, Rk	R	PE
7. *Dendrelaphis tristis*	1,2	245-650	S	VR	OR
8. *Elaphe helena*	1,4,7,10-13	245-1010	S, Sf	UC	OR
9. *Elaphe hodgsoni*	12,19,24,25	1010-2720	Tf	UC	PE
10. *Lycodon aulicus*	1,4,5,7,11,12	245-1010	S, Rd Sf	VC	OR
11. *Lycodon striatus*	7,12,14	610-1650	Sf, Rd	UC	OR
12. *Oligodon arnensis*	1,4,5,7,12	245-1010	Rd,S	UC	OR
13. *Psammophis leithi*	5	290	S	VR	OR
14. *Psammophis schokari*	10	950	Sf	VR	PE
15. *Ptyas mucosus*	1-16, 18, 19, 25	245-2680	S, N Sf, Rd, Tf	VC	OR
16. *Sibynophis sagittarius*	1,4	245-275	S	UC	OR
17. *Sphalerosophic arenarius*	4,14	275-1650	S, Tf	UC	OR
18. *Sphalerosophis diadema diadema*	1,4,6,7,11,12	245-1010	S,T f St, Rd	C	OR

Contd...

Name of the species	Places of collection (Station Nos.)	Altitudinal range in meters	Habitat	Status	Affinities
	1	2	3	4	5
19. *Sphalerosophis d. atriceps*	1,7,15	245-1670	S, Tf Sf, Rd	UC	OR
20. *Xenochrophis piscator*	1,4,7, 9-12, 14 18, 20	245-2000	A,S Sf	VC	OR
Family : Elapidae					
1. *Naja naja naja*	1,4,5,7	245-610	S, Rd	C	OR
2. *Naja naja oxiana*	1,4,5,7	245-610	Rd, S	C	PE
3. *Bungarus caeruleus*	1,4,7,9,12,14,17	245-1828	S, Rd Sf	C	OR
Family : Viperidae					
1. *Vipera russelli*	1-7, 10-12, 14, 15	245-1670	S, Rd Rk	VC	OR
2. *Vipera lepetina*	19,22	1895-2680	Tf	VR	PE
3. *Echis carinatus*	1-7, 9-12	245-1010	S, Rk Rd	C	PE
4. *Agkistrodon himalayyanus*	8,12,14,15,19	800-1895	Tf	UC	PE

Abbreviations

A – Aquatic
Af – Alpine forest
C – Common
OR – Oriental
PE – Palearctic
R – Rare
Rd – Ruderal
Rk – Rocky
S – Sandy
St – Subtropical
Tf – Temperate forest
UC – Uncommon
VR – Very rare

Unlike lacertilian fauna majority of the ophidian species (71.9%) are Palearctic and the rest (28.1%) Oriental in their affinities.

In summering up, the amphibians of Jammu and Kashmir state are few and are derived equally from Palearctic and Oriental regions. The turtle fauna is impoverish and entirely Oriental in origin and restricted to 710 m only. These forms require immediate attention and scientific planning for their conservation. The lizard fauna is dominated by Paleatropical species from the northwestern areas of Jammu and Kashmir state. The snake fauna is; however, dominated by Oriental species of the Indian sub-region. Interesting no lizard of Oriental affinities is able to cross Outer Himalayas to reach Middle Himalayas, whereas amphibian of the Oriental affinities (*R. cyanophlyctis* and *M. ornata*) and about ten species of snakes (*T. porrectus, E. conicus, A. stolata, B. trigonata, B. multifasciata, L. striatus, S. arenarius. S. d. atriceps, X. piscator* and *P. mucosus* have crossed the Outer Himalayas to reach Middle Himalaya. The ubiqutous *P. mocosus* has managed to reach upto Greater Himalaya.

Two anuran species (*B. latestei, S. occidantalis*) and seven lacertilian species (*P. reticulatus, P. theobaldi, A. himalayana S. ladecense, C. lawderanus, C. stoliczkai* and *C. monntium salsorum*) are restricted to alpine areas of Kashmir Himalayas and of Palearctic affinities. Interestingly, all the endemic herptiles are from the alpine zones of Kashmir Himalayas which are geologically quite recent, having been established only after the Pleistocene glaciation (Kachroo, 1978; Mani, 1979).

References

Das, S. M. 1967. Palearctic elements in the fauna of Kashmir. Nature 212 (5068): 1327–1330.

Dubois, A. 1978. Une espece nouvelle de *Scutiger* Theobald 1868 de l 'Himalaya occidental. (Anura: Pelobatidae) Senckenbergina Biol. 59 (3–40): 163–171.

Duda, P. L., S. P. S. Dutta and D. Sahi 1977. Some tortoises from Jammu. Geobios 4: 263–264.

Duda, P. L. and D. N. Sahi 1978. *Cyrtodactylus himalayanus:* a new gekkonid species from Jammu, India. Herpetol. 12(3): 351–354.

Jayram, K. C. 1974. Ecology and distribution of freshwater fishes, amphibia and reptiles. In: Ecology and Biogeography in India. M. S. Mani (ed.). Dr. Junk b. v. Publ. Hague, Netherlands pp. 517–584.

Kachroo, P. 1978. A reflection on the flora of Kashmir. Bio Vigyanam. 4 : 13–28.

Mani, M. S. 1978. Ecology and phytogeography of high altitude plants of the North-West Himalayas. Oxford and IBH Publ. Co., New Delhi.

Minton, S. A. 1966. A contribution to the herpetology of west Pakistan. Bull. Amer. Mus. Nat. Hist. 134(2): 27–184.

Sahi, D. N. and P. L. Duda. 1978. Gedragvan *Natrix piscator* in gerangenschap. Lacerta (Netherlands), 36(1) : 119.

Sahi, D. N. and P. L. Duda. 1983. Notes on little known lizards from Jammu and Kashmir state. J. B. N. H. Society 90(1): 227–229.

Sahi, D. N. and P. L. Duda, 1985. A checklist and keys to the herptiles of Jammu and Kashmir State, India. Bull. Chi. Herp. Soc. 29 (32–34): 86–97.

Sahi, D. N. and P. L. Duda 1986. Occurrence of *Ablepharus pannonicus* Fitzinger (Lacertilia: Scincidae) in Jammu: an addition to the India reptilian fauna. Bull. Chi. Herp. Soc. 21 (21–34): 92–93.

Sharma, B. D. and T. Sharma 1975. Some new records of snakes (Reptilin: Serpentes) from Jammu and Kashmir State. Curr. Sci. 44 (17) : 646–647.

Smith, M. A. 1931. Fauna of British India, including Ceylon and Burma (Reptilia and Amphibia). Vol. 1. Loricata and Testudines. Taylor and Francis, London xii, 185 pp.

Smith, M. A. 1935. Fauna of British India, including Ceylon and Burma (Reptilia and Amphibia) Vol. II. Sauria. Taylor and Francis, London xii, 440 pp.

Smith, M. A. 1943. Fauna of British India, including Ceylon and Burma. (Reptilia and Amphibia) Vol. III. Ophidia. Taylor and Francis, London xii, 583 pp.

Chapter 25

Shrinking Wetlands of India

*S. K. Chadha**

Introduction

Mankind is heading towards environmental degradation and it is earnestly desired to ponder over the problem failing which the very existence of life will be in danger. Wetlands are areas of marshes, moors, fens, peatlands of water whether natural or artificial, fresh or saline, dynamic or static, perennial or temporary, etc. The very characteristic phenomenon about the wetlands is their closeness to the sub-surface water. In certain instances a little scratch makes the sub-surface water to appear on the earth. Mangroves, ponds and estuaries are important components of wetlands.

Wetlands are transitional bodies between marine bodies and the terrestrial ecosystems. They are the repository of fish varieties and organic materials. According to a survey nearly two third of the global fish harvest is from the wetlands, moors and ponds. On account of great diversity in climatic, physiographic and topographic conditions the country has substantial number of wetlands. Ecologically significant and biological productive wetlands are facing a serious threat due to enormous increase of the pollutants, weeds sediments and silt. The ever increasing rate of debris is due to deforestation. Every year about two million hectares of forests disappear and the total forest area has been reduced to about 35 million hectares from the original 80 million hectares. About 12,000,000,000 tonnes of fertile top soil is washed every year. Fairly a good proportion of the debris enters the wetlands, moors, marshes and ponds. The fauna and flora in the wetlands is adversely affected by the silt and sediments.

Poisonous gases that are released through nuclear explosions, industrial smokes and chemical weapons also enter the wetlands and destroy the organic life. Wetlands are under a great threat from the illegal encroachment, developmental activities and human interferance. The continued neglect and deterioration of the wetlands is a reflection of underestimation of the significance of these by the State and Central government. It is only in the present decade that some attention has been paid to the improvement of the marshes, moorlands, ponds, lakes and

* Controller of Examinations, University of Jammu, Jammu–180004.

wetlands. The World Wild Fund for Nature has promised to support financially for the rehabilitation and improvement of the wetlands.

Wetlands Under Threat

Among the wetlands of India, those which are under great environmental threat are:

(a) The Chilka Lake in Orissa.

(b) The Dal Lake in Kashmir.

(c) The Wular Lake in Kashmir.

(d) The Vembanad Lake in Kerala.

(e) The Loktak Lake in Manipur.

The mounting threat to all these wetlands is the enormous rise of silt/sediments, growth of weeds, illegal human encroachment and the influx of organic pollutants. In all there are 16 wetlands in the country.

The Chilka lake is certainly turning into a lagoon due to excessive sedimentation and this in turn is causing threat to millions of organisms that are present in the lake. Excessive weeding is also responsible for the shrinking of the lake. The outer channel that joins the lake with the Bay of Bengal is almost choked and has virtually blocked the mouth of the lake. This has also disallowed the entry of the sea water into the lake and has greatly damaged the estuarine fauna and flora. This has also affected the supply of the catch of the fish and also the economy of the natives. The limonologists

Eutrophicating Mansar Lake in Jammu Siwaliks (Photo courtesy: B.L. Kaul)

are of the opinion that the environmental threat has decimated some of the rare species of organisms. The harmful weeds are spreading around the Chilka lake and is damaging the aesthetic set up of the lake area. The growth of the unwanted weed coupled with other factors has reduced the northern portion of the lake into marsh. In order to conserve this wetland it is necessary that the work of deweeding, desiltation and dredging be initiated. The wetlands are the most fragile component of the environment and their conservation needs priority. From tourist's point of view also, it is desirable to monitor the upkeep of the lake. The economic stability of the fishermen depends purely on the survival of the lake.

Under the new action plan 16 wetlands need immediate attention. Of these the wetlands of Gujarat (Nalsarovar), Himachal Pradesh (Renuka), Punjab (Kanjali & Harike), Kashmir (Wular), Orissa (Chilka), Andhra Pradesh (Kolleru), Manipur (Lokat), Madhya Pradesh (Bhoj), Rajasthan (Sambar & Pichola) Maharashtra (Ujni), Kerala (Kotta), Bihar (Kabar), Chandigarh (Sukhna) etc. These wetlands have been selected for conservation by the National Wetland Committee.

The Wular is the biggest fresh water lake in Asia and has a water storage capacity of about 300,000 cubic feet. Bohnar, Madhumati and Erin join it but it receives bulk water supply from the Jhelum. The river Jhelum leaves it at the South-Western corner near Sopore and the lake acts as a bowl for refuse, dust and garbage. The night soil of the Srinagar city, hotels and industrial effluents has greatly polluted the water of the lake. The insecticides that are sprayed for agricultural crops and fruits finally find their destination in the lake. The excessive growth of weeds and accummulation of silt in the lake has further damaged its ecology. The area of the lake is shrinking due to illegal encroachment and/or reclamation. The haphazard and unplanned plantation of the willow has further damaged the geographical extent of the lake. The polluted water has reduced the fish supply as their mortality rate has greatly accellerated. The local species such as carp and *Schizothorax,* etc; have greatly been reduced. The Mahaseer a fish of great economic value which used to visit the lake during summer for breeding has stopped migrating from West Punjab due to construction of Mangla dam on the river Jhelum in Pakistan Occupied Kashmir. The migratory birds do not like to stay here on account of polluted conditions. The quantity of water nut has gone down and other species of lacustrine fruits and flora have become extinct.

The deterioration in the ecological conditions of the Keoladeo Ghana Sanctuary in Rajasthan has greatly affected the stay of migratory birds from Siberia. In the same way the industrial effluents entering the Vedaranyam swamp on Tamil Nadu coast which supports the famous bird sanctuary at point calimere has adversely affected the stay of the migratory birds in the area. The Southern Petrochemicals Industries Corporation which is producing caustic soda is greatly damaging the ecology of the swamp. The swamp serves as spawning, rearing and nursury area providing fin fish, shell fish, shrimps, lobesters and clams. If this wetland is properly monitored it can provide livelihood to millions of natives. The wetlands provide ideal sites for inland fisheries, fertilizers, foods and fuel. The wetlands control marine floods and natural sewage system. They are in fact the vital components of aquaculture. The wetland of Calcutta annually yield 8000 tonnes of fish and provide employment to over 20,000 fishermen, farmers, ragpickers and salt manufacturers. Same is true of Bombay, Madras, Goa and many other swamps of the country.

Conclusion

Reclaiming the wetlands is the biggest threat and if due measures are not taken now there will be a great loss to aesthetic set up, topography and aquaculture. Biologically deweeding is undesirable as it may adversely affect the ecological balance and efforts should be made to control it rather then eradicating it. Desiltation can be checked by afforestation in the watersheds and conserving the soils from erosion. Dredging can also improve the health of the moorlands and wetlands. Monitoring of the pollutants and checking the entry of industrial effluents can check the ecology of the wetlands.

References

Bedlock D. (1990). Agriculture and Habitat Loss in Europe. WWF/EFP.

Collins M. (1990). The Last Rain Forests. Mitchell Beazlay.

Dugan P. J. (1990). Wetland Conservation: a review of current issues and required action. IUCN, Gland, Switzerland.

Finalayson M. & M.E. Moser (1991). Wetlands. Facts on Fate, U. K.

Hoskins J. & M.E. Moser (1991). Wetlands. Facts on Fate, U. K.

Hoskins J. & A.W. Hopkins (1991). The Mekong: a river and its people. Post Publishing, Bangkok.

Jamsai S. (1988). Naga: Cultural origins in Siam and the West Pacific.

Kaul B. L. (1992). Eco-degradation of Dal Lake. In Himalayan Environment, Man and Economic Activities, Ed. J. L. Raina, Pointer, Jaipur.

Lee D. B. (1990). Okevango Delta: Old Africa's last refuge. Nat. Geog. Dec. 38–69.

Patrick Dugan (1993). Wetlands In Danger, Mitchell Beazlay & IUCN.

Pearce F. (1991). The rivers that won't be tamed. New Scientist, 13, April, pp 38–41.

Scott D. A. (1989). A Directory of Asian Wetlands, IUCN, Gland, Switzerland.

Struzik E. (1992). The rise and fall of Wood Bufflo, National Park, Birealis 10: 11–23.

Chapter 26

Ecology and Status of Wildlife in Ladakh

*B. L. Kaul**

Introduction

Ladakh is the land of passes (La–pass, dakh-land). Comprising of the districts of Leh and Kargil, it is the largest in area of the three main regions of Jammu and Kashmir state viz. Jammu, Kashmir and Ladakh. Its most striking features are the mountain ranges which stretch from the south-east to the north-east. Ladakh stretches between 32° – 36° N and 76° – 79° E. Although most of Ladakh is mountainous yet there are many valleys lying in the lap of the mountain ranges, which include the great Himalayan range, Zanaskar range, the Ladakh range and the Korakoram range. Like all the high altitude mountainous regions, Ladakh is sparsely populated being inhabited by about 1.5 lakh people. Although life is difficult at a high altitude of 2900–5900 metres yet both man and beast live there happily. The rich variety of wildlife in Ladakh bears testimony to this fact.

The landscape of Ladakh includes snow clad mountains, treeless arctic deserts and a few narrow fertile valleys, some of them breathtakingly beautiful. With the advent of summer a miraculous change occurs when life-sustaining vegetation lends colour to an otherwise bleak landscape. Ladakh is a large barren but beautiful region of intense sunlight and clear unpolluted air. The temperature in summer rises to 35 degrees to 37 degrees celcius but drops to - 30 degree to - 40 degrees celcius in winter. Dras in Kargil district experiences the lowest winter temperatures. There is very low rainfall in the region, about 92 mm. because it is situated in the rain shadow of the Himalayas. Most of the water available in Ladakh is in the form of frozen snow or ice.

Despite the extremely harsh environment and sparse, vegetation wildlife in Ladakh thrived before independence. This was mainly due to the Buddhist philosophy of non-violence. After independence, Ladakh became more approachable. Construction of roads for defence purposes and the development of the region ironically provided greater opportunities for killing the wildlife. The defence personnel also had a share in the killing because of their occasional indulgence in shikar, a legacy of the Raj days. In short the opening up of Ladakh to rest of the world has exposed its wildlife.

* Department of Zoology, Govt. College for Women, Parade, Jammu-180001.

Fauna of Ladakh

The rich fauna of Ladakh is a mixture of Palaearctic and oriental regions. It comprises mostly of mammals and birds. The region is poor in reptiles and amphibians. The mammalian population consists mainly of ungulates, rodents and carnivores. The avifauna include birds of prey, finches, plovers, larks, geese, grebes, pigeons, snowcocks, patridges, gulls, ducks and the famous "Tung tung" or black necked crane.

(1) Mammals

The most notable animal of the Ladakh, Tibet and adjacent parts of Central Asia is the wild yak (*Bos gruniens*). It is a large animal weighing about a tonne. It has long black hair which is tinged with grey at the muzzle. The horns are curved and long which may grow up to 75 cm in length. The wild yak which moves in herds spends summers at great heights and descends down to lower altitudes in winter.

Lama Yuru Ladakh

Another pride of Ladakh is nyan also called as Tibetian sheep (*ovis ammon*) found at a height of about 4500 m. A male may weigh upto a quintal. Dar (1983) has estimated their number to be around 200 in eastern Ladakh. A close relative of nyan is shapo (*Ovis orientalis*). It is smaller than the nyan, a large one being around 80 kg. in weight. It prefers heights between 3000 to 4000 meters. The bharal or blue sheep (*Pseudovis nahoor*) is a strange looking creature, half-sheep and half-goat. It differs from the goat in the absence of a beard. Likewise it differs from the sheep in the complete absence of glands on the face, (Gergan, 1962). The bharal are found at an altitude of 6000 metres. In summer they graze in large herds on rich and abundant grass of the meadows. Ladakh and its neighbourhood is also rich in wild goats. The

Asiatic ibex (*Capra ibex*) is the most beautiful of all. It supports two long scimiter-shaped horns marked with bold transverse knots on the front surface. They may be over a metre long. The mature males have a long beard and weigh over a quintal. They move in herds of 10 to 16 and prefer black precipitous rocks and cliffs. In winter they descend to the lower altitude for food and shelter.

The Tibetian antelope (Chiru) and gazelle (Goa) are two small but most valuable creatures which inhabit high altitudes in Ladakh. The chiru (*Pantholops hodgsoni*) is slender but stout looking due to the thick coat of interwoven pale-brown hair. The colours vary from season to season and from individual to individual. They roam the eastern fringe of Ladakh and inhabit Aksaichin and Tibetian plateau above 5000 metres. Their horns measure upto 70 cm. The animals are slaughtered in large numbers for the sake of under fleece-Shahtush - from which the famous shahtush shawls are made. Hardly a pound of under fleece is obtained from one animal. Among the other herbivores of Ladakh are the mouse hare, tibetian wooly hare, marmot and many other rodents. Marmots are common rodents found above the tree line. They live in burrows and produce shrill ringing calls. Marmots and the hare are a favourite food item of snow leopard, lynx, fox and large birds of prey. These herbivores hibernate during the winter.

In Ladakh one comes across the Himalayan bear (*Solenarctos thibetanus*) as well as the brown bear (*Ursus arctos*). The animal lives upto 4500 metres while the brown bears live higher up. The bears are omnivorous and feed upon anything available, plant or animal. The wolves and foxes are also found in Ladakh. There are two varieties of wolves. One has light brown fur while the other has darker fur. These move in pairs and predate on domestic animals, a fact for which they are hated by Ladakhis. However, some of them are trapped and killed for their skins. Foxes inhabit thorny bush lands and willow growths near to streams. They have a thick golden fur. They predate on marmots, mouse hare and partridges etc.

The most precious predators found in Ladakh are the snow-leopard or ounce and Tibetian lynx. The snow leopard (*Uncia uncia*) lives at great heights. It has a soft woolly coat. The upper side has a coat which is grey with cream or buff tinge and the underwool is dark grey. On the body there are irregular large rosettes with dark grey, while on the head and neck are solid black spots. The head is clear grey with white around the eyes and upper lip. The backs of the ears are black with a white patch. The forespaws are spotted while the hind paws are spotless. The snow leopard inhabits rocky hill sides above the tree line. Its natural food is ibex, shapo, bharal, markhor, marmot and hares. Occasionally it also kills domestic goat, sheep and ponies of nomads ets. Unless wounded a snow leopard never attacks a man. The skins of ounce fetch high prices - a lakh rupees per good piece, a fact which provides poachers enough attraction for killing them. Another predator, the tibetian lynx (*Lynx lynx isabellinus*) is an attractive animal. It is found at heights above 4400 metres. It is reported that there are several species of lynx found in J & K state of which the Tibetian lynx inhabits Ladakh.

The tibetian wild ass (*Equus kiang*) is a relative of the wild ass of the Rann of Kutch. It has a wide distribution. Besides Ladakh it is found in Tibet, Turkistan and Mangolia. Kiang is shy but an inquisitive animal. It inhabits plain and rocky grounds

Asiatic Ibex *(Capra Ibex)*

Nyan *(Ovis ammon hodgsoni)*

and can gallop at a good speed. Solitary animals or groups of 10 to 12 animals can be seen. The colouring of the upper portion of the body is chestnut or red and the underparts are white. There is a dark brown dorsal stripe extending from the stiff black name to the tuft of the tail

(2) Birds

Despite its inhospitable climate and high and dry terrain Ladakh has a varied avifauna most of which are migrants. There are a few residents also, the most spectacular of them being the snowcocks and patridges. The most well known migrant bird found in Ladakh is the black-necked crane (*Grus nigricollis*) called "Tung-Tung" in Ladakhi. These are large graceful birds which inhabit open marsh-lands, wet plains and prairies. The body is pale grey with black head, neck and feet. The bill is greenish and the legs are large. Black-necked cranes visit Ladakh during the summer months for nesting. Hussain (1985) has reported sighting of 3 breeding pairs in Ladakh region. The nest is built on a mound inthe centre of a marsh from marsh-grass and weeds. Two dull white or brown eggs are laid in the nest which is well protected by the parents from predators.

The Himalayan snow cock (*Tetragallus himalayensis*) inhabits stony slopes between 4000 to 5500 metres. It is a large majestic bird and descends to 2400 metres in extreme cold. These birds are distributed throughout Ladakh and Kashmir but their population has very much declined due to excessive hunting. The chukor patridge (*Alectoris chukor pallasus*) are colourful birds found widely in Ladakh and Kashmir. They breed at a height of above 5000 metres. 5 to 7 eggs are laid in a nest made under the bushes.

Some other birds of Ladakh are:-

1. Eagle (*Cuncuma leucorypha*)
2. Vulture (*Gyps himalayanus*)
3. Kite (*Milvus migrans lineatus*)
4. Owl (*Tyto alba javanica*)
5. Kestrel (*Falco tinnuculus*)
6. Chugh (*Pyrrhocoax pyrrhocoax*)
7. Mongolian plover (*Charadrius mongolus*)
8. Red Shank (*Triga totanus*)
9. Great Rosefinch (*Carpodacus rubicilla*)
10. Barheaded geese (*Anser indicus*)
11. Black necked grebe (*Podiceps nigricollis*)
12. Horned lark (*Eremophila alpestris*).

Threat to Wildlife

According to Dar (1995) the population of the most of the wildlife in Ladakh has dwindled greatly in the recent past. Ecodegradation of Ladakh Himalaya due to vagaries of nature together with unsuitable conditions created by man have resulted in a serious situation. Many of the species are now facing danger of extinction. Due to difficult terrain, extreme cold and low oxygen content in the air very few people have ventured to make a serious integrated study of the present status of wildlife in Ladakh Himalaya. However, such a study is necessary because it will throw some light on the causes responsible for dwindling populations and enable us to arrive at some conclusions in order to seek remedial measures to help preserve wildlife in this ecologically important part of our country. The region is also interesting from biogeographical and evolutionary points of view because it includes animals which are a mixture of palaearctic and oriental regions. The study of the distribution and evolution of fauna in the region is closely linked with the evolution and rise of the great Himalayas.

References

Atul and Punam (1989). "Potentials of Agroforestry in Himachal Himalaya" in *Environmental Hazards in Himalayas*, pp. 37-56, S. K. Chadha (Ed.) Pointer Publishers.

Chadha, S. K. (1989). *Environmental Hazards in Himalayas*, pp. 1-13, Pointer Publishers, Jaipur.

Dar, Moti (1983). "Ladakh" in *Wildlife of J & K*, Sanctuary Publications.

Dar, Moti (1995). "Wild-life of Ladakh" in Ecodegradation of Himalayas. (Ed. B. L. Kaul) Yak Book Channel Jammu.

Ganher J. N. (1979). *The wild-life of Ladakh*. Harmukh Publications, Srinagar.

Gergan S. S. (1962). "Big game of Jammu and Kashmir" A Govt. of J & K Publication.

Kaul B. L. (1987). *Ecology and wildlife of Kashmir Himalaya.*

Kaul B. L. (1989). "Ecology and Wildlife of Ladakh" in *Ecological Hazards in the Himalayas* pp. 66-74. S. K. Chadha (Ed.).

Chapter 27

On the Freshwater Chelonian Fauna of Jammu and Kashmir State

D. N. Sahi and Anil K. Verma

Introduction

Jammu and Kashmir water bodies are the least investigated for herptelian elements and therefore, the freshwater chelonian fauna of the state is unknown and no specific information on these animals can be found in the literature save for (Duda *et al.*, 1977; Sahi and Duda, 1985). As a part of studies on Herpeto fauna of Kashmir Himalayas, supported by D. O. E., New Delhi, a survey of freshwater chelones of J & K state was undertaken during the years 1988-92. The riverine systems of Indus, Jhelum, Chenab and Tawi; the perennial water bodies; lake Mansar, Surinsar and Dal lake; the ponds and ditches along the course of rivers, associated reservoirs and nearby village ponds were surveyed for chelones.

During present investigations, authors discovered 8 species of freshwater chelones contained in 5 genera and belonging to 2 families representing the freshwater chelonian fauna of the state.

In reviewing these chelones; we have surveyed the literature pertaining to Indian turtles (Boulenger, 1890; Smith, 1931; Moll, 1984 & Das, 1985). The representative forms of the collected species have been preserved and deposited in the Museum of the Department of Biosciences, University of Jammu, Jammu.

Study Area

J & K state displays a blend of climate and ecology ranging from sub-tropical plains to temperate hills between 36° 58' N and 32° 17' N latitude, and from 80° 20' E to 76° 20' E longitude. The physiography and topography of the state has been explained elsewhere in detail (Sahi & Duda, 1985). Chelones were discovered only in the Jammu province of the state and the localities wherefrom chelones were recovered/discovered are already recorded in J & K Gazetteer. The latitude, longitude and elevations of the collection sites are at best only approximations.

Methodology Followed

A full description of the techniques by which collection of the chelones was done

* Department of Bio-sciences, University of Jammu, Jammu, J & K State-180001, India.

has been described elsewhere (Verma, 1992). Nomenclature followed is that of Smith, 1931; Moll, 1984; Das, 1985 and Iverson 1986. CL = Carapace length; CW = Carapace width; PL = Plastron length; PW = Plastron width; SH = Shell height; W = Weight of chelone; SnL = Snout length.

Results and Discussion

A total of 72 waterbodies (31 in Jammu; 7 in Kathua; 10 in Udhampur; 6 in Rajouri; 4 in Poonch and 6 in Doda districts of Jammu Province; and 3 in Srinagar; 2 in Kargil and 3 in Leh district of Kashmir province) were extensively surveyed to explore possible presence of chelonian fauna in the waterbodies of above mentioned places. In all about 345 specimens of chelones, 337 live and 5 dead were collected, besides 3 chelone shells picked from fields and ditches.

Altogether, 8 species of chelones *(Table 1)* were recorded during the surveys; three hard shells viz., *Geoclemys hamiltoni; Kachuga tectum tectum* and *K. smithi;* and remaining five viz., *Lissemys punctata punctata; Trionyx gangeticus; T. hurum; Trionyx* sp. and *Chitra indica,* the soft shells.

During present investigations, following eight species of freshwater chelones inhabiting various lotic as well as lentic water bodies of the state were recovered/discovered from the Jammu province. The colour, measurements and other descriptive data given here under are from live specimens.

Family: Emydidae

1. Spotted pond turtle *Geoclemys hamiltoni* (Gray)

Material examined: (2; 1 juvenile and 1 adult male). 1 juvenile from Satwari, Jammu; 1 adult male from R. S. Pura, Jammu.

Measurements: CL 125 - 224 mm; CW 108-152 mm; PL 91-160 mm; PW 84-135 mm; SH 52-86 mm.

Table 1 : Showing Freshwater Chelonian Fauna of J & K State

	Turtle species	*Common name*	*Collection site*
Family: Emydidae			
1.	*Geoclemys hamiltoni* (Gray)	Spotted pond turtle	3, 6
2.	*Kachuga smithi* (Gray)	Brown Roof turtle	1-4, 6-7, 10
3.	*K. tectum tectum* (Gray)	Indian Saw back turtle	1-4, 6
Family: Trionychidae			
4.	*Lissemys punctata punctata* (Bonnaterre)	Indian flap shell turtle	1-2, 6, 8-9
5.	Trionyx gangeticus Cuvier	Indian soft shell turtle	1, 6, 8-9
6.	*T. hurum* (Gray)	Indian Peacock soft shell turtle.	6
7.	*Trionyx* sp.	Soft shell plather (proposed)	3, 6
8.	*Chitra indica* (Gray)	Narrow headed soft shell turtle	7

Contd...

Collecting site (s)		No.	Collecting site (s)		No.
i.	Gho-Manhasan	1	ii.	Gajansu	2
iii.	Gadigarh	3	iv.	Miran Sahib	4
v.	Sehi	5	vi.	R. S. Pura	6
vii.	Kathua	7	viii	Mansar	8
ix.	Surinsar	9	x.	Satwari	10

Description: Carapace tricarinate, oblong in outline widest at the posterior end of second vertebral shield, width 70-80 per cent of its length, strongly arched and highest a little ant. to mid-point, shell height about 38-41 per cent of carapace length, margin slightly serrated posteriorly. Plastron rigidly joined to carapace, deeply notched behind; plastron shields in full complement, upper jaw slightly projected, snout as long as orbit; nostril terminal, edges of jaws smooth, lower with central cusp; alveolar surface of both jaws well expanded; skin or hinder part of head divided into shields; fore limb with strong scales in transverse rows; hind limbs less scaly; digits webbed to bases of claws.

Colouration in life: Carapace black with a lattice work of yellow; plastron shields brown with radially arranged spots and striae, head black with lemon yellow spots, throat spotted but less intensely; limbs gray to black, sparsely spotted.

Distribution: J & K state: river Tawi in Jammu district of the Jammu province. Elsewhere: northern to north-eastern India (Das, 1985).

2. Brown Roof Turtle *Kachuga smithi* (Gray)

Material examined: (136; males 46, females 90). Collected from Gho-Manhasan 117, Gajansu 5, Satwari 3, Gadigarh 1, Miran Sahib 3, R. S. Pura 5 in Jammu district; and Kathua town 2, in Kathua district of Jammu province.

Measurements: CL 99-236 mm; CW 85-208 mm; PL 85-219 mm; PW 78-187; SH 54-85 mm; W 95-1327 g.

Description: Carapace ovoid, depressed with interrupted central keel, carapace width 80-92% of its length, widest just anterior to hind limbs, strongly arched, highest at or around the middle; shell height 31-39% of carapace length, Plastron nearly as long as the carapace, truncate anteriorly, posterior end widely notched, head moderate, snout shorter than orbit, nostril sub-terminal, edges of both jaws finely serrated, large terminal cusp on lower jaw; skin of hinder part of head divided into shields; limbs with transversely enlarged scales; digits webbed to beyond bases of nails.

Colouration in life: Carapace olive brown, plastral shields black with yellow border, head olive to greyish yellow, limbs pale-grey, webb dull yellow.

Distribution: J & K state: water bodies of Jammu & Kathua districts of Jammu province. Elsewhere: Northern to eastern India (Das, 1985).

3. Indian Saw back Turtle *Kachuga tectum tectum* (Gray)

Material examined: (133; 47 males and 86 females). Collected from Gho-Manhasan 114, Gajansu 5; Gadigarh 3; Miran Sahib 5 and R. S. Pura 6, all in Jammu district.

Measurement: CL 100-215 mm; CW 79-195 mm; PL 76-180 mm; PW 74-165 mm; SH 40-84 mm; W 90-1050 g.

Description: Carapace elevated, shell height 36-47% of carapace length, hinder margin feebly serrated; first central lamina variable in shape; second central lamina normally longer than third, obtusely pointed behind; third pentagonal, pointed anteriorly as a short blunt spine; the fourth central longer than all other centrals; fifth central broader than long and broadest of all other centrals. Remaining morphometric details as in *K. smithi*.

Colouration in life: Carapace green with small black dots and an orange central stripe; with age shell gets darker and turns olive. Plastron yellow to orange with irregular black spots; head blackish, neck with narrow pale-green stripes, limbs dark grey and yellow spotted.

Distribution: J & K State: As in the case of *K. smithi* with which it is found sympatric except that it is less common in rivers. Elsewhere: Northern to north-eastern India (Das, 1985).

Family: Trionychidae

4. Indian Flapshell Turtle *Lissemys punctata punctata* (Bonnaterre)

Material examined: (40; 16 males and 24 females). Collected from R. S. Pura 11; Gajansu 10; Gho-Manhasan 7 Mansar lake 9 and Surinsar lake 3.

Measurements: CL 184-310 mm; CW 175-268 mm; PL 168-265 mm; PW 130-205 mm; SH 54-113 mm; W 652-2900 g.

Description: Carapace circular to oval in outline widest just anterior to hind limb; width of disc 87-95% of its length, carapace moderately arched, shell height 25-36% of carapace length, margin of carapace smooth, slightly flared posteriorly. Plastron nearly as long as (85-95%) carapace, mostly cartilagenous and with 7 callosities. Head moderate to large proboscis short and stout; edges of jaw smooth, claw large and heavy.

Colouration in life: Carapace olive green with numerous yellow spots of varying shapes and sizes. Plastron creamish-yellow. Head and limbs grey with light yellow spots on head and neck.

Distribution: J & K State: The species is found in irrigation channels, streams, distributaries of the river Chenab and ponds of R. S. Pura, Gadigarh, Gajansu, Golgujral of district Jammu; Kathua district; and lake Mansar and Surinsar at an altitude of 665 metres in district Udhampur of the Jammu province. Elsewhere: Peninsular India, Andaman Islands (Das, 1985).

5. Indian Soft Shell Turtle *Trionyx gangeticus* Cuvier.

Material examined: (21; 8 males, 13 females). Collected from R. S. Pura 4; Gho Manhasan 3; Surinsar lake 3 and Mansar lake 11.

Measurements: CL 440-530 mm; CW 349-403 mm; PL 329-391 mm; PW 302-370 mm, SH 91-110 mm.

Description: Carapace oval in adult, widest near the mid of the body; arch of carapace low; vertebral region slightly depressed in adults; carapaceal plates Hyoplastron, Hypo-plastron and Xiphiplastron strongly rugose and coarsely pitted in adult; plastron length 72-75% of carapace length; head broad and massive with long proboscis; edges of jaws smooth, snout on the skull nearly as long as the diameter of the adult.

Colouration in life: Carapace of young adult olive-green with black reticulation, plastron whitish in colour; head and limbs greenish with a black longitudinal streak from between the eyes on to the nape and three oblique black stripes on either side diverging from it, another streak starting from behind the eyes.

Distribution: J & K State: River Tawi and its tributaries, R. S. Pura in Jammu and lake Mansar and lake Surinsar in Udhampur district; and Kathua district. Elsewhere: Northern to eastern India (Das, 1985).

6. Indian Peacock Soft Shell Turtle *Trionyx hurum* Gray.

Material examined: 1 collected from R. S. Pura, Jammu district.

Measurement: CL 250 mm; CW 208 mm; PL 228 mm; PW 215 mm.

Description: This turtle species resembles *Trionyx gangeticus* in nearly all essential characters excepting that carapace in *T. hurum* has 4 well developed ocelli. Head and neck lacks the longitudinal streaks; yellow and black spots present, angles of mouth yellow, two large spots on temple. Inner margins of mandible not raised, a groove in the middle.

Colouration in life: Like that of *T. gangeticus* except that carapace has dull reticulations and plastron is brownish grey in colour.

Distribution: J & K State: R. S. Pura of Jammu district, Jammu province. Elsewhere: Northern to north-eastern India (Das, 1985).

7. *Trionyx* sp.*

Material examined: (3; 1 male and 2 females). Collected from Gadigarh stream 1; Sehi stream (R. S. Pura) 2, in Jammu district.

---

* For lack of good sample size the designation of separate species status has been withheld for the time being.

Measurements: CL 116-178 mm; CW 87-131 mm; *SL 38-50 mm; *SW 21-27 mm; SnL 7-8.5 mm.

Description: Carapace disc-like, head large, oblong with a large proboscis, interorbital space considerably narower than diameter of the eye; carapace subcircular, coarsely tuberculate with tubercles in longitudinal series though not well marked. Disc finely reticulated with 4 black conspicuous ocelli. Longitudinal black-streaks from between the eyes runs posteriorly onto the nape, two oblique black streaks diverging from it, another black streak starts from behind the eyes to end on the temporal region, super-ciliary regions marked with black spots, two spots on interorbital space and one on each side of temple.

Colouration in life: Carapace dull coloured with coarse tubercles, head and neck olivaceous with definite black markings; limbs dark olive green, underpart grey.

Distribution: J & K State: Jammu district of the Jammu province. Elsewhere: unknown (?).

8. Narrow headed Soft Shell Turtle *Chitra indica* Gray.

Material examined : 1 Collected from a stream at Kathua.

Measurements: CL 206 mm; CW 151 mm; PL 155 mm; PW 143 mm; SH 46 mm.

Description: The specimen a young one possessed the features as described by Das (1985), carapace low and disc-like or oval in outline, bearing wavy reticulations. Head small and narrow or elongated, snout round; eyes and nostrils closely associated; a V-shaped mark present on the neck.

Colouration in life: Carapace dull olive or grey in colour with a diagnostic prominent wavy pattern. Plastron is creamish or white in colour.

Distribution: J & K State: Kathua district of Jammu province. Elsewhere: Northern to eastern India (Das, 1985).

Summary and Conclusion

Interestingly no freshwater chelones could be located in Jhelum and Indus river systems and its tributaries or other perennial water bodies of Kashmir and Leh regions of J & K state. No testudine have ever been sighted in any parts of the state inspite of our several years survey. All the freshwater chelones have been recorded from the various districts of Jammu province only and that too from the height below 700 m.

In the present survey, 8 species in 5 genera belonging to 2 families have been recorded (Table 1).

* SL = Skull length

* SW = Skull width

Four species namely, *Kachuga smithi, Chitra indica, Trionyx hurum;* and *Trionyx* sp. were recorded from the lotic water bodies i.e., river Tawi and river Chenab drainage system. Whereas, remaining four species viz., *Lissemys punctata, punctata; Trionyx gangeticus; Geoclemys hamiltoni* and *Kachuga tectum tectum* were found inhabiting both the lentic as well as lotic water bodies.

Kachuga tectum tectum and *K. smithi* were found sympatrically in the lotic water bodies (Gho-Manhasan stream and Gajansu stream) as were also seen *Lissemys punctata punctata* and *T. gangeticus* in the lentic water bodies like lake Mansar and lake Surinsar. *Lissemys punctata puncata* was also recovered along with the hard-shells - *K. t. tectum* and *K. smithi* in the Gho-Manhasan stream and marshes in Jammu district.

Geoclemys hamiltoni, Trionyx hurum; Trionyx sp. and *Chitra indica* have been recovered in very small numbers and that too in the restricted sites in the district Jammu (R. S. Pura) and Kathua.

Hence, the freshwater chelonian fauna of J & K state, as it emerges out from the present investigation is distributed in Jammu province and is constituted by 8 species which constitutes about 36.3% of the total freshwater chelonian fauna reported from rest of the India (Das, 1985).

Acknowledgement

Authors are grateful to the Department of Environment and Forests, Govt. of India, New Delhi for providing financial assistance under the research project, "Herpeto-fauna of Kashmir Himalayas etc.."

References

Boulenger, G. A. 1890. The Fauna of British India including Ceylon and Burma. Reptilia and Amphibia (Taylor & Francis): 541.

Das, I. 1985. Indian Turtles. A Field Guide. World Wildlife Fund India, Eastern Region: 1-119, Ms. Bonani Kakkar (Publ.)

Duda, P. L., Dutta, S. P. S., and Sahi, D. N. 1977. Some tortoises from Jammu. *Geobios*, **4**: 263-64.

Iverson, J. B. 1986. Checklist with distribution maps of the turtles of the world. Paust Printing, Richmond, Indiana.

Moll, E. O. 1984. Freshwater turtles in India: Their Status, conservation and Management. *Hamadryad*, **9** (3): 9-17.

Sahi, D. N., and Duda, P. L. 1985. Checklist and keys to the herptiles of Jammu & Kashmir State. *Chic. Herp. Soc.* *20* (3-4): 86-97.

Smith, M. A. 1931. The Fauna of British India, Including Ceylon and Burma. Reptilia and Amphibia 1. Loricatas, Testudines. Taylor Francis (Eds.), London.

Verma, A. K. 1992. Studies on the Male Reproductive Cycles of Some Emydid Turtles of Jammu. Ph. D. thesis, University of Jammu, Jammu.

Chapter 28

Threatened Wildlife Habitat in Kashmir Himalayas

B. L. Kaul

Introduction

Pandit Jawaharlal Nehru's love of nature animated his very soul, reflecting a poet's dream. Mountains, valleys, trees, flowers and animals symbolized for him the multifarious and innumerable aspects of creation. Nehru's first visit to Kashmir after his marriage in 1916 cast a spell on his mind and soul which can be discerned in these words: "Like some supremely beautiful woman whose beauty is almost impersonal and above human desire, such was Kashmir in all its feminine beauty of river and valleys and lake and graceful trees." His love for the mighty Himalayas and wildlife is also well known, as he himself remarked that life would be dull and colourless without magnificent birds and animals.

Unfortunately in a span of four decades after India gained freedom our hills and mountains wear a desolate look as a mute testimony to the virtual desecretion of nature's charm and beauty. Much of the forest cover that used to be the pride of the lofty Himalayas and adorn them with rupturous charm, has disappeared, thanks to the unchecked activities of some unscrupulous people who through an indiscriminate felling of trees denuded the mountains and hills rendering them ecologically useless and counterproductive. As a natural corollary to the ruthless deforestation, many species of wildlife in the Himalayas in general and in the Kashmir Himalayas in particular, are facing the threat of extinction.

The author would suggest that a serious study be made to co-relate the ecodegradation of a wide range of natural habitats to the problems connected with the conservation of a variety of fascinating but endangered species of wildlife in Kashmir Himalayas. The study may prove beneficial in evolving steps towards the preservation of our precious wildlife heritage, besides being a humble homage to the memory of such a humane and many splendoured personality as Pandit Jawaharlal Nehru whose birth centenary year has just passed.

Kashmir Himalayas

The three main constituent regions of the state of Jammu and Kashmir are the sub-Himalayan Jammu, the Himalayan Kashmir and the Trans-Himalayan Ladakh.

* Department of Zoology, Government College for Women, Parade Jammu-180001.

These three regions can be likened to the three steps of a terrace, the lower step being the Jammu region, the middle one as the Kashmir valley and the upper one as Ladakh. In the Jammu region we have a range of hills known as Siwaliks and to the north of these Siwaliks are the middle mountains called as the lesser Himalayas or the Pir Panjal range. The Kashmir valley is enclosed by the Pir Panjal range in the south and by the Greater Himalayas in the north. Ladakh is the land of mountain passes with the mountain ranges constituting the most striking features of the land. These ranges stretch from the south-east to the north-west and include the Greater Himalayan range as well besides the Zanskar range, the Ladakh range and the Korakoram range. Many valleys nestle in the laps of these formidable mountain ranges.

Forests and Vegetal Cover

The type of vegetation found in Jammu and Kashmir is typical of western Himalayas, of which the state is a part. A largely mountainous terrain dominates the twin provinces of Jammu and Kashmir and rainfall is sufficient to favour growth of vegetation including trees. But in Ladakh region the cold and arid climate does not favour the growth of forests.

The outer slopes of Siwaliks in the Jammu region are covered with evergreen bushes and bear an open scrub forest. The slopes of the middle mountains are also clothed in temperate monsoon forests. Important species of trees found in the region are deodar, chir pine, blue pine, silver fir and oak. The mountain ranges surrounding the valley of Kashmir abound in deodar, blue pine, silver fir, birch trees and alpine brush wood of *Juniper* and *Rhododendron*. Besides these types of trees, willows and poplars also abound in the plains of the valley.

Before 1947, Jammu and Kashmir state boasted of a well managed forest wealth. However, after independence the forests which till then had been protected from the ravages of deforestation became victims of vandalism through an indiscriminate felling of trees. According to the figures available, the state has now 1.95 lakh hectares of degraded forests of which 1.10 lakh hectares are culturable and the rest i.e. 0.85 lakh hectares constitute a "dead loss". This is without doubt a matter of great concern, not only in terms of economy but also from the ecological point of view.

With the fast disappearance of forests, wild animals too have fallen an easy prey to poachers and the trigger-happy hunters. The situation has already assumed alarming proportions, and some of the most beautiful and rare wild species like Hangul, Snow-leopard, Spotted deer, Musk deer and Himalayan Monal are threatened with extinction in their respective habitats. Biologically it is a well known fact that a close relationship exists between an organism and its environment and any tampering with the habitat, automatically results in upsetting the natural life of the organism. Besides playing havoc with the natural habitats of wild animals deforestation in Jammu and Kashmir has also brought in its wake floods as a result of soil erosion on the mountain slopes. Frequent land-slides on the Srinagar- Jammu and Srinagar-Leh highways are also a direct consequence of eco-degradation of the Kashmir Himalayas which has been going on unchecked for quite some time now.

Disappearing Wildlife

In the not so distant past, the state of Jammu and Kashmir boasted of a rich and varied wildlife. Unfortunately due to rapid depletion of the forest cover wildlife has been rendered vulnerable to reckless hunting and poaching. A stage has now come where some of the species of wild animals have come to the verge of extinction thereby finding inclusion in the red book of International Union of Conservation of Nature (I. U. C. N.). Unless immediate remedial steps are taken on a priority basis, we shall soon lose some of the finest wildlife species in our part of the Himalayas. Jammu and Kashmir state has the unique distinction of having the "Hangul" (*Cervus elephas hanglu*) found nowhere else in the world and it has already become an endangered species. Similarly, snow-leopard (*Panthera uncia*), Markhor (*Capra falconeri*) Musk deer (*Moschus moschiferous*) are other rare and beautiful animals facing danger of extinction.

Due to the varying type of terrain and climate, ranging from the plains of Jammu and alpine forests and shallow water bodies of Kashmir, to the lunar type mountains of Ladakh a rich bird fauna abounds in Kashmir Himalayas. These birds include the beautiful pheasants, partridges and water birds. Partridges and pheasants remain confined to dense forests ranging in altitude from 3000 to 15000 ft. above the sea level. Some of these birds like Western Tragopon (*Tragopon melanocephalus*) and cheer pheasant (*Catreus wallichi*) have been listed among endangered species in the red book of I. U. C. N.

What Can Be Done?

Although fragmentary accounts are available on the geomorphology vegetation, fauna and wildlife in Kashmir Himalayas, yet so far no attempt has been made to have a realistic and comprehensive overview of the situation as it exists today. The available data are scattered in different journals and periodicals and there is no single publication to which a layman or even an expert can refer. Further, there has been no serious attempt to co-relate the ecodegradation with the existence of wildlife in Kashmir Himalayas. As such the author suggests that the following aspects of the problems of conservation of wildlife be studied in detail:

1. Assessing the eco-degradation of natural habitats pertaining to the wildlife of Kashmir Himalayas.
2. Collection of information related to present status and distribution of endangered species of mammals and birds.
3. Study of predator -prey distribution in natural ecosystems and the role played by man in disturbing this relationship.
4. Collection of data regarding viable breeding populations of endangered species of wildlife.
5. An assessment of the role played by man and natural agents in eco-degradation of the habitats of wildlife.
6. Cataloguing the rich zoogeographic diversity of wildlife in Kashmir Himalayas, a combination of oriental and palaearctic fauna.
7. Suggesting remedial measures for restoration of habitats and conservation of endangered species of mammals and birds.

References

Atul and Punam (1989). "Potentials of Agroforestry in Himachal Himalaya" in *Environmental Hazards in Himalayas* pp. 37-56, S. K. Chadha (Ed.) Pointer Publishers, Jaipur.

Chadha, S. K. (1989). *Environmental Hazards in Himalayas*, pp. 1-13, Pointer Publishers, Jaipur.

Dar, Moti (1987). "Ladakh" in *Wildlife of J & K*, Sanctuary Publications.

Dar, Moti (1995). "Wildlife of Ladakh" in Ecodegradation of Himalayas (Ed. B. L. Kaul) Yak Book Channel, Jammu.

Granhar, J. N. (1979). *The Wildlife of Ladakh*, Harmukh Publications, Srinagar.

Gergan, S. S. (1962). "Big game of Jammu and Kashmir" A Govt. of J & K Publication.

Kaul, B. L. (1987). *Ecology and Wildlife of Kashmir Himalaya*, S. K. Chadha (Ed.)

Kaul, B. L. (1989). "Ecology and wildlife of Ladakh" in *Ecological Hazards in the Himalayas*, pp. 66-74. S. K. Chadha (ed.)

Kaul, B. L. (1994). "Call of the wild" In the *Hindu*, Madras (May 1, 1994).

Chapter 29

Role of Gastropods in Trematode Transmission Among Herptiles
Part - I Amphibia: A Numerical Analysis

*Anil K. Verma and P. L. Duda**

Trematodiasis afflicts farm and wild animals in equal measure. Snails are important vectors in the dissemination and transmission of several such diseases in amphibians and reptiles like they do so in other vertebrates. Paramphistomiasis, Opisthorchiasis and Dicrocoeliiasis are a few of the many such most dreaded diseases. Snail constitutes the essential first intermediate host, sometimes the only intermediate host of many digenetic trematodes. The presence of susceptible snail host is thus a primary requirement for the establishment of trematode infection. In India, only a few workers (Sewell, 1922; Annandale and Rao, 1925; Mohandas, 1974, 1984) have reported larval trematodes from snail host of a particular locality or from a new snail host.

The digeneans exhibit many variations in life cycle patterns (Heyneman, 1960). In a study on the life cycles of digeneans of birds and mammals, Ewers (1964) found that in the life cycles of 279 spp. of digenetic trematodes, the molluscs belonging to families viz., Lymnaeidae, Physidae, Planorbidae, Thiaridae and Hydrobiidae were by for the most important hosts.

The present contribution is a numerical analysis of the data available in literature (Dawes, 1956; Yamaguti, 1975; Erasmus, 1972; Olsen, 1974) on the life-cycles of the digenea of Amphibia. This includes 13 genera of snails belonging to 6 families and 2 subclasses of the class Gastropoda, Phylum Mollusca. Wide range of variations in the percentage of infection in total snail population is available from different regions.It has been observed that the population of snail spp. and frequency of infection varies from place to place and also with the season (Cross, 1971; Erasmus, 1972). Incidence of infection in snails with larval trematodes is higher in ponds than in any other habitat (Das Mahapatra, 1983; Majumdar, 1983).

Most cercarial spp. are host specific and are restricted to one molluscan host,

* Department of Biosciences, University of Jammu, Jammu - 180001, India.

* Paper presented at "First International Conference of the IUCN /SSC-ISRAG" at Utkal University, Bhubaneshwar (Orissa) 23-25 Feb., 1992.

most presumably due to the selection mechanism exhibited by the Miracidium. It has been analysed that of the different families of digenean parasites of amphibians, *Helisoma* and *Radix* harboured larval stages of 5 families (45.45%) followed by *Lymnea* (36.36%) in 4 families; *Planorbis* and Physa (27.27% each in 3 families) *Bulinus, Gyraulus* and *Coretus* (18.18% in 2 families) leaving hardly a few gastropods to harbour a particular cercarial type of a family with incidence rate being just 9.09% (Table 1).

Interestingly, majority of these vector-snails (*Physa* and *Bithinia* excepted) belong to two families of snails viz., Lymneaidae and Planorbidae. Information on various Gastropods involved in the life cycles of digenetic trematodes of amphibians belonging to different families is tabulated in Table 2. A perusal of the Table 2 indicates that family Planorbidae which includes the genera *Helisoma, Planorbis, Gyraulus, Bulinus, Anisus, Planorbula* and *Coretus* is involved in the transmission of trematodes pertaining to as many as 8 digenean families; followed in order of Lymnaeidae (including *Lymnea, Radix* and *Pseudosuccinia*) in 8 families; Physidae *(Physa)* which carry larvae of 3-families of digeneans and Hydrobiidae (Bithinea) involved in the dissesmination and transmission of only 1 family of flukes (i.e., Lecithodendriidae) among amphibians. The different Gastropod combinations and patterns of multiple infection in amphibians harbouring digeneans has been tabulated in Table 3.

Table 1: Showing percentage of infection catered by snails in the life-cycle of digenetic flukes of amphibians

S. No.	*Snail (s)*	*No. of families of flukes involved*	*Percentage of infection catered by snails*
1.	*Helisoma*	5	45.45
2.	*Radix*	5	45.45
3.	*Lymnea*	4	36.36
4.	*Planorbis*	3	27.27
5.	*Physa*	3	27.27
6.	*Bulinus*	2	18.18
7.	*Gyraulus*	2	18.18
8.	*Coretus*	2	18.18
9.	*Pseudosuccinea*	1	9.09
10.	*Anisus*	1	9.09
11.	*Planorbula*	1	9.09
12.	*Bithinia*	1	9.09

Table 2: Showing different families of digenetic trematodes of amphibians, vector-snails and % age of total snail genera involved in their life cycles

S. No.	*Family of Digenea*	*Gastropod genera involved (n = 12)*	*Number of gastropod families*	*% age of total snails involved*
1.	Hematolechidae	*Lymnea, Planorbis Gyraulus, Anisus, Helisoma, Planorbula, Coretus & Pseudosuccinia*	2	53.33
2.	Plagiorchiidae	*Buliinus, Radix, Physa, Coretus, Planorbis*	3	33.33
3.	Macrodendriidae	*Lymnea, Helisoma, Physa, Physella*	3	26.26
4.	Lecithodendriidae	*Lymnea, Bulinus, Bithinia.*	3	20.00
5.	Hemiuridae	*Radix, Physa and Helisoma*	3	20.00
6.	Pharamphistomidae	*Gyraulus, Planorbis and Radix*	2	20.00
7.	Omphalometridae	*Lymnea & Radix*	1	13.33
8.	Gorgoderidae	*Helisoma**	1	6.66
9.	Cephalogonimidae	*Helisoma*	1	6.66
10.	Telorchiidae	*Radix*	1	6.66
11.	Brachycoelidae	*Land slugs & snails*	–	–

Molluscan foraying done by the authors in and around Jammu city and some places within the geographical limits of Doda district of Jammu Province of J & K state, from May, 1991 till-date provided over 500 specimens of Gastropods belonging to 4 families (Terrestrial forms excepted) namely, Planorbidae and Lymnaeidae among Pulmonates and Viviparidae and Melaniidae (=Thiaridae) among prosobranchiates. The various gastropod genera recorded so far under present investigation includes, *Planorbis, Anisus, Gyraulus, Lymnea, Vivipara* and *Melanoides (=Thiara)* excluding the terrestrial and amphibious forms.

Table 3 : Showing gastropod host, combinations and patterns of multiple infections in amphibians world-over

S. No.	*Gastropod families*	*Families of digenea*	*No. of families*
1.	Planorbidae + Lymnaeidae+ Hydrobiidae	Lecithodendriidae	1
2.	Planorbidae + Lymnaeidae+ Physidae	Plagiorchiidae, Hemiuridae, Macrodendridae	3
3.	Planorbidae + Lymnaeidae	Hematolechidae & Paramphistomidae	2
4.	Planorbidae	Cephalogonimidae & Gorgoderidae	2
5.	Lymnaeidae	Telorchiidae & Omphalometridae	2

It has been found that the amphibians particularly the Anurans in Jammu province of the state harbour a number of digenetic trematodes namely *Diplodiscus* Diesing, 1936; *Haematolechus* Looss, 1899; *Plagiorchis* Luhe, 1890; *Ganeo* Klein 1905; Prosotocus Looss, 1899; *Pleurogenoids* Travassos, 1921; *Loxogenes* Stafford, 1905; *Phyllodistomum* Brunn, 1899; Astiotrema Loss, 1900 and *Mehrorchis* Srivastava, 1934. These trematodes involve 1 or more gastropod as intermediate host in the completion of their life cycle. The snail-vector involved in the life cycles of these trematodes of amphibians from Jammu has been tabulated in Table 4.

It is worthwhile to mention that *Radix, Coretus, Physa, Bithinia and Bulinus* have not been collected so far from Jammu during the present survey and therefore, some alternative intermediate host may be involved in the life cycle of the flukes viz., *Prosotocus, Astiotrema, Loxogenes* and *Pleurogenoids*. Investigations are under progress to unravel the mystery by exploring the snail host involved in the completion of the life cycle of three most prevalent digeneans viz., *Ganeo, Phyllodistomum* and *Mehrorchis* of the anurans. These flukes afflicts damage to a vast population of frogs inhabiting the diverse habitats in various places world-over.

Table 4 : Showing various digenetic flukes of amphibians from Jammu together with the vector-snails involved in their life cycles

S. No.	*Trematode parasite of amphibians*	*Snails involved in their life-cycle*	*Number of snails Genera involved*
1.	Diplodiscus	*Planorbis, Gyraulus Anisus*	3
2.	Hematolechus	*Planorbula, Gyraulus, Lymnea, Coretus, Planorbis Pseudosuccinea*	6
3.	Plagiorchis	*Radix, Bulinus, Physa, Planorbis*	4
4.	Prosotocus	*Bithinia*	1
5.	Astiotrema	*Coretus*	1
6.	Loxogenes	*Bulinus*	1
7.	Pleurogenoids	*Bulinus (?)*	–
8.	Ganeo	*	–
9.	Phyllodistomum	*	–
10.	Mehrorchis	*	–

* The life-cycles of these flukes have not been worked out so far.

Accumulating sufficient knowledge about the vector-snails involved in the life cycle of the amphibians, latter can be saved from their fluke parasites, by putting a check on the population of the vector-snails involved in the life cycles of their flukes.

Management of Trematode Diseases of Amphibians

As per data available (Yamaguti, 1975) about 394 spp. of digenea belonging to 66 genera and 19 families are known so far from amphibians. Out of these, knowledge (Publ.) exists for 61 spp. (15.5%) of 24 genera (36.4%) from 11 families (57.9%) and the life cycle of the remainder of the species numbering 333 belonging to 42 genera and 8 families of digenea are still unknown.

It is thus obvious (Table 2) that the trematodes belonging to family; Hematolechidae possesses a very broad spectrum of intermediate hosts covering as many as 53.33% of the gastropod spp. known to be involved as vector of trematode larvae in amphibians. On the contrary, Telorchiidae and Cephalogonimidae have a relatively narrower host-spectrum and have resricted their choice to just a single gastropod genus i.e., Radix and Helisoma, respectively. Evidently, management of Hematolechid infection is rather difficult in comparison to the Telorchiids and cephalogonimids in which case population can be considerably reduced by checking the population of just a single genus of the vector-gastropod. However, at the present state of the art, when *Radix* is yet to be recovered from Jammu, the management of Telorchiids and Cephalogonimids shall have to await further investigations.

As the parasites depends on the movement of vector-snail or definite host for dissemination or dispersal, the management of the parasitic diseases and the parasite would thus have to depend on the main objective of the interruption of transmission of infection and dissemination or dispersal, through control of vector.

Apparently, the role played by individual gastropod spp. will be related to the population of adult parasites harboured by the amphibians associated with that particular habitat. The restriction of the larval trematodes to certain molluscan hosts may be related to the feeding behaviour of the final host and partly to the ecology of the intermediate host.

Studies on larval infection, their seasonal dynamics and rhythmic behaviour would be most necessary in any snail-vector control programme. Because of the economic, sociological as well as scientific difficulties, snail control is the best means of control of trematode infections at the present.

Acknowledgement

The senior author (AKV) is thankful to the D. O. E., Government of India, New Delhi for financial assistance under the project No. MAB/RE/14/89.

References

Annandale, N. and Rao, H. S. 1925. Material for revision of recent Lymnaeidae (Mollusca: Pulmonata). Rec. Indian Mus., 27: 137-189.

Croll, N. A. 1971. Ecology of Parasites – Heinemann Publ. London, +127 pp.

Das Mahapatra, S., Das Gupta, B. R. and Chowdhury, A. 1982 *Lymnea*, a vector of digenetic trematodes of West Bengal. Proc. Symp. Host. Environ., Z. S. I. India, pp. 65-68.

Dawes, B. 1956. The trematode with special reference to British and other European forms XVI +644 pp. Cambridge University Press.

Erasmus, D. A. 1972. The Biology of Trematodes. +212pp Edward Arnold (Publ.) Ltd; London.

Ewers, W. H. 1964. An analysis of the molluscan hosts of the trematodes of birds and mammals and some speculation on host-specificity. Parasitology, *54*: 571-8.

Heyneman, D. 1960. On the origin of complex life-cycles in the Digenetic flukes. Sobretino/del Libro Homenaje al Dr. Eduordo Caba Caballero Y Caballero Mexico, 1960: 1933-52.

Khar, H. K. 1984. The morphology and ecology of certain trematode parasite larvae from some selected freshwater environments from Jammu (Unpublished Ph. D. thesis) University of Jammu, +386pp.

Majumdar, S. S., Das Gupta, B. and Chowdhury, A. 1983. *Indoplanorbis exustus* as the intermediate host of *Furcocercous cercariae* in West Bengal. Proc. Symp. Host Environ. Z. S. I. India, pp. 17-22.

Mohandas, A. 1974. Studies on freshwater *Cercariae* of Kerala. Biology. III Emergence, behaviour and viability of *Cercariae*. Proc. Nat. Acad. Sc. India, 44 (B) III: 139-44.

Olsen, W. O. 1974. Animal parasites. Their life cycles and Ecology + 562pp. University Park Press, Baltimore.'

Sewell, R. B. S. 1922. *Cercariae indicae*. Indian J. Med. Res., Calcutta, 10: 1–370.

Yamaguti, S. 1975. A synoptical review of life histories of Digenetic trematodes of Vertebrates. +590 pp. Keigaku Publ. Co., Tokyo, Japan.

INDEX

Ablepharus, 291
Acanthodactylus, 291
Accessory pneumectomy, 51
Advanced fry, 195
Agama tuberculata, 291, 292
Agamidae, 294
Agkistrodon, 292
Alevin phase, 196
Amia, 102
Amiurus, 132
Amphibians in Kashmir Himalaya, 290
Anisoptera, 74
Anisus, 322
Aquatic odonata, 69
Arionidae, 230, 231
Ariophantidae, 231
Asiatic Ibex, 305
Asphyxiation, 51
Aspidoparia, 152
Asplanchnidae, 191
Avifauna, 283
Barbus, 4
Bar headed geese, 280, 307
Basomatophora, 231
Beasts of prey, 283
Bee, 201
Beej, 201
Belostomatidae, 78
Benthic biomass, 239
Benthic macro-invertebrates, 135
Benthic sampling methods, 137
Bharat, 304
Bhoj wetland, 301
Biochemical studies of tissue, 31
Biological control agents, 70
Biosphere reserves
    Overa-aru, 287
    Gulmarg, 288
Biotype, 135
Bithinia, 231, 322
Bithyniidae, 231
Black necked crane, 280, 304
Black necked grebe, 307
Blue sheep, 304
Boidae, 295
Boiga, 292
Botine type kidneys, 78, 95, 100, 101
Botia birdi, 88
    kidneys, 95, 117
    ureters, 95, 119
    urinary bladder, 95, 119
Bowman's Capsule, 106, 110, 112, 115, 117, 121
Bos gruniens, 304
Brachionidae, 188
Breeding habits, 4
Breeding seasons, 4
Brown bear, 305
Bufo, 291
Bufonidae, 293
Bulinus, 322
Calotes, 291
Capra ibex, 305
Capra falconeri, 319
Carapace, 270
Carp, 301
Carpine type kidneys, 87, 100, 101
Carassius, 132
Catla , 35, 65, 100

Caudal fin stage, 197
Caudal peduncle, 3
Cervus elephas hanglu, 319
Chalapona, 200
Channa punctatus, 31
Cheer pheasant, 280, 319
Chelonia, 293
Cheloniidae, 266
Chironomous, 70
Chilka lake, 300
Chiru, 305
Chitra, 310
Chorionic gonadotropins, 65
Choriogin, 65
Chugh, 307
Cirrhina kidney, 100
Cladoceran zooplankton, 203, 249
Clarias kidney, 51
Cleavage stages, 5, 198
Coelenterata, 248
Collecting tubule, 108, 111, 113, 115, 119
Coluber, 292
Colubridae, 295
Colurellidae, 189
Comparative indices, 145
Copepoda, 249
Coretus, 322
Corixidae, 77
Corpuscles of stannius, 90, 94, 95, 99
Crossocheilus
- head kidney, 94, 115
- kidney, 90, 115
- ureters, 95, 117
- urinary bladder, 95, 117

Cyprinus
- head kidney, 94, 112
- kidney, 94, 112
- Ureters, 94, 95
- Urinary bladder, 94, 95

Cyprinine type kidney, 87, 94, 100, 101
Cyrtodactylus, 291
Dachhigam National Park, 286
Dal lake, 300
Danio, 132
Daphnia, 70
Decapod crustaceans, 61
Dendralaphis, 292
Dermochelyidae, 266
Destructive insects, 71
Devikoppa tank, 187
Diatoms, 61
Differential leucocyte count, 54
Dim, 201
Diplodiscus, 324
Diptera, 61
Distal convoluted tubule, 108, 111, 113, 115, 119, 121
Diversity indices, 141
Eagle, 307
Echis, 292
Ecogeographical distribution of Herptiles, 298
Ecobiochemical adaptation, 31
Egg ratio, 38
Elapidae, 296
Eleuteroembryonic phase, 195
Embryo, 198
Embryonic development, 4
Emydid turtles, 265
Emydidae, 266, 293, 310
Enidae, 231
Ephemeroptera, 61
Eryx, 292
Estrovest, 65
Eublepharus, 292
Eutrophication of lake Mansar, 235
Eutrophichthyes kidney, 100
Exogenous feeding, 4
Equns kiang, 305
Farooq Nagar pond, 209

Fertility of rotifers, 47
Fierasfer, 102
Filinidae, 191
Fingerling stage, 17, 196, 199
Fish haematology, 54
Frequency of rotifer distribution, 38
Fry stages, 7, 20, 194, 198
Game animals, 283
Ganeo, 324
Gastropods of Jammu, 229, 321
Gastrulation stage, 5
Gazelle, 305
Gekkonidae, 293
Geoclemys, 310
Giant water bugs, 71
Glomerulus, 106, 110, 112, 115, 117, 121
Glyptothorax, 95
    head kidney, 95, 121
    kidney, 95, 121
    ureter, 99, 123
    urinary bladder, 99, 132
Golden eagle, 280
Great rosefinch, 307
Green algae, 61
Gyraulus, 232
Hangul, 319
Hatching of egg, 7, 11, 12, 13-17
Haemoglobin contents, 54
Haemotological values, 51
Haematological studies, 52
Haematolechus, 324
Hare, 283
Harike lake, 301
Helisoma, 322
Hemiptera, 61, 69, 76
Hemis National Park, 286
Hemidactylus, 292
Hepatectomy, 51
Heteropneustes fossilis, 35, 61
Himalayan snowcock, 280
Himalayan bear, 305
Horned lark, 307
Hydrometridae, 77
Hypophysine forte, 65
Hypophysectomy, 51
Ibex, 283
Ictalurus, 128
Ichthyofauna of Mansar lake, 221-227
Ichthyofauna of Kaveri, 155
Indicator organisms, 139
Induced spawning, 65
Insects as food of fish, 72
Insects of Surinsar lake, 249
Intermediate tubule, 108, 110, 113, 115, 119
Interspecific competition in cladocera, 203
Jammu and Kashmir, 289
Juniper, 318
Juvenile, 196
Juxtaglomerular cells, 132
Kabar wetland, 301
Kachuga tectum, 268, 310, 312
Kachuga smithi, 269, 310
Kaleej, 280
Kanjalee wetland, 301
Kestrel, 307
Kiang, 282, 305
Kishtwar National Park, 286
Kite, 307
Kolleru wetland, 301
Kotta wetland, 301
Labeo kidney, 100
Lacertelia, 293
Lacertidae, 294
Lake Mansar, 221
Lapidocephalichthyes, 100
Larva, 194, 198
Larval development, 7
Larval periodization, 3, 4
Lateral-line melanophores, 18
Late post-lavra, 195

Lemna, 209
Lepidogaster, 102
Lepisosteus, 102
Lissemys, 291
Loktek lake, 300
Lophius, 128
Loxogenes, 324
Lycodon, 292
Lymnea, 322
Lymnaeidae, 231
Lynx, 305
Mabuya, 291
Macrobenthos biomass densities, 239
Macrobenthic fauna of Mansar lake, 235
Macrobra Chium, 63
Macrovegetation, 61
Mahseer, 3
Major carps, 65
Management of wildlife in J & K, 286
Mangla Dam, 301
Mansar lake, 300
Marmot, 283
Markhor, 319
Mastacembelus, 100
Maxillary barbels, 3
Mechanical stress on early embryos, 151
Mehrorchis, 324
Melenophores, 3, 18
Meso-larva, 197
Meso-gastropoda, 231
Mesonephros, 102
Meta larva, 198
Mesovellidae, 76
Microhyla, 291
Microhylidae, 293
Microphis, 130
Monal, 280
Mongolian plover, 307
Mrigal, 65
Myotome stage, 6
Myonocephalus, 128
Mystus kidney, 100
Musk deer, 319
Nalsarovar wetland, 301
Nandus, 170
Naucoridae, 79
Neck segment, 106, 110, 113, 115, 119, 121
Nepidae, 78
Notonectidae, 78
Notopterus kidney, 100
Notopterine kidney, 87, 101
Nyan, 304
Odonata, 61, 73
Oligodon, 292
Onchorhynchus, 124
Ophiosops, 291
Ophidia, 294
Opsanus, 128
Oreinus, 90, 105
    head kidney, 90, 112
    kidney, 90, 110
    ureters, 90, 112
    urinary bladder, 90, 112
Ostracoda of Surinsar lake, 249
Oviprim, 65
Ovis ammon, 304
Ovis orientalis, 301
Owl, 307
Pantholeps hodgsoni, 305
Panthera uncia, 319
Pelobatidae, 293
Pelvic bud stage, 197
Pelvic ray stage, 198
Performance indices, 146
Philodimidae, 191
Phrynocephalus, 291
Phylodistomum, 324
Physa, 322, 324
Placoglossus, 124, 130
Plagiorchis, 322

Planobidae, 231
Planorbis, 322
Planorbula, 322
Planktonic rotifers, 37
Planktonic food, 4
Plastron, 270
Pleidae, 77
Pleuroganoids, 324
Poang, 201
Poecilia, 157
Population indices, 141
Population structure, 47, 245
Post-cleavage stages, 6, 20, 198
Post-fry stages, 11, 24, 26, 198
Post-fingerling stages, 28
Post-flexion phase, 197
Post-hatching stage, 22
Post-larva, 195
Preflexion phase, 197
Pre-juvenile, 196
Pro-larva, 194
Prosotocus, 324
Pronephros, 102
Protopterygio larva, 196
Protozoa, 248
Proximal convoluted tubule, 106, 110, 113, 115, 119, 121
Pseudovis nahoor, 304
Pseudosuccinia, 322
Pterigio larva, 197
Ptyas, 292
Puntius, 157, 222
Python, 292
Radix, 322, 324
Ramphotyphlops, 292
Rana, 291
Ranidae, 293
Rasbora
  kidney, 100
  relative abundance, 222
Red shank, 307
Renal organs
  morphology, 87-102
  histology, 105-132
Renuka wetland, 301
Reptiles of Kashmir Himalya, 291
Rhododendron, 318
Rotifer fauna of Dharwad, 187
Rotifer populations, 37, 248
Rohu, 65
Rostral barbels, 3
Sac fry, 194
Sambar wetland, 301
Salvelinus, 124
Saprophilic species, 139
Saprobian system, 135
Saprophobic species, 139
Saproxenic species, 139
Schizothorax, 301
Schizothorax esocinus
  head kidney, 88, 110
  kidney, 88, 106, 108
  ureter, 90, 108
  urinary bladder, 90, 108
Scincidae, 294
Scinecella, 291
Scutiger, 291
Seasonal succession of zooplankton, 245, 249
Seed, 201
Sexual maturity in Emydid turtles, 265
Shahtush, 305
Shapo, 304
Sibynophis, 292
Silurine type kidney, 87, 95, 100, 101
Sitana, 291
Snow leopard, 305, 319
Solenarctos thibetanus, 305
Sphalerosophis, 292
Stage, 40, 194, 198
Status of herptiles in Kashmir, 289

Status of wildlife in J & K, 279
Stylomatophora, 231
S ıkhna lake, 301
Surinsar lake, 245
Tail length, 270
Terminology of fish larval stages, 193, 194
Testudinidae, 266
Tibetian sheep, 304
Tor, 3
Tor kidney, 100
Total erythrocytic count, 54
Total lencocyte count 54
Tragopon, 319
Transitional larva, 196
Trematode transmission, 321
Trichoceriodae, 191
Trionychidae, 266, 291, 293, 310
Trionyx, 310, 313
Typhlops, 292
Typhlopidae, 294
Ujni wetland, 301
Uncia uncia, 305
Ureter histology, 108, 110, 113, 117, 119, 123
Ureter morphology, 90, 94, 95, 99
Urinary bladder morphology, 90, 94, 95, 99
Urinary bladder histology, 108, 112, 117, 119, 123
Ursus arctos, 305
Varanidae, 294
Varanus, 292
Vellidae, 77
Vembanad lake, 300
Veronicellidae, 231
Viviparidae, 231, 296
Vulture, 307
Water scorpion, 71
Western tragopon, 280
Wetlands, 299
Wild ass, 305
Wildlife sanctuaries
  Jasrota, 287
  Nandini, 287
  Overa, 287
  Ramnagar, 287
  Surinsar, 287
  Trikuta, 287
Wild Yak, 282
Wolffia, 209
Wular lake, 300
Xenochrophis, 292
Xenentodon cancilla
  embryonic development, 161-165
  larval development, 165-169
  periodization, 159
Yak, 304
Yolk plug stage, 5
Yolk-sac larva, 194
Zoarces, 102
Zoobenthic density, 239
Zooplankton, 38, 247
Zygoptera, 73

35
SR

FS-1